2004

Univer

SO-AIS-776

3 0301 00213610 5

Mathematical Modelling with Case Studies

Mathematical Modelling with Case Studies

A differential equation approach using Maple

Belinda Barnes

School of Mathematical Sciences
Australian National University
Canberra

and

Glenn R. Fulford

School of Mathematics and Statistics
The University of New South Wales (ADFA)
Canberra

LIBRARY
UNIVERSITY OF ST. FRANCIS
JOLIET, ILLINOIS

London and New York

First published 2002
by Taylor & Francis
11 New Fetter Lane, London EC4P 4EE

Simultaneously published in the USA and Canada
by Taylor & Francis Inc,
29 West 35th Street, New York, NY 10001

Taylor & Francis is an imprint of the Taylor & Francis Group

© 2002 Taylor & Francis

Publisher's Note
This book has been prepared from camera-ready copy provided
by the authors.
Printed and bound in Great Britain by
TJ International Ltd, Padstow, Cornwall

All rights reserved. No part of this book may be reprinted or reproduced
or utilised in any form or by any electronic, mechanical, or other means,
now known or hereafter invented, including photocopying and recording,
or in any information storage or retrieval system, without permission in
writing from the publishers.

Every effort has been made to ensure that the advice and information
in this book is true and accurate at the time of going to press. However,
neither the publisher nor the authors can accept any legal responsibility
or liability for any errors or omissions that may be made. In the case of
drug administration, any medical procedure or the use of technical
equipment mentioned within this book, you are strongly advised to
consult the manufacturer's guidelines.

British Library Cataloguing in Publication Data
A catalogue record for this book is available from the British Library

Library of Congress Cataloging in Publication Data
A catalog record for this book has been requested

ISBN 0–415–29803–2 (HB)
ISBN 0–415–29804–0 (PB)

511.8
B261

70.93

1367

1/14/64

Contents

Part IV   Models with Partial Differential Equations   353

Part V   Appendices   381

Preface

Aims and objectives

This book aims to provide the student with some basic modelling skills, which will have application to a wide variety of problems.

The focus is on those mathematical techniques which are applicable to models involving differential equations. We consider models in three different areas: growth and decay processes, interacting populations and heating/cooling problems. The main mathematical technique used is that of solving differential equations. However, the range of applications and mathematical techniques presented should provide a broad appreciation of the scope of this type of modelling, while the skills developed are transferable to many other areas.

Layout

The book is divided into three general sections: compartmental models, population models and heat transfer problems.

Within each of the modelling sections the process of constructing a model is carefully considered and presented in full detail. Because this is one of the most difficult and crucial aspects of modelling, we emphasise that the process followed is: start simple, evaluate and extend.

One very important skill is that of converting a seemingly complex problem into a much simpler one. To illustrate this, there is an emphasis on the development of simple models, where some reality may be sacrificed in order to gain a basic understanding of the problem being modelled. Flow diagrams and word equations are constructed, both to aid in the model building process and to develop the mathematical equations. The assumptions made and consequent limitations are discussed together with their ramifications in the interpretation of results. This structured and detailed approach has proved a successful learning strategy.

Analysis using a variety of techniques with theoretical, graphical and computational tools follows. Understanding the behaviour of the model under changing conditions (such as changes in the parameters and initial conditions) are part of this process. Graphical methods are encouraged and it is the intention of this course to provide an interactive use of software. To this end `Maple` code and ensuing results are included within the text.

Validation of the model is discussed, using observations or prior knowledge, and extensions to the models are suggested and developed or incorporated into the exercises at the end of each chapter. There is an emphasis on identifying and recognising the strengths and limitations of a model, as well as its relevance in context.

A more detailed overview of the layout is presented in Section 1.2.

Features

Applications and case studies are integral to this text. Case studies have been included throughout to provide the reader with relevant applications of the skills learned and an understanding of how such skills are used and interpreted in current research. We have aimed to include models from a diverse range of disciplines, environmental, biological, archaeological, etc. The authors strongly encourage students to read some of the research papers, for which references are provided.

Another feature of this text is the use of marginal notes throughout, to help the student follow the main ideas of the text and find important information easily.

Furthermore, certain models are developed and extended in sequential chapters of the book, allowing students to concentrate initially on formulation, then in later chapters, to analyse and extend them as the theory is developed. This approach provides continuity.

Prerequisites

Basic calculus, such as a first year modelling/calculus course or advanced school mathematics, as well as fundamental computing skills, are required to make optimal use of this text. Many details of the mathematics required for differential equations and their analysis are covered in the Appendices, and thus a prior knowledge of these equations and their solution is not necessary. Since much `Maple` code is included, prior detailed knowledge of any software package is not a prerequisite and basic computing skills should suffice.

Course suggestions

This book contains far more than was given by the authors in any one course. Part I, Compartmental Models, provides a good introduction for students with little previous background in modelling. This could be followed by a choice in emphasis on either Population Models (Part II) or Heat Transport (Part III).

The authors suggest the inclusion of a project, to complement the course and contextualise the modelling process. This may involve an examination in depth of one of the case studies or other relevant research papers which incorporate the theory learned.

Software integration

The analysis of these models is carried out using a analytical approach as well as the integrated use of a software package, which provides both numerical solutions and an excellent means by which graphics are displayed. It allows us to obtain solutions to differential equations, which are not always possible analytically, and provides a valuable tool in the analysis process. `Maple` code and output (based on `Maple`V Release 5.1, and tested on Release 6) are provided in the text so that students may reproduce the results exactly. Other software packages could replace `Maple` (such as `Mathematica` or `Matlab`) where students are more familiar with these, as the principles are the same and their use would in no way detract from the thrust of the book. Many universities have a site licence for `Maple`, or one of the other packages, with cheap student versions available. (For further information on obtaining `Maple`, see `http://www.maplesoft.com` or `info@maplesoft.com`.)

Acknowledgements

There are many friends and colleagues who have had an input into this book, with typing, technical support, suggestions and encouragement, and we would like to offer them all our warmest thanks.

We are extremely grateful to Jill Smith, Sharon Eldridge and Annette Hughes at the Australian National University, and Annabelle Lippiatt from University College, ADFA, who contributed significantly to the typing of this document. And to Tracey Hansen who undertook the copy editing. We would also like to thank James Gifford and Nick Guoth for the valuable technical advice they provided, as well as Harvinder Sidhu for his `Maple` code to generate the bifurcation diagram in Section 3.6. Furthermore, we thank Scott Rabuka and Waterloo `Maple` for their support.

One of the students taking the course, Cezary Niewiadomski, constructed many of the diagrams incorporated into the book, and we thank him warmly for this generous contribution. There are many other students who pointed out errors and provided suggestions, to all of whom we are most grateful. Willow Hart spent a large number of tedious hours compiling the index and updating files, to our great appreciation.

We are very grateful for information on alcohol and drug absorption which was so readily supplied to us by Kelly FitzGerald of Ainslie Pharmacy (Canberra, Australia) as well as the Alcohol and Other Drugs Council of Australia. Thanks are also due to Duck Australia who provided details of Australian standards for hot water heaters. Many thanks also to Tony Richings for his background information on land mines; we greatly admire your devastating task of clearing Cambodia of mines.

We extend our thanks to those colleagues who read and commented on sections of the book, particularly Harvinder Sidhu, David Rowland, Daryl Daley and Joe Gani. Your comments were a highly valued contribution in the construction of the text.

Finally, we would like to thank our families for their support and encouragement, as such an enterprise always has an impact on family life. Our warmest thanks for this, which has made the experience all the more enjoyable for us.

Belinda Barnes and Glenn Fulford

Belinda.Barnes@anu.edu.au
glenn.fulford@ga.gov.au

Chapter 1

Introduction to mathematical modelling

Mathematical modelling has a diversity of applications and thus a range of possible approaches. Together with an outline of the contents, aims and scope of the book, this chapter provides an introduction to the cyclic nature of the modelling process, a brief description of some modelling approaches and a discussion of the wider ramifications of model application.

1.1   *Mathematical models*

uses of
models

Models of systems have become part of our everyday lives: they range from global decisions having a profound impact on our future, to local decisions about whether to cycle to university based on weather predictions. Together with their provision of a deeper understanding of the processes involved, this predictive nature of models, which aids in decision-making, is one of their key strengths.

diverse ap-
plications

In particular, many processes can be described with mathematical equations: that is, by mathematical models. Such models have a use in a diverse range of disciplines.

aesthetic
uses of
mathe-
matics

There is an aesthetic use, for example, in constructing perspective in paintings or etchings such as is seen in the paradoxical work of Escher. The proportions of the golden mean and the Fibonacci series of numbers, occurring in many natural phenomena such as the arrangement of seed spirals in sunflowers, has been applied to methods of information storage in computers. This well-known mathematical series is also applied in models describing the growth nodes on the stems of plants, as well as in aesthetically pleasing proportions in painting and sculpture and the design of musical instruments. From a philosophical perspective, mathematical logic and rigour provide a model for the construction of argument.

practical
and
analytical
applica-
tions

In a more practical and analytical mode there is a plethora of applications. Mathematical optimisation theory has been applied in the clothing industry to minimise the required cloth for the maximum number of garments, and to the arrangement of odd-shaped chocolates in a box to minimise the number required to give the impression that the box is full! The mathematics of fractals has allowed the successful development of fractal image compression techniques, requiring little storage for extremely precise images; it has now been adopted by Microsoft Encarta.

further ap-
plications

Some other areas of application include the physical sciences (such as astronomy), medicine (such as the absorption of medication), and the social sciences (such as patterns in election voting). Mathematical models are used extensively in biology and ecology to examine population fluctuations, water catchments, erosion and the spread of pollutants, to name just a few. Fluid mechanics is another extensive area of research, with applications ranging from the modelling of evolving tsunamis across the ocean, to the flow of lolly mixture into moulds. (Mathematicians were consulted to establish the best entry points for the mixture to the mould in order to ensure a filled nose for a Mickey Mouse lollypop!)

The use and versatility of such mathematical models has been heightened by the power of computers. This trend is likely to continue, as such modelling results in an efficient and economical way of understanding, analysing and designing processes. Furthermore, the diverse range of applications, where mathematical models are used, implicates an appeal to many areas of mathematics as well as to many types of models. Some models can be extremely accurate in their predictions, whereas others may be more susceptible to a range of interpretations, particularly in the case of large systems with many interacting mechanisms. *use of computers*

The use of information from this modelling process to reach decisions is now very much in the public view, particularly in the case of environmental issues exemplified in the 1998 international meeting on Global Warming in Kyoto. We would hope that by encouraging the responsible use of models, we may improve public confidence in the role of science, which has been declining over recent years (Beck et al. (1993)). It is well to remember that models are not reality and should not be promoted as such. They are considered by the authors to be a valuable and integral part of a multi-disciplinary approach towards decision-making on issues we face today and we encourage future modellers to adopt and promote this perspective. *decision-making*

1.2   An overview of the book

This course aims to provide the student with some basic modelling skills which have application to a wide variety of problems. There is an emphasis on the initial development of simple models, which are then extended and improved by incorporating further features. As well, we have focused attention on the importance of declaring any assumptions made in the modelling process, and bearing these in mind when interpreting the predictions. We discuss various theoretical approaches as they arise, but the emphasis is on the modelling itself and the interpretation of the results. *overview*

Layout of the book

The analysis of these models is carried out using a theoretical approach as well as the integrated use of `Maple` (a software package), which provides *Maple software* both numerical solutions and an excellent means by which graphics are displayed. Where used `Maple` code and output are presented in the text.

case
studies

Furthermore, case studies have been included throughout to provide the reader with relevant applications of the skills learned and an understanding of how such skills are used and interpreted in current research. We have aimed to include models from a diverse range of subjects, environmental, biological, archaeological, etc. The authors strongly encourage students to read some of the research papers, for which references are provided.

types of
math-
ematics
used

In general, with such a diversity of modelling applications it is not surprising that many different areas of mathematics are used. These include differential and integral calculus, differential equations, matrices, statistics, probability, to name just a few.

differential
equation
models

This course focuses on those mathematical techniques which are applicable to differential equation models, in particular ordinary differential equations (ODEs) although partial differential equations (PDEs) are also discussed. We consider models in three different areas: decay processes, interacting populations and heating/cooling problems. The main mathematical technique used is that of solving differential equations. However, this range of applications and the mathematical techniques presented should still provide an appreciation of the scope of this type of mathematical modelling, because the skills developed are transferable to many other areas.

book
sections

The book is divided into three general parts: compartmental models, population models and heat transfer problems.

modelling
process

Within each of the modelling sections, the process of constructing a model is carefully considered, because this aspect is one of the most difficult and crucial parts of modelling. One very important skill is that of converting a seemingly complex problem into a much simpler one. There is an emphasis in this book on the development of simple models where some features of reality are sacrificed in order to gain a basic understanding of the underlying problem. With each model type the text covers this building process in detail, with a critical evaluation of the assumptions made and their subsequent limitations on interpretation. Flow diagrams aid the understanding of this process, and clarify any division of subsections within a model. Then 'word equations' are developed describing the process under consideration, and from these the differential equations follow. This structured approach has proved a successful learning strategy.

Once a model has been established it is analysed using a variety of *model analysis* techniques with theoretical, graphical and computational tools. Understanding the behaviour of the model under a variety of conditions (changes in the parameters and different initial conditions) is discussed, with an awareness of possible limitations on the interpretations (or misinterpretations) of the results. Graphical methods are encouraged, and it is the intention of this course to provide an interactive use of `Maple` software. To this end `Maple` code and ensuing results are included within the text; they are easy to use and applicable to a variety of problems. Simple analytical techniques are developed and applied to solve the differential equations; typically this is confirmed with `Maple`. Again, all relevant code is included within the text and a list of commonly used procedures can be found under the `Maple` entry in the index.

Validation of the model is discussed, using observations or prior knowl- *model validation* edge. Extensions are suggested and developed or incorporated into the exercises at the end of each chapter. Thus the process followed is to start simple, evaluate and extend. Subsequent interpretations are critically evaluated in terms of the assumptions made, with an emphasis on the strengths, limitations and relevance of the models constructed.

We now outline two typical problems discussed in this book to provide *outline of typical approach* an indication of how this process of simplifying a complex problem into its basic parts is approached in the text.

A typical problem in population growth

A major issue in the farming industry is dealing with insect pests which *pest control* infest and destroy crops. To this end we examine an example from the Citrus Industry in America which was threatened by an insect pest (introduced from Australia). One obvious approach is to spray the crops with a pesticide, such as DDT, to eradicate the insects. However, this is not necessarily the most efficient approach in that complicated ecological interactions between the pest and other animals or plants may be affected in unforeseen ways.

To simplify the problem we consider this population of aphid-like insects *simplified problem* (the pest) which interact with a type of ladybird beetle. The interaction is a predator–prey relationship since the aphids are eaten by the beetles. Thus the aphid population and the beetle population interact in that the aphid numbers are reduced by the presence of the beetles and the beetle population is enhanced by the presence of the aphids. Furthermore, our model needs to incorporate the effect of the pesticide on both the aphid and beetle populations.

coupled
equations

We formulate two differential equations: one describing the aphid population as a function of time and the other describing the beetle population as a function of time. Solving the differential equations gives us an unexpected result: the effect of the pesticide actually increases the aphid population, which is not what was expected or desired. This problem illustrates how simple models can provide improved insight into the way that populations interact.

other
similar
problems

Other interacting populations which we examine in detail include competing species models, as well as models of populations in combat and the spread of epidemics.

A typical problem in cooling

a typical
problem

Mathematical models can be useful in domestic situations. To illustrate this, we consider how to determine the optimal amount of insulation to wrap around a water pipe to prevent its freezing on a cold morning. As heat is lost from the water in the pipe to the outside, the temperature of the pipe will drop, and if the water freezes it may burst the pipe. The placement of a layer of insulating material around the pipe is designed to reduce the rate of heat loss and thus prevent the water freezing as quickly. The aim is to determine the rate of heat loss, as a function of the thickness of the insulating material, in order to determine the specifications for sufficient and effective coverage of the pipe.

simplified
models

To produce a mathematical model, we envisage first a simplified situation. We consider an infinite pipe where the heat loss is in the radial direction from the centre to the outside of the pipe. We can also simplify the problem by ignoring the metal pipe and just considering the heat transfer through the insulation itself.

differential
equation
approach

We aim to develop a differential equation which predicts the temperature at any point in the insulation, and use this to find the rate of heat loss. Again a surprising result emerges from the mathematics. It turns out that for some (not all) insulating materials there can be more heat loss than if there were no insulation at all.

other
similar
problems

Similar models examined are those of insulating the wall of a house and the effect of double glazing in windows. Further problems include determining the optimal length of heat fins which are used to reduce the temperature of computer chips, and establishing criteria to avoid spontaneous combustion in storage bins.

1.3   Some modelling approaches

There are many modelling approaches, and while this book is mainly *overview* concerned with mathematical modelling with differential equations it is essential to recognise that there may be other approaches with alternative conceptual frameworks which are more appropriate for different problems. A combination of approaches may provide the 'best' understanding, each approach providing insight into some different aspect.

Below we have briefly outlined some of the approaches that can be taken when formulating a mathematical model. In order to clarify each method we apply it to the problem of determining a population size as a function of time.

Empirical models

The empirical approach is the most basic but also the least useful. The *guessing a* idea is to fit a curve through a set of data and then to use this curve in *suitable* order to predict outcomes where there are no data. *curve*

In terms of the population example, if t denotes time and $P(t)$ denotes *example* the population at time t then we might try curves of the form *empirical model of a population*

$$P(t) = a + bt + ct^2 \qquad \text{and} \qquad P(t) = ae^{bt}$$

and use the method of least squares[1] to estimate the parameters a, b and c. The disadvantage with this approach is that we cannot be confident that the formula applies outside the range of data considered and the parameters a, b and c have no specific meaning. Thus we cannot extend the model easily to a related problem.

This method is often used to create a continuous data set for input to an- *typical ap-* other model, such as completing a temperature profile for a water column *plication* to be used as data in an ecological model predicting water velocities.

Stochastic models

Another approach to modelling is the probabilistic or stochastic approach *probabilis-* ('stochastic' comes from the Greek word to guess). Using this method we *tic* try to estimate the probability of certain outcomes based on the available *approach* data. In terms of the population example we would aim to establish a formula for the probability of a population changing size on the basis

---

[1]In the method of least squares the total error is given as the sum of the squares of the difference between the curve and the data points and the unknowns are estimated by minimising this error.

of assumed formulae for the probability of population births or deaths. These assumed formulae may in turn be based on probability distributions. Quantities of interest to be calculated would be standard statistics such as the mean and the standard deviation of the population size.

typical application

These models can by extremely complicated, although this is not necessarily the case. They do have the advantage of incorporating a degree of uncertainty within them, and ideally should be used when there is a high degree of variability in the problem. This method is typically used for models of small populations when reproduction rates need to be predicted over a time interval. They also have valuable application in many other areas such as economic fluctuations, insurance problems, telecommunications and traffic theory, and biological models.

Simulation models

programming a set of rules

In a simulation model one writes a computer program which applies a set of rules, or possibly even physically builds a scale model. It is intended to produce a set of data which mimic a real outcome including extreme events. In a population model, the rules would apply to each member of the population. Usually a simulation will involve some random components. The computer program can be run many times and statistical information gained in the process.

typical application

Typically, such models are used in engineering applications as an aid to identifying problems which may arise during use or construction. They also provide a useful means by which data sets can be generated. One such example is rainfall data, which may be required as input to a model forecasting phosphorus concentrations in a stream network.

complexity versus simplicity

It might be argued that this simulation approach provides the most realistic models, but this does not mean it provides the best models. The best models are usually those which are simplest yet still provide results which are useful.

Deterministic models

ignore random variation

Another approach to mathematical modelling is the deterministic approach. This is the approach we adopt in this book. Modelling in this manner we ignore random variation and try to formulate mathematical equations describing the basic fundamental relationships between the variables of the problem.

example

For instance, in the population model we would aim to obtain an equation relating birth rates and death rates which themselves are related through equations to the population size at any given time. In a sense we are constructing a model made of several empirical and inter-related sub-models, linking these sub-models together and then using the whole system to predict the outcome from a set of initial conditions.

This process is widely used and can be extremely accurate, such as in the case of predicting satellite orbits. It has the drawback that in other cases it is not possible to establish all the component mechanisms of a process or, even if this were possible, that including all known relationships renders the model unwieldy. However, as we shall see, even under these conditions this modelling technique can provide valuable insights into a process. *typical application*

Statistical models

Statistical models concern the testing (referred to as hypothesis testing) of whether a set of empirical data is from one or another category. These categories are assumed to have particular distributions (with associated means and standard deviations) and the results suggest the data are drawn from one such category. This distribution can then be used to predict the outcome of further trials. In terms of our population such a model may test whether a sample of heights was taken from a population with an average height of six feet (say category A), or one with an average height of five feet and ten inches (say category B). Our results would indicate a percentage error margin with which we have made this prediction. *testing data sets*

Statistical testing is used widely in psychology, paleontology and the biological sciences. One example application would be to establish spatial distributions of some ancient species from sets of fossilised teeth found in different locations. *typical application*

1.4 The cyclic process

The main processes involved in constructing a mathematical model are discussed below. When you start to build models you may find it productive to return to this section, as the process can be deceptively complex. *overview*

The modelling cycle

Designing a useful and appropriate model may be a challenging task; whilst the model needs to incorporate the major features of the phenomena it describes, it is also essential to keep it simple to allow for its *start simple* meaningful analysis and application. So, start simple. Use the large volume of knowledge that has been tested already and from which you can draw. You will need to make assumptions, that is conditions under which the model equations are valid, and then the model should be tested to ensure that it complies with known results. It may need adjustments, criticism or even total rejection if it does not perform adequately.

refine the model

This process may be followed again in an iterative manner until its performance is acceptable for the required task. So the process is cyclic. Much can be learned about a problem in going from a simple representation to a more complex one, far more than if one attempted to go directly to the more complex representation. Most sophisticated models in use today have been constructed this way: starting simple.

a cyclic process

A mathematical model is a simplification of a complex real world problem which is cast into the form of mathematical equations. There are several stages to the construction of a mathematical model. Figure 1.4.1 shows a schematic diagram of the process.

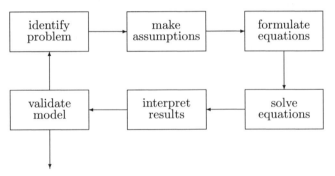

Figure 1.4.1: Schematic diagram of the modelling cycle. A process for constructing and continually improving a mathematical model.

The steps in the cycle

step 1: identify the problem

The first step in constructing a mathematical model is to identify the problem. This is not always as trivial as it sounds. For the mathematical model to be useful it is important to have a goal to work towards, and it can sometimes take great skill and experience to pose the right questions when called upon as a consultant to model a process or system.

step 2: make simplifying assumptions

Next, make simplifying assumptions about the problem which render it manageable and easy to understand. For example, to model accurately the effect of spraying pesticides on crops to reduce the population of aphids, it is important to model the interaction of the aphids with many other species. However, one might first look at how they interact with just one species, or indeed just look at the effect of the pesticide on the aphids alone, ignoring any interaction with other species.

The following procedure is the formulation of equations relating the variables of interest. At this stage we might also have to make additional simplifying assumptions, not made in step 2. To ease the difficulty of this step we introduce the concept of a word equation which relates quantities expressed in words before assigning mathematical symbols to them.

step 3: formulate equations

It may turn out that we cannot solve the equations with the techniques available, in which case it would be necessary to return to the previous step and simplify the equations in such a way as to be able to solve them.

step 4: solve the equations

The solutions obtained can then be applied to try and answer the questions posed in the first step. This is done by graphing results and considering how a change in one variable may cause changes in another. For example, in a model for insulating a water pipe, we would investigate how the rate of heat loss varies with respect to the thickness of the insulation. Sometimes we may gain far more from the solutions than merely the answers to the questions we originally posed.

step 5: interpret the solutions

Finally, the model may not behave in a way that is consistent with observations. In this case, we would return to step 2 and relax some of the simplifying assumptions we made. Depending on the context, a simple model may be adequate. Alternatively, a more complicated model under a particular set of conditions might be required for the best information. It is up to the modeller to make these judgements and refine the model accordingly. In doing so, the process provides us with an improved insight into the system we are modelling.

step 6: repeat the process

When to stop refining

While repeating this process to provide the best possible model, the modeller should bear in mind that models need to remain simple so as to allow for meaningful analysis, but should describe as accurately as possible the main phenomena they attempt to describe. Equations which are very complicated may be too complex to solve, and thus interpretation becomes difficult. On the other hand, models which are too simplistic do not represent the process well, and may not be very reliable in their interpretation. So a reasonable balance needs to be found. Furthermore, the modeller should remain aware of, and account for, aspects not included in the model which may affect the conclusions.

balance simplicity with predictability

uses

A model is an attempt at mimicking the real world and cannot describe it completely. However, while some models are at best a reasonable estimate of a process others can provide extremely accurate and reliable information. We emphasise the value of using more than one model in a given situation. A more philosophical discussion of how models are chosen and used is given in Section 1.5.

1.5   Modelling for decision-making

overview

Currently modelling, and in particular mathematical modelling, provides a means by which many political and management planning decisions are made, both locally and globally. Such models have huge economic clout and wield social and environmental ramifications. There is a need to use these models wisely.

models are not infallible

Modelling is an extremely powerful tool, a framework for research, debate and planning, which provides a valuable source of information for decision-making. However, while models aim to incorporate the main mechanisms of a process and have the capacity not only to describe and explain but to predict as well, they are not infallible, and should not be considered in isolation particularly in the light of the above remarks. Decision makers need to be made aware of the modelling process and its context-specific strengths and limitations, in order to make best use of the interpretations.

objectives: under-standing and prediction

The main objectives of any model are twofold: first, to provide a deeper understanding of a system or process, and second, to use this knowledge for predictions and decision-making. In the first instance, mathematical modelling can be extremely successful with, in many cases, a combination of models providing the best insight into a problem. Of course, any model is, at best, merely a means of mimicking reality and as such at each stage of the process, there is the possibility of error or introduced limitations. This does not render the process useless, but rather suggests a need for an understanding of the complexities involved.

different interpret-ations of outcomes

No part of the process is necessarily simple. For example, in the decision-making process individuals invariably have different interpretations of the results, and differ in opinion as to what might most benefit a situation. They may disagree on a 'safe' level of pollution in a local river, or how 'green' the Sydney Green Olympics should be, or what constitutes an endangered species, or even whether a population facing extinction warrants concern at all.

Furthermore, the model needs to be sufficiently simple so that the available equipment is adequate for sensible and useful predictions. If it took until next week to predict tomorrow's weather we might as well not bother: clearly accurate predictions are only useful before the event! Certain physical processes, such as the path of a satellite, can be accurately described while others, such as ecological systems or the process of global warming, are far more complex and require modelling and interpretation of a very different sort. Thus the modelling process is not necessarily simple and it needs to be well understood, with all its strengths, simplifications, assumptions and limitations to utilise the results optimally.

best utilisation of results

When involved in decision-making processes, to take the results of any single model as absolute would be short-sighted. It constitutes one part of a multi-dimensional field, far more complex in most cases than our single mathematical model can hope to represent. But, for their part, decision makers need to be made aware of the limitations of the model predictions, as well as their strengths. They need to recognise the assumptions on which the model is based and the rationale behind them. They should consider a variety of interpretations. They should be as well-informed as possible.

recognition of limitations

Furthermore, beyond the mathematics there may be, for example, historical, legal or cultural issues involved which are integral to the planning process and should be acknowledged. It is the modellers' responsibility to provide the best information possible, not to sell a 'perfect' model. It is well to remember that, ultimately, political and management decisions have an impact on us all.

other considerations

In many instances, the reticence of scientists to work together with those in other 'softer' disciplines, and *vice versa*, has emphasised the distrust between them and the lack of respect for each others' research. There are many who ignore the imperfect nature of models, and emphasise only the accuracy of mathematics. There are those who deride quantitative research. The authors consider the contributions from each discipline to carry value, and encourage their collaboration and cooperation in order to realise the full and powerful potential of modelling.

respect for other disciplines

With a combination of the qualitative and the quantitative, a recognition of the complexity of systems and the true nature of models, we might aim to capture (to the best of our ability) more than just the physics of the system. If we are willing to recognise and acknowledge the complexities and limitations involved, cooperative and informed modelling has a truly enormous potential.

combination of qualitative and quantitative

further   For further discussion on certain of the issues raised in this section, see
reading   Gani (1972), Gani (1980) and Beck et al. (1993).

Part I

Introduction to Compartmental Models

Chapter 2

Compartmental models

The compartmental model framework is an extremely natural and valuable means with which to formulate models for processes which have inputs and/or outputs over time. In this chapter, we model radioactive decay processes, pollution levels in lake systems and the absorption of drugs into the bloodstream, using compartmental techniques.

2.1   Introduction

overview

Many processes may be considered as compartmental models: that is, the process has inputs to and outputs from a 'compartment' over time.

Compartmental diagram

compartmental concept

One example of this compartmental notion is the amount of carbon dioxide in the Earth's atmosphere. The compartment is the atmosphere. The input of CO_2 occurs through many processes, such as burning, and the output occurs through processes such as plant respiration. This is illustrated in Figure 2.1.1 in a diagram known as a *compartmental diagram*.

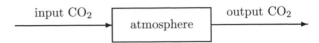

Figure 2.1.1: Input–output compartmental diagram for CO_2.

quantity of interest

What we are interested in is the amount of CO_2 in the compartment at any given time, or whether over long periods the levels are increasing or decreasing. Currently, the input of CO_2 into the Earth's atmosphere appears to exceed the output, so that there is an overall increase in the level of CO_2 in the compartment.

Balance law

word equation

Suppose we are modelling the amount of CO_2 as above, or the size of a population which is changing with time. We can think of the amount of a substance (CO_2 or the population) as occupying the compartment and the rate of change can be considered as the 'rate in' minus the 'rate out'. Thus we have what is called the *balance law*. In words this is

$$\left\{ \begin{array}{c} net\ rate \\ of\ change \\ of\ a\ substance \end{array} \right\} = \left\{ \begin{array}{c} rate \\ in \end{array} \right\} - \left\{ \begin{array}{c} rate \\ out \end{array} \right\}. \tag{1}$$

It could describe, for example,

- the decay process of radioactive elements,
- births and deaths in a population,
- pollution into and out of a lake or river, or the atmosphere,
- drug assimilation into, and removal from, the bloodstream.

In the following sections, we use this balance law approach to formulate mathematical models of differential equations which describe such processes.

2.2   Exponential decay and radioactivity

The process of dating aspects of our environment is essential to the un-   overview
derstanding of our history. From the formation of the earth through the
evolution of life and the development of mankind historians, geologists,
archaeologists, paleontologists and many others use dating procedures to
establish theories within their disciplines.

General compartmental model

While certain elements are stable, others (or their isotopes) are not, and   radioactive
emit α-particles, β-particles or photons while decaying into isotopes of   decay
other elements. Such elements are called *radioactive*.

The decay, or disintegration, of one nucleus is a random event and so for   modelling
small numbers of nuclei one might apply probability functions. However,   radioactive
when dealing with large numbers of nuclei we can be reasonably certain   decay
that a proportion of the nuclei will decay in any time interval and thus
we can model the process as continuous with some fixed rate of decay.
We can consider the process in terms of a compartment without input
but with output over time, as in Figure 2.2.1.

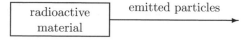

Figure 2.2.1: Input–output compartmental diagram for radioactive nuclei.

We make some assumptions and then, based on these, develop a model   model as-
to describe the process.   sumptions

- We assume that the amount of an element present is large enough
  so that we are justified in ignoring random fluctuations.
- We assume the process is continuous in time.
- We assume a fixed rate of decay for an element.
- We further assume there is no increase in mass of the body of
  material.

The first step is to determine an equation describing this disintegration   starting
process and, as with each subsequent model building in this book, we   with a
start with a *word equation* and then replace each 'word-term' with math-   word
ematical symbols. In a word equation we can express the above decay   equation
process as

$$\left\{\begin{array}{c} \text{amount of} \\ \text{radioactive} \\ \text{material} \\ \text{at time } t + \Delta t \end{array}\right\} = \left\{\begin{array}{c} \text{amount of} \\ \text{radioactive} \\ \text{material} \\ \text{at time } t \end{array}\right\} - \left\{\begin{array}{c} \text{amount of} \\ \text{radioactive} \\ \text{material} \\ \text{decayed in time } \Delta t \end{array}\right\}. \qquad (1)$$

Formulating the differential equation

terms into
symbols

To convert the terms in the word equation into symbols let $N(t)$ be the number of radioactive nuclei at time t and let Δt be a small change in time. We know that the change in the number of nuclei is proportional to the number of nuclei at the start of the time period. Hence (1) translates to

$$N(t + \Delta t) = N(t) - kN(t)\,\Delta t \qquad (2)$$

where k is a positive constant of proportionality indicating the rate of decay per nucleus in unit time. We assume k to be fixed for each element although it will have a different value for different elements/isotopes. Note that $N(t)$ is an integer (number of nuclei) while $-kN(t)\,\Delta t$ may not be. However, if instead we consider N as a mass in grams (from which we can find the number of nuclei) then we can overcome this problem. Note that we then have an equation for $N(t)$, which assumes that $N(t)$ is a continuous, differentiable function rather than an integer. This is an appropriate approximation, particularly when the number of nuclei is large.

letting
$\Delta t \to 0$

We let $\Delta t \to 0$ in (2) and take the limit. This gives

$$\lim_{\Delta t \to 0} \frac{N(t + \Delta t) - N(t)}{\Delta t} = -kN(t). \qquad (3)$$

From the definition of the derivative, equation (3) becomes the *differential equation*

$$\frac{dN}{dt} = -kN(t). \qquad (4)$$

alternative
formula-
tion

Alternatively, we can use the balance law and obtain the differential equation directly. In this case we have the mass of nuclei of the element in a 'compartment'. There is no input, only an output as the sample decays. Thus from the words of the balance law to symbols we have

$$\left\{ \begin{array}{c} rate\ of\ change \\ of\ mass \\ of\ nuclei \end{array} \right\} = - \left\{ \begin{array}{c} rate\ of \\ mass \\ decaying \end{array} \right\} \qquad (5)$$

Setting the left hand term to dN/dt and the right hand term to $-kN(t)$ gives the same differential equation.

initial
value
problem

Given a sample of a radioactive element at some initial time, say n_0 nuclei at t_0, we may want to predict the mass of nuclei at some later time t_1. We require the value of $k > 0$ for the calculations: it is usually found through experimentation. Then, with known k and an initial condition $N(t_0) = n_0$ we have an initial value problem (IVP)

$$\frac{dN}{dt} = -kN, \qquad N(t_0) = n_0. \tag{6}$$

Solving the differential equation

We can solve this equation theoretically using the separation of variables technique (see Appendix A.3) as in the following example.

Example 1: *Solve the initial value problem (IVP) in equation (6).*

Solution: *Since the differential equation is separable,*

$$\int \frac{1}{N} \frac{dN}{dt} \, dt = \int -k \, dt,$$
$$\int \frac{1}{N} \, dN = -k \int dt,$$

and hence, since N is a positive quantity,

$$\ell n \, N = -kt + C,$$

where C is an arbitrary constant. Taking exponentials of both sides we obtain

$$N(t) = Ae^{-kt}, \qquad where \quad A = e^C,$$

and note that $N \geq 0$.

Using the initial condition $N(t_0) = n_0$ we get

$$n_0 = Ae^{-kt_0} \qquad and \quad A = n_0 e^{kt_0}.$$

Thus

$$N(t) = n_0 e^{-k(t-t_0)}, \tag{7}$$

which is the solution of the IVP.

Note that above we have considered the indefinite integral, when tech- *definite or* nically we should specify the integral over a particular interval $[0, t]$. In *indefinite* the next example we show that this has no impact on the final solution, *integrals* and thus for the many integrations carried out in this book the interval is often not given explicitly, but understood from the context.

Example 2: *Solve the IVP in equation (6) on the interval $[0, t]$.*

Solution: *Since the differential equation is separable,*

$$\int_0^t \frac{1}{N}\frac{dN}{ds}\,ds = \int_0^t -k\,dt,$$

$$\int_{n_0}^N \frac{1}{N}\,dN = -k\int_0^t dt,$$

and hence, since N and n_0 are positive,

$$\ln N - \ln n_0 = -kt + 0,$$

$$\ln\frac{N}{n_0} = -kt.$$

Taking exponentials of both sides, we obtain

$$N(t) = n_0 e^{-kt},$$

which is the solution of the IVP.

Maple solution We can also obtain the analytic solution of the differential equation using Maple, or some other software package. We include many inserts of Maple code throughout the book, which are listed in the index, and which allow the reader to reproduce the analytic results and all illustrated graphs. For a brief introduction to Maple see Appendix C.

Maple code The code to establish the analytic solution is:

```
> restart;
> de := diff(N(t),t) = k*N(t);
> dsolve({de, N(t0)=n0}, N(t));
```

graph of solution Consider a graph of the solution $N = n_0 e^{-k(t-t_0)}$. First, since $t - t_0 > 0$, we have that $e^{-k(t-t_0)}$ is a negative exponential. We also have that

$$\lim_{t\to\infty} n_0 e^{-k(t-t_0)} = 0$$

and $N' = -kN < 0$ (since N and k are positive). This implies that the function $N(t)$ is monotonically decreasing and has no turning points. This information allows us to construct a graph of the solution illustrated in Figure 2.2.2.

graph using Maple Equally, Maple could have been used to construct the graph with the following code:

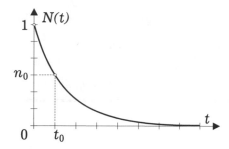

Figure 2.2.2: Graph of solution (7) showing exponential decay.

```
> restart:
> de := diff(N(t),t)=k*N(t);
> soln := dsolve({de, N(t0)=n0}, N(t));
> k := -.5; n0 := 1; t0:=0;
> plot({rhs(soln), N(t)}, t=0..15);
```

Often experimental data provides information concerning $N(t)$ at a given time t_1 and what is required is to estimate the constant k and thus predict future or past values of N. This can be done algebraically, or using the substitution procedure of Maple (see Question 1 of the exercises) with code as follows:

finding the constant of proportionality

```
> restart:
> de := diff(N(t),t)=k*N(t);
> t0:=0; n0:=1;
> soln := n0->dsolve({de, N(t0)=n0}, N(t)):
> ans := soln(n0);
> get_k := solve(subs(t=2, N(2)=1/4, ans), k);
> evalf(get_k);
```

The *half-life* τ of the radioactive nuclei can be used to determine k, where τ is the time required for half of the nuclei to decay. The half-life τ is more commonly known than the value of the rate constant k for radioactive elements.

half-life

Example 3: *If the half-life is τ then find k in terms of τ.*

Solution: *Using the solution (7), and setting $N(t+\tau) = N(t)/2$, we have*

$$\frac{N(t+\tau)}{N(t)} = \frac{1}{2} \tag{8}$$

which gives

$$\frac{n_0 e^{-k(t+\tau-t_0)}}{n_0 e^{-k(t-t_0)}} = \frac{1}{2}.$$

This simplifies to

$$e^{-k\tau} = \frac{1}{2}.$$

Taking logarithms of both sides gives $\ell n(1/2) = -k\tau$. *Hence*

$$k = \frac{\ell n(2)}{\tau}. \tag{9}$$

Note that both τ and k are independent of n_0 and t_0.

Radiocarbon dating

applying
the theory

We can apply the above theory to the problem of dating artifacts and paintings by considering the decay process of certain radioactive elements in each.

radioactive
carbon

The following description and example are adapted from Borelli and Coleman (1996). All living organisms absorb carbon from carbon dioxide (CO_2) in the air, and thus all contain some radioactive carbon nuclei. This follows since CO_2 is composed of a radioactive form of carbon ^{14}C, as well as the common ^{12}C. (^{14}C is produced by the collisions of cosmic rays (neutrons) with nitrogen in the atmosphere and the ^{14}C nuclei decay back to nitrogen atoms by emitting β particles.) Nobel Prize winner Willard Libby, during the late 1940s, established how the known decay rate and half-life of ^{14}C, together with the carbon remaining in fragments of bones or other dead tissue, could be used to determine the year of its death. Because of the particular half-life of carbon, internationally agreed upon as 5568 ± 30 years for ^{14}C, this process is most effective with material between 200 and 70 000 years old.

how
carbon
dating
works

Carbon dating depends on the fact that for any living organism the ratio of the amount of ^{14}C to the total amount of carbon in the cells, is the same as that ratio in the surroundings. Assuming the ratio in air is constant, then so is the ratio in living organisms. However, when an organism dies, CO_2 from the air is no longer absorbed although ^{14}C within the organism continues to undergo radioactive decay.

Lascaux
Wall
Paintings

In the Cave of Lascaux in France there are some ancient wall paintings, believed to be prehistoric. Using a Geiger counter the current decay rate of ^{14}C in charcoal fragments collected from the cave was measured as approximately 1.69 disintegrations per minute per gram of carbon. In comparison, for living tissue in 1950 the measurement was 13.5 disintegrations per minute per gram of carbon.

Example 4: *How long ago was the radioactive carbon formed and, within an error margin, the Lascaux Cave paintings painted?*

Solution: Let $N(t)$ be the amount of ^{14}C per gram in the charcoal at time t. We apply the model of exponential decay, $N' = -kN$. We know $\tau = 5568 \pm 30$ years (the half-life of ^{14}C). From equation (9) $k \simeq 0.0001245$ per year.

Let $t = t_0 = 0$ be the current time. Let T be the time that the charcoal was formed, and thus $T < 0$. For $t > T$, ^{14}C decays at the rate

$$\frac{dN}{dt} = -kN \quad \text{with} \quad N(0) = n_0$$

and

$$N(T) = n_0 e^{-kT}$$

or

$$T = -\frac{1}{k} \ell n \left(\frac{N(T)}{n_0} \right).$$

But we do not know $N(T)$ or n_0. However,

$$\frac{N'(T)}{N'(0)} = \frac{-kN(T)}{-kN(0)} = \frac{N(T)}{n_0}$$

and we do have $N'(T)$ and $N'(0)$. Thus

$$T = -\frac{1}{k} \ell n \left(\frac{N(T)}{n_0} \right) = -16690 \pm 94 \, years.$$

The accuracy of this process depends on having prior knowledge of the exact ratio of ^{14}C to the total C in the atmosphere, but this has changed over the years. There is a basic sinusoidal variation with an 8000-year period (approximately) to consider. Further, volcanic eruptions and industrial smoke emit only ^{12}C (from tissue older than $100\,000$ years) into the atmosphere and decrease the ratio, while nuclear testing has resulted in a 100% increase of the ratio in certain parts of the northern hemisphere! These are now factored into dating calculations (see Borelli and Coleman (1996)). *some limitations*

As previously mentioned radiocarbon dating is inaccurate for the recent past and for very long time periods. Thus the lead isotope with a relatively short half-life ($\simeq 22$ years) is used for the dating of paintings in our recent history, while at the other extreme, radioactive substances (such as Uranium-238) with half-lives of billions of years, are used to date the earth. *other dating methods*

LIBRARY
UNIVERSITY OF ST. FRANCIS
JOLIET, ILLINOIS

Skills developed here:

- Understand the concept of modelling with compartments and the balance law.
- Find the constant of proportionality from the half-life for radioactive elements.
- Generate a solution, using `Maple` or analytical methods, to simple separable differential equations.

2.3   Case Study: Detecting art forgeries

This case study is based on an article in Braun (1979).

After World War II, Europe was in turmoil. Some well-known artworks had disappeared, others had turned up miles from the galleries to which they belonged, and some new ones had appeared! The Dutch Field Security uncovered a firm which, acting through a banker, had been involved in selling artworks to the Nazi leader, Goering. In particular in the case of a painting by the seventeenth century artist Jan Vermeer, "Woman taken in adultery", the banker claimed that he had been acting on behalf of Van Meegeren, a third rate painter, who was duly imprisoned for the crime of collaboration. However, shortly after his arrest Van Meegeren announced that he had never sold this painting to Goering and that it was in fact a forgery. He claimed that, together with "Disciples at Emmaus" and several others, he himself had painted them. In order to prove this claim he began to forge a further work by Vermeer while in prison. When he learned that his charge had been changed from that of collaboration to one of forgery, he refused to finish the work as it would provide evidence against himself and reveal the means by which he had 'aged' the paintings.

A team of art historians, chemists and physicists was employed to settle the dispute. They took X-rays to establish what paintings might lie underneath, analysed pigments in the paints and looked for chemical and physical signs of old age authenticity. They found traces of modern pigment in the cobalt blue, although Van Meegeren had been very careful, in all other cases, to use pigments which the artists themselves might have used. Further examination indicated Van Meegeren had scraped old canvases which he had used as a base for the forgeries. The discovery of phenoformaldehyde, which he had cleverly mixed into the paints and then baked in the oven so that it would harden to

*bakelite as was typical of old paint, finally sealed his fate and he was convicted
in October 1947 and sentenced to a year in prison. (He died of a heart attack
while in prison in December of that year.)*

*But the controversy was not over. Art lovers claimed that "Disciples at Em-
maus" was far too beautifully painted to be a forgery and was in quite a
different class from the other forgeries. They demanded more scientific evidence to prove it a fake. Besides, a noted art historian A. Bredius had
previously certified it as authentic and it had been bought by the Rembrandt
Society for a sizeable sum of money. Reputations and money were at stake
here.*

*A scientific study was undertaken with the results based on the notion of
radioactive decay. The atoms of certain elements are unstable, or radioactive,
and within a given time (which depends on the element involved) a number
of the atoms disintegrate. This rate of disintegration has been shown to be
proportional to the number of atoms present at that time, $N(t)$. So, starting
from $N(t_0) = n_0$, where n_0 is the original amount of material,*

$$\frac{dN}{dt} = -\lambda N, \qquad \text{with} \qquad N = n_0 e^{-\lambda(t-t_0)},$$

where λ, the decay constant, is chosen positive.

*From here, if the initial formation of the substance was at time t_0 then its age
is $(1/\lambda)\ell n(n_0/N)$. N can be measured, but n_0, the original amount of the
material, is harder to find.*

*A useful quantity to know is the half-life of an element, the time taken for the
number of atoms present at a particular time to halve. Now if $N = \frac{1}{2}n_0$ then,
rearranging the above gives*

$$\text{half-life} = \frac{\ell n\, 2}{\lambda} \simeq \frac{0.69}{\lambda}.$$

*The half-lives of many radioactive substances are well-known: for Uranium-
238 it is 4.5 billion years, for Carbon-14 is 5568 years, for Lead-210 it is 22
years and for Polonium-214 it is less than 1 second.*

*All paintings contain small amounts of radioactive Lead-210 which we know
to decay with a half-life of 22 years, as this is contained in lead white which
has been used by artists as a pigment for over 2000 years. Lead white contains
lead metal (which is smelted from rocks) which in turn contains small amounts
of Lead-210 and extremely small amounts of Radium-226 (mostly removed in
the smelting process). There is no further supply of lead due to the absence
of Radium-226 which decays to Lead-210 in a series of steps, and so Lead-210
initially decays rapidly. Radium-226 has a half-life of 1600 years and since it
exists in very small amounts in lead white, eventually the process of lead decay
stabilises when the amount disintegrating balances with the amount produced
by radium disintegration.*

Let $N(t)$ be the amount of Lead-210, and then, as before,

$$\frac{dN}{dt} = -\lambda N(t) + R(t), \qquad n_0 = N(t_0),$$

where $R(t)$ is the rate of disintegration of Radium-226 in minutes per gram of white lead. Since we are interested in a time period of 300 years at the most, and the half-life of Radium-226 is 1600 years, and the amount present is so small, we can consider $R(t) = R$ to be a constant. Solving the equation, using an integrating factor of $e^{\lambda t}$, we conclude that

$$N(t) = \frac{R}{\lambda}\left(1 - e^{-\lambda(t-t_0)}\right) + n_0 e^{-\lambda(t-t_0)}.$$

R and N can be measured easily and λ is known, however n_0 is more difficult and is dealt with in the following way. The above equation can be rearranged to give

$$\lambda n_0 = \lambda N e^{\lambda(t-t_0)} - R(e^{\lambda(t-t_0)} - 1)$$

where λn_0 is the disintegration rate when the white lead was extracted from the ore at time t_0. A good estimate of a reasonable size for this disintegration rate can be obtained by taking current measurements for such ore, and they provide a range from 0.18 to 140 disintegrations per minute per gram. In exceptional circumstances the rate has been measured as 30 000 disintegrations (per minute per gram). This gives a very wide range and thus excludes any possibility of predicting the age of the painting with much accuracy. However, an exact date is not required. What is needed is to establish if the painting is old (about 300 years old) or modern (painted within the last 30 to 50 years).

Suppose it were ±300 years old, then λn_0 as expressed above should provide an original (at t_0) disintegration rate of much less than 30 000. Alternatively, if it is a modern forgery then λn_0 should be ridiculously large, or far greater than 30 000 disintegrations per minute per gram.

Using the half-life formula to find λ ($\lambda = (\ell n\, 2)/22$) it follows that $e^{300\lambda} = 2^{150/11}$. The current disintegration rate can be measured, and since it is the same as for Polonium-210 after a few years, and as the latter is easier to measure in white lead, it is taken as 8.5 disintegrations $(\text{min}^{-1}\text{g}^{-1})$ (measured from ^{210}Po) for the painting "Disciples of Emmaus". Now, with $R = 0.8$, the equation gives

$$\lambda n_0 = 8.5 \times 2^{150/11} - 0.8(2^{150/11} - 1) > 98\,000.$$

This painting was definitely not painted by Vermeer in the seventeenth century.

2.4   Case Study: Pacific rats colonise New Zealand

The case study which follows, emphasises the care with which such models should be adopted when applying this dating procedure as 'proof' of a proposed theory.

This case study is based on an article by Anderson (1996).

===

Opinions are divided as to when humans first occupied New Zealand. Some, due to changes in the vegetation, favour an arrival date of 1400BP (where BP means before present) or earlier, while others argue, based on archaeological evidence, a date of 900BP or later. Anderson (Anderson (1996)) discusses the dating of Pacific rats (rattus exulans) as an indicator of human arrival as they would have arrived as 'stowaways' with the first Polynesians. The advantage of this means of establishing an arrival time is that the rat population would have expanded rapidly and thus evidence of them should be easy to find.

To ensure the least contamination of the material by other carbon sources, clean hard bones were carefully selected from dry shelters and dunes in the Shag River Mouth. The idea was to use the bone protein (collagen) for obtaining the age as it would hold no inbuilt age, a potential problem associated with marine reservoir effects. (Some shell exposed only to water from deep in the ocean, is exposed to a very different ratio, a lower ratio, of carbon to ^{14}C than that in air. Thus different dates can result. These are marine reservoir effects.)

The AMS (accelerator mass spectroscopy) radiocarbon dating method was used since it only requires very tiny amounts of carbon and the Institute of Geological and Nuclear Sciences, Ltd. (IGNS) produced a range with dates as early as 2000BP. This suggested that the rats were in existence in NZ at least a millennium earlier than previously thought. By 1996 this date had been accepted as 'fact'.

Anderson proposes that these 'facts' and dates are questionable. Such dates are not consistent with the carbon chronologies associated with moa eggshell, charcoal and bones using non-protein (non-collagen) samples. He suggests that there are further factors which contribute to carbon levels. There is a considerable likelihood and evidence that the rats ate insects living in rotting wood, in which case the radiocarbon activity in the rats may have been depleted to levels approaching the age of the old logs—possibly a thousand or thousands of years older than the rats.

Thus, in the case of the NZ rat bones, the radiocarbon dating procedure was in fact dating these ancient NZ logs and not the rats which, 1000 or so years later, arrived with the first New Zealanders and set about eating the insects which were eating the rotten old logs! All this in spite of the extreme care taken in avoiding any possible contamination of the samples.

2.5   Lake pollution models

overview   First we consider the simple example of the concentration of salt in a tank of water. We then apply this method to model the changing concentration of pollution in a lake system.

Background

predicting pollution change over time

Pollution in our lakes and rivers has become a major problem particularly over the last 50 years. In order to improve this situation in the future it is necessary to gain a good understanding of the processes involved. Some way of predicting how the situation may improve (or decline) as a result of current management practices is vital. To this end we need to be able to predict how pollutant amounts or concentrations vary over time and under different management strategies.

General compartmental model

single compartment

This problem can be considered as a compartmental model with a single compartment, the lake, as is illustrated in Figure 2.5.1. Applying the balance law there is an input of polluted water from the river(s) flowing into the lake, or due to a pollution dump into the lake, and an output as water flows from the lake carrying some pollution with it.

Figure 2.5.1: Input–output compartmental diagram for pollution in a lake.

word equation

This leads us to the word equation, for the mass of pollutant in the lake,

$$\left\{ \begin{array}{c} rate\ of\ change \\ of\ mass \\ in\ lake \end{array} \right\} = \left\{ \begin{array}{c} rate \\ mass \\ enters\ lake \end{array} \right\} - \left\{ \begin{array}{c} rate \\ mass \\ leaves\ lake \end{array} \right\}. \tag{1}$$

Before developing the differential equations describing this process, we examine the simple model of a salt solution in a tank.

An example problem of salt dissolved in a tank

(The following example is adapted from Borelli and Coleman (1996).) example
As an example problem involving concentrations let us consider how to problem
model the concentration of salt dissolved in a tank of water. The solution
of water and salt is sometimes called brine. A large tank contains 100
litres of salt water. Initially s_0 kg of salt is dissolved. Salt water flows
into the tank at the rate of 10 litres per minute, and the concentration
$c_{in}(t)$ (kg of salt/litre) of this incoming water-salt mixture varies with
time. We assume that the solution in the tank is thoroughly mixed and
that the salt solution flows out at the same rate at which it flows in: that
is, the volume of water–salt mixture in the tank remains constant.

Example 1: *Find a differential equation for the amount of salt in the tank
at any time t. (Note that concentration can be defined as the mass of salt per
unit volume of mixture.)*

Solution: *The tank is the 'compartment'. Let $S(t)$ be the amount (mass)
of salt in the tank at time t. $S(t)$ is what we want to find. First we draw a
diagram of the situation as in Figure 2.5.2.*

Figure 2.5.2: Input–output compartmental diagram for salt solution in tank.

*Then, applying the balance law to the mass of salt, this can be described by
the word equation*

$$
\left\{\begin{array}{c} \text{rate of change} \\ \text{of mass of salt} \\ \text{in tank} \end{array}\right\} = \left\{\begin{array}{c} \text{rate} \\ \text{mass of salt} \\ \text{enters tank} \end{array}\right\} - \left\{\begin{array}{c} \text{rate} \\ \text{mass of salt} \\ \text{leaves tank} \end{array}\right\}. \tag{2}
$$

*The rate at which the mass of salt is added is the flow rate 10 (litres) multiplied
by the concentration of the incoming mixture $c_{in}(t)$ (in kg/litre),*

$$
\left\{\begin{array}{c} \text{rate} \\ \text{salt} \\ \text{enters tank} \end{array}\right\} = 10c_{in}(t).
$$

*The rate at which the salt is leaving the tank is the flow rate out multiplied
by the concentration of salt in the tank, $S(t)/100$, so*

$$
\left\{\begin{array}{c} \text{rate} \\ \text{salt} \\ \text{leaves tank} \end{array}\right\} = 10 \times \frac{S(t)}{100} = \frac{S(t)}{10},
$$

again measured in kg/min.

So the differential equation describing the rate of change in the mass of salt in the tank is given by

$$\frac{dS}{dt} = 10c_{\text{in}}(t) - \frac{S(t)}{10}.$$

solving the IVP To solve the differential equation we also need an initial condition. At time $t = 0$ we assume the amount of salt dissolved in the tank is s_0 (measured in kg). Hence we are required to solve IVP

$$\frac{dS}{dt} = 10c_{\text{in}}(t) - \frac{1}{10}S(t), \qquad S(0) = s_0. \tag{3}$$

This is a linear, first-order ODE (ordinary differential equation). It is not a separable equation, but we can solve it using the technique of integrating factors (see Appendix A.4 on integrating factors). We obtain a solution for an arbitrary function for input concentration $c_{\text{in}}(t)$ and then later investigate different functional forms of $c_{\text{in}}(t)$.

Example 2:   *Using the technique of integrating factors, solve the IVP* (3) *on the interval* $[0, t]$, *in terms of an arbitrary function* $c_{\text{in}}(t)$.

Solution: *Multiplying equation* (3) *by the integrating factor (IF)* $e^{\int_0^t \frac{1}{10}\, ds} = e^{(t-0)/10} = e^{t/10}$ *we get*

$$\frac{d}{dt}\left(S(t) \times e^{t/10}\right) = 10c_{\text{in}}(t)e^{t/10},$$

whence, integrating over $[0, t]$, *gives*

$$S(t)e^{t/10} - s_0 e^{0/10} = 10\int_0^t c_{\text{in}}(s)e^{s/10}\, ds.$$

Hence,

$$S(t) = s_0 e^{-t/10} + 10e^{-t/10}\int_0^t c_{\text{in}}(s)e^{s/10}\, ds. \tag{4}$$

definite or indefinite integrals Note that the initial condition $S(0) = s_0$ was included when applying the boundaries of the given interval of integration. Alternatively, for the more general case of indefinite integrals see Appendix A.4.

specifying the input concentration The original question was: How much salt is in the tank at time t? To answer this we need to know $c_{\text{in}}(t)$, the concentration of the incoming salt solution. Suppose $c_{\text{in}}(t) = c_1$, a constant, then (4) gives

$$S(t) = s_0 e^{-t/10} + 100c_1(1 - e^{-t/10}). \tag{5}$$

Suppose, instead, c_{in} is a sinusoidal function, say $c_{in}(t) = 0.2 + 0.1 \sin t$, then evaluating the integral (either by tables or by using `Maple`) gives

$$S(t) = s_0 e^{-t/10} + 20 + \frac{10}{101}(\sin t - 10 \cos t + 192 e^{-t/10}). \qquad (6)$$

Note that in equations (4), (5) and (6), as $t \to \infty$ the effect of the initial condition on the solution decreases and becomes negligible. Also, these solutions have two parts: one which is the response to the initial data and one which is the response to the input. We discuss this further in Section 2.10 where we cover some general theory about first-order linear differential equations. *(two parts to the solution)*

A lake pollution model

We return now to the problem of pollution in a lake. We apply the above theory to investigate the changing concentration of a pollutant. *(lake pollution)*

As usual we need to make some assumptions while developing the model and to this end we assume that the lake has a constant volume V, and that it is continuously well mixed so that the pollution is uniform throughout. *(assumptions)*

Let $C(t)$ be the concentration of the pollutant in the lake at time t. Let F be the rate at which water flows out of the lake in m^3/day. Since the volume is constant we have *(word equation)*

$$\left\{ \begin{array}{c} flow \ of \\ mixture \\ into \ lake \end{array} \right\} = \left\{ \begin{array}{c} flow \ of \\ mixture \\ out \ of \ lake \end{array} \right\} = F.$$

Applying the balance law to the mass of the pollutant $M(t)$ we can describe the process in words

$$\left\{ \begin{array}{c} rate \ of \\ change \ of \\ mass \ of \\ pollutant \\ in \ lake \end{array} \right\} = \left\{ \begin{array}{c} rate \ at \\ which \ the \\ pollutant \ enters \\ the \ lake \end{array} \right\} - \left\{ \begin{array}{c} rate \ at \\ which \ the \\ pollutant \ leaves \\ the \ lake \end{array} \right\}.$$

This translates into the differential equation for the changing mass *(formulating the equation)*

$$M'(t) = F c_{in} - F \frac{M(t)}{V}, \qquad (7)$$

where c_{in} is the concentration (in units of mass per unit of volume, such as g/m^3) of the pollutant in the flow entering the lake.

the initial value problem

Now, since $M(t) = C(t)V$ we have that $M'(t) = C'(t)V$ (since V is constant) and hence that $C'(t) = M'(t)/V$. With this change of variable the differential equation (7) for the mass is transformed to a differential for the concentration of the pollutant in the lake,

$$\frac{dC}{dt} = \frac{F}{V}c_{in} - \frac{F}{V}C. \qquad (8)$$

solving the equation

If the flow rate F is constant with time then we can use the technique of separating the variables to solve the equation, as in the following example.

Example 3:   Solve the differential equation (8) with the initial condition $C(0) = c_0$.

Solution: Separating the variables,

$$\int \frac{1}{c_{in} - C} dC = \int \frac{F}{V} dt.$$

Integrating gives

$$-\ell n\,|c_{in} - C| = \frac{F}{V}t + K \quad (K \text{ an arbitrary constant}),$$

and thus

$$C(t) = c_{in} - e^{-K}e^{-Ft/V}.$$

Using the initial condition $C(0) = c_0$, we obtain

$$C(t) = c_{in} - (c_{in} - c_0)e^{-Ft/V}. \qquad (9)$$

two parts to the solution

Note again that the solution can be divided into two parts: $c_0e^{-Ft/V}$ which is the contribution from the initial data, and $(c_{in} - c_{in}e^{-Ft/V})$ which is the contribution from the pollution inflow to the system. Also note that as $t \to \infty$, then $C(t) \to c_{in}$. That is,

$$\lim_{t\to\infty}(c_{in} - (c_{in} - c_0)e^{-Ft/V}) = c_{in}$$

and the concentration in the lake increases/decreases steadily to this limit.

Maple solution

Maple could have been used to establish the solution and plot the results. To this end the following code provides the time-dependent plot for a variety of initial pollution concentrations where the safety threshold for the level of pollution is illustrated with a line:

```
> restart: with(plots):
> cin:=3;V:=28;F:=4*12;threshold:=4;init_c:=10;
> de1:=diff(c(t),t)=(F/V)*(cin-c(t));
> soln:=c0->dsolve({de1,c(0)=c0},c(t),numeric):
> plot1:=c0->odeplot(soln(c0),[t,c(t)],0..8):
> list1:=seq(plot1(i/2),i=1..12):
> line1:=plot([[0,threshold],[8,threshold]]):
> display({list1,line1});
```

(See Appendix C for a screen dump of similar code.)

Suppose we start with a polluted lake, with initial concentration c_0, but subsequently only fresh water enters the lake. We are interested in calculating the time for the pollution level in the lake to reduce to a specified value. *using the solution*

Example 4: *How long will it take for the lake's pollution level to reach 5% of its initial level, if only fresh water flows into the lake.*

Solution: *From the solution* (9), *and setting* $c_{in} = 0$, *we have*

$$C(t) = c_0 e^{-Ft/V}. \tag{10}$$

Rearranging to solve for t,

$$t = -\frac{V}{F} \ell n \left(\frac{C}{c_0} \right).$$

Since $C = 0.05c_0$, *we find* t *given by*

$$t = -\ell n(0.05) \times \frac{V}{F} \simeq \frac{3V}{F}.$$

Consider two American lakes where Lake Erie flows into Lake Ontario (Mesterton-Gibbons (1989)). Lake Erie has $V = 458 \times 10^9 \, \text{m}^3$ and $F = 480 \times 10^6 \, \text{m}^3/\text{day}$. Thus if $t_{0.05}$ is the time it takes for the pollution level to reach 5% of its current level, then from above $t_{0.05} \approx 7.8$ years. Lake Ontario has $V = 1636 \times 10^9 \, \text{m}^3$, $F = 572 \times 10^6 \, \text{m}^3/\text{day}$ and $t_{0.05} \approx 23.5$ years. *application to American lakes*

Maple provides a method of substitution which can be used to find this time. The following code fragment, with $c_0 = 10$ in the above example, is appropriate: *Maple solution*

```
> restart:
> de1:=diff(C(t),t)=(F/V)*(cin-C(t));
> soln:=c0->dsolve({de1,C(0)=c0},C(t)):
> cin:=0;c0:=10:
> ans:=soln(c0):
> get_t:=solve(subs(C(t)=(0.05*c0),ans),t):
> evalf(get_t);
```

some
limitations
We have made some assumptions in this model. First, we assumed the lake to be well mixed. However, if the lake were not well mixed then the time taken to flush the lake is likely to be longer since poor mixing will prolong the process. We have also assumed the flow rate F to be constant.

seasonal
flow rate
Suppose that the flow rate into and out of the lake, $F(t)$, varied seasonally. We could let $F(t) = F_0(1 + \epsilon \cos(2\pi t))$ where F_0 is the mean in/out flow and $|\epsilon| < 1$ so that $F \geq 0$. Let us look at the equation again:

$$\frac{dC}{dt} = -\frac{F_0}{V}\left(1 + \epsilon \cos(2\pi t)\right) C.$$

Separating the variables and integrating gives

$$\ln |C(t)| = \frac{-F_0}{V}t + \frac{F_0}{V}\frac{\epsilon}{2\pi}\sin(2\pi t) + K,$$

with K some arbitrary constant, and therefore

$$C(t) = c_0 e^{-F_0(t - \frac{\epsilon}{2\pi}\sin(2\pi t))/V}.$$

Since $\sin(2\pi t) \leq 1$, and ϵ is small, we have

$$C(t) \leq c_0 e^{-F_0(t - \frac{\epsilon}{2\pi})/V}$$
$$\leq c_0 e^{-F_0(t - \frac{1}{2\pi})/V}. \tag{11}$$

For large t, there is little difference between the predictions from this solution and that of the simpler model as given in (10), concerning the time taken for the lake to purify. Thus, from (11) and (10), we can be reasonably sure that the times we estimated applying the previous very simple model, will provide a lower bound for the purification time.

further
extensions
This model can easily be extended to encompass a system of lakes (one lake flowing into another) by setting the inflow in the model of the downstream lake equal to the outflow of the upstream lake. Clearly, there may be several lakes/rivers feeding into the downstream lake at any time, and a single upstream lake may only be a fraction of the flow feeding the lower lake.

Skills developed here:

- Apply the concept of modelling with compartments and the balance law.
- Extend the model developed in this section to a system of lakes, or the addition of further pollution sources into a single lake.
- Sketch graphs of pollution levels over time.
- Generate solutions, using `Maple` or analytical methods, to the model and its extensions.

2.6 Case Study: Lake Burley Griffin

The information for the case study is adapted from the research paper Burgess and Olive (1975).

===

Lake Burley Griffin in Canberra, the capital city of Australia, was created artificially in 1962 for both recreational and aesthetic purposes. In 1974 the public health authorities indicated that pollution standards set down for safe recreational use were being violated and that this was attributable to the sewage works in Queanbeyan upstream (or rather the discharge of untreated sewage into the lake's feeder river).

After extensive measurements of pollution levels taken in the 1970s it was established that, while the sewage plants (of which there are three above the lake) certainly exacerbated the problem, there were significant contributions from rural and urban runoff as well, particularly during summer rainstorms. These contributed to dramatic increases in pollution levels and at times were totally responsible for lifting the concentration levels above the safety limits. As a point of interest Queanbeyan (where the sewage plants are situated) is in the state of New South Wales (NSW), while the lake is in the Australian Capital Territory, and although they are a ten-minute drive apart the safety levels/standards for those who swim in NSW are different from the standards for those who swim in the Capital Territory!

In 1974 the mean concentration of the bacteria faecal coliform count was approximately $10^7 \, m^{-3}$ at the point where the river feeds into the lake. The safety threshold for this faecal coliform count in the water is such that for contact recreational sports no more than 10% of total samples over a 30-day *period should exceed* $4 \times 10^6\,\mathrm{m}^{-3}$. *(That is, if you are in the Capital Territory and not NSW!)*

Given that the lake was polluted it is of interest to examine how, if sewage management were improved, the lake would flush out and if and when the pollution levels would drop below the safety threshold.

The system can be modelled, very simply, under a few assumptions. Flow (F) into the lake is assumed equal to flow out of the lake and the volume (V) of the lake will be considered constant and is approximately $28 \times 10^6\,\mathrm{m}^3$. *Further, the lake can be considered as well mixed in the sense that the pollution concentration throughout will be taken as constant. Under these assumptions a suitable differential equation model for the pollutant concentration is*

$$\frac{dC}{dt} = \frac{F}{V}c_{\text{in}} - \frac{F}{V}C, \tag{1}$$

where c_{in} *is the concentration of the pollutant entering the lake. With the initial concentration taken as* c_0 *the solution is*

$$C(t) = c_{\text{in}} - (c_{\text{in}} - c_0)e^{-Ft/V}. \tag{2}$$

With only fresh water entering the lake $(c_{\text{in}} = 0)$, *with a mean monthly flow of* $4 \times 10^6\mathrm{m}^3$ *and with the initial faecal coliform count of* 10^7 *parts/m*^{3} *(as was measured in 1974) the lake will take approximately 6 months for the pollution level to drop below the safety threshold. However, pure water entering the lake is not a very realistic scenario with three sewage plants and much farmland upstream, and so including the entrance of polluted river water into the lake model is essential.*

From the solution (2), as time increases so the concentration of a pollutant in the lake will approach the concentration of the polluted water entering the lake. This level is independent of the initial pollution level in the lake and if $c_0 > c_{\text{in}}$ *then the level of pollution in the lake decreases monotonically to* c_{in}, *while if* $c_0 < c_{\text{in}}$ *then the level increases steadily until it reaches* c_{in}. *Thus with the faecal coliform entering the lake at a count of* 3×10^6*parts/m*^{3} *the concentration of the pollutant in the lake will approach this level with time. This is evident in Figure 2.6.1, where* Maple *has been used to solve the differential equation (1) and plot the results, for a number of different initial concentrations.*

However, the model is still simplistic. Flow rates change over the year with on average a seasonal pattern, while the amount of pollution reaching the lake will itself be seasonal. Assuming a sinusoidal pattern over the year, a rough estimate of the concentration from the available data in the 1970s is taken as $c_{\text{in}}(t) = 10^6(10+10\cos(2\pi t))$ *and for the flow,* $F(t) = 10^6(1+6\sin(2\pi t))$. *Note here that the concentration of the pollutant increases when the flow is low. We need to re-solve the differential equation (1) with* F *no longer constant.*

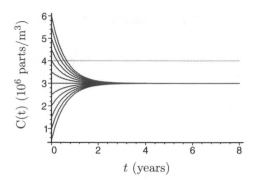

Figure 2.6.1: Pollution levels in Lake Burley Griffin over time with constant rates of flow and pollution. A range of initial concentrations has been used. The grey line is the pollution threshold level.

Using `Maple` *to determine the solution and plot the concentration of pollution in the lake over time, the results are displayed in Figure 2.6.2.*

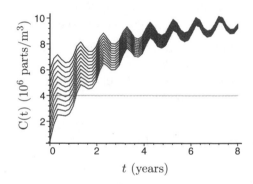

Figure 2.6.2: Pollution levels in Lake Burley Griffin over time with seasonal incoming flows and pollutant concentrations. A range of initial concentrations has been used. The grey line is the pollution threshold level.

It should be noted that this model for Lake Burley Griffin is still simplistic in its assumption of a well mixed body of water. If the concentration decreased from the point of river entry to the point of outflow then the flushing time could take considerably longer. Further, in most lakes there is a main channel of water flow which flushes regularly, and adjacent to this channel are areas of trapped water which flush less frequently and through a very different process. The process is that of diffusion which operates at a microscopic level and is extremely gradual. Thus pockets of the lake may have a much higher (or lower) pollution concentration than others and these may also be the protected bays where swimming is most likely to take place.

2.7   Drug assimilation into the blood

overview   We investigate two simple models of cold pill assimilation into the blood-
            stream which are adapted from Borelli and Coleman (1996). In the first
            model we consider a single cold pill and in the second a course of cold
            pills. The basic idea of using a compartmental diagram to build the
            models is used.

Background

drug   We readily take pills or drink alcohol, without necessarily having a good
warnings   understanding of how these drugs are absorbed into the bloodstream or
            for how long they have an effect on us. The warnings on the packaging list
            some of the effects and are intended to ensure safety for all users. In the
            following example and case study we see that different drugs are absorbed
            into, and extracted from, the blood at very different rates. Some may
            affect us for many hours after the medication has ceased. Further, in the
            case study on alcohol absorption, we see how different body mass and
            the sex of an individual can radically modify the effects of alcohol.

what   The drug dissolves in the gastrointestinal tract (GI-tract) and each ingre-
happens   dient is diffused into the bloodstream. They are carried to the locations
to the
drug   in which they act and are removed from the blood by the kidneys and
            the liver. The assimilation and removal may occur at different rates for
            the different ingredients of the same pill.

General compartmental model

a two   We can consider this problem as a compartmental model with two com-
compart-   partments, corresponding to the GI-tract and the bloodstream. This is
mental
model   illustrated in Figure 2.7.1. The GI-tract compartment has a single input
            and output and the bloodstream compartment has a single input and
            output.

Figure 2.7.1: Input–output compartmental diagram for drug assimilation.

This application of the balance law gives us two word equation, one for each compartment. We have

$$\left\{\begin{array}{c} rate\ of\ change \\ of\ drug \\ in\ GI\ tract \end{array}\right\} = \left\{\begin{array}{c} rate\ of \\ drug \\ intake \end{array}\right\} - \left\{\begin{array}{c} rate \\ drug \\ leaves\ GI\text{-}tract \end{array}\right\},$$

$$\left\{\begin{array}{c} rate\ of\ change \\ of\ drug \\ in\ blood \end{array}\right\} = \left\{\begin{array}{c} rate \\ drug \\ enters\ blood \end{array}\right\} - \left\{\begin{array}{c} rate \\ drug \\ leaves\ blood \end{array}\right\}. \tag{1}$$

We let $x(t)$ be the amount of a drug in the GI-tract at time t and $y(t)$ the amount in the bloodstream at time t. We consider two models: a single cold pill where there is no ingestion of the drug except that which occurs initially, and a course of cold pills where the drug intake is assumed to occur continuously.

The common cold remains without a cure. However, there are pills which   cold pills
can be taken to relieve some of the congestion and symptoms, such as watering eyes and a running nose, through the action of a decongestant and an antihistamine. The cold pills we consider in the following two models consist of these two drugs.

Model I: Case of a single cold pill

In the GI-tract we consider the pill to have been swallowed, and so after   formulat-
this event (over subsequent time) we have nothing more entering the   ing the
GI-tract. The pill dissolves and diffuses into the bloodstream from the   equations
GI-tract. So, for the GI-tract there is only an output term. Assuming the output rate is proportional to the GI-tract drug concentration $x(t)$, then

$$\frac{dx}{dt} = -k_1 x, \qquad x(0) = x_0, \tag{2}$$

where x_0 is the amount of a drug in the pill, our initial condition, and $k_1 > 0$ is the coefficient of proportionality (the rate constant).

In the bloodstream the initial amount of the drug is zero. The level increases as the drug diffuses from the GI-tract and decreases as the kidneys and liver remove it. Thus

$$\frac{dy}{dt} = k_1 x - k_2 y, \qquad y(0) = 0, \tag{3}$$

with $k_2 > 0$. The cold pill is made up of a decongestant and an antihistamine, and the coefficients of proportionality, k_1 and k_2, are different for the different component drugs in the pill.

solving the differential equations

The differential equations (2) and (3) are not coupled because the first equation is totally independent of y. So we can solve it independently of the second equation. The solution to this first differential equation is (see Exercises, Question 10)

$$x = x_0 e^{-k_1 t}.$$

Using this solution the second differential equation (3) becomes

$$\frac{dy}{dt} = k_1 x_0 e^{-k_1 t} - k_2 y.$$

This equation is a first-order linear differential equation whose solution when $k_1 \neq k_2$ is (see Exercises, Question 10)

$$y = \frac{k_1 x_0}{k_1 - k_2} \left(e^{-k_2 t} - e^{-k_1 t} \right).$$

(Note that a different solution is valid when $k_1 = k_2$.)

interpretation

As expected, as t increases both x and y approach zero although the rate at which this occurs depends on the coefficients k_1 and k_2 associated with each drug. Some values are given in Table 2.7.1 from Borelli and Coleman (1996).

Table 2.7.1: Values of constants for decongestant and antihistamine in cold pills.

	Decongestant	Antihistamine
k_1	1.3860/hr	0.6931/hr
k_2	0.1386/hr	0.0231/hr

Using these values we can graph the changing values for x and y in each of the two 'compartments'. With `Maple`, we can easily explore the effects of the antihistamine in the bloodstream as k_2 increases/decreases and determine how sensitive the system is to changes in the parameter k_2. See Figure 2.7.2 with the associated `Maple` code:

```
> restart: with(plots): with(DEtools):
> k1:=1.386; k2:=0.1386; k3:=0.6931; k4:=0.0231;
    I1:=0; I2:=0; hours:=15; initx:=1;inity:=0;
> de1 := diff(x(t),t)=I1-k1*x(t);
> de2 := diff(y(t),t)=k1*x(t)-k2*y(t);
> pair1 := [de1,de2]:
> plot1 := DEplot(pair1, [x,y], t=0..hours, {[0,initx,inity]},
    stepsize=0.1, scene=[t,x], linecolor=gray, arrows=NONE):
```

```
> plot2 := DEplot(pair1, [x,y], t=0..hours, {[0,initx,inity]},
     stepsize=0.1, scene=[t,y], linecolor=gray, arrows=NONE):
> de3 := diff(w(t),t)=I1-k3*w(t);
> de4 := diff(z(t),t)=k3*w(t)-k4*z(t);
> pair2 := [de3,de4]:
> plot3 := DEplot(pair2,[w,z], t=0..hours, {[0,initx,inity]},
     stepsize=0.1, scene=[t,w], linecolor=black, arrows=NONE):
> plot4:=DEplot(pair2,[w,z],t=0..hours, {[0,initx,inity]},
     stepsize=0.1, scene=[t,z], linecolor=black, arrows=NONE):
> display(plot1,plot3);display(plot2,plot4);
```

Figure 2.7.2: Single cold pill. The graph on the left represents the GI-tract and that on the right the bloodstream. The levels of antihistamine are illustrated with black lines ($k_2 = 0.0231$ thick line, $k_2 = 0.08$ thin line) and the decongestant with a grey line. The other parameter values are given in Table 2.7.1, with typical initial values for the antihistamine and decongestant, $4\,\text{mg}$ and $60\,\text{mg}$.

The coefficients also depend on the age and health of the person involved, and the concentration of a drug may also depend on the persons' body mass, which means that for some people the doses may peak higher and/or faster than for the average person, and of course this may be potentially dangerous.

Model II: Case of a course of cold pills

In reality, particularly for a cold, we take a course of pills rather than just one. In this way there is a continuous flow of drugs into the GI-tract. We adjust our model of the previous section to give

$$\frac{dx}{dt} = I - k_1 x, \qquad x(0) = 0,$$
$$\frac{dy}{dl} = k_1 x - k_2 y, \qquad y(0) = 0,$$

(4)

formulating the equations

where I is a positive constant representing the rate of ingestion of the drug.

the
solutions

Once again we have a sequence (cascade) of linear equations which are not coupled and thus can be solved sequentially. For

$$\frac{dx}{dt} = I - k_1 x$$

the solution is (see Exercises, Question 11)

$$x(t) = \frac{I}{k_1}(1 - e^{-k_1 t}).$$

Then solving for $y(t)$ in the second equation gives (see Exercises, Question 11)

$$y(t) = \frac{I}{k_2}\left[1 - \frac{1}{k_2 - k_1}\left(k_2 e^{-k_1 t} - k_1 e^{-k_2 t}\right)\right].$$

(Note again that this solution is valid only if $k_1 \neq k_2$.)

solutions
over time

As usual we are interested in what happens over time; in this case as $t \to \infty$ then $x \to I/k_1$ and $y \to I/k_2$. Now we draw the graphs of these functions to establish the accumulating levels of the decongestant and antihistamine in the GI-tract and the bloodstream over a time period where pills are taken each hour and the value of I is constant.

Figure 2.7.3: Course of cold pills. The graph on the left represents the GI-tract and that on the right the bloodstream. The levels of antihistamine are illustrated with black lines and the decongestant with a grey line. The parameter values are given in Table 2.7.1, with typical initial values for the antihistamine and decongestant, 0.5 mg and 15 mg.

interpret-
ation of
graph

Note that the level of the antihistamine (which makes you sleepy) is slow to rise, but settles to a high level, while that of the decongestant is faster to rise but settles to a lesser level with time. This is illustrated in Figure 2.7.3. (The Maple code is as provided above but with the initial condition zero and I constant and non-zero.)

Our assumption of I constant holds in the case where drugs are embedded   extension
in resins which dissolve at varying rates, allowing the drug to be released
slowly and evenly over a period of hours. In reality some pills dissolve
quickly and thus $I(t)$ should be a pulsing function, possibly a sinusoidal
function, or some other function of t providing an initial and substantial
boost to the drug level and then very little during the remaining time
period before the next dose is taken. We do not consider those cases
here, but what is required in the model is the replacement of I with an
appropriate time dependent function.

Skills developed here:

- Be able to model a sequence of processes using the compartmental technique.
- Understand how to find solutions, with `Maple` and analytically, in each of the sequence of compartments.
- Sketch graphs of solutions in each compartment over time.
- Establish how changes in parameters impact on these solutions.

2.8 *Case Study: Dull, dizzy or dead?*

This case study has been adapted from Oakley and Ksir (1993).

Alcohol is unusual in that it requires no digestion and can be absorbed ex-
tremely rapidly from the stomach into the bloodstream (particularly if the
stomach is empty of food or other liquids), and diffuses even more rapidly
from the intestines. The vapour can also be absorbed through the lungs. Thus,
unless alcohol is heavily diluted or taken with food, very little metabolism oc-
curs in the GI-tract (gastrointestinal tract) and all the alcohol diffuses into
the bloodstream. Alcohol is distributed freely to all body fluids and the con-
centration of alcohol in the brain rapidly approaches that in the blood. Most
of the alcohol (90–98%) is oxidised through the liver and excreted while the
remainder leaves the body through the lungs, urine, saliva and sweat.

The state of drunkenness (or 'happy feeling') experienced, is a measure of
the blood alcohol level (BAL) or the alcohol concentration in the blood. It
is measured in grams/100 ml of blood. Thus 100 g/100 ml corresponds to

a BAL of 100 (grams/100 ml), and 100 mg/100 ml measures a BAL of 0.1 (grams/100 ml). As is common practice we will not stipulate the units of BAL on each occurrence, but assume that grams/(100 ml blood) is understood.

Australian law prohibits the driving of vehicles (including boats and horse or camel-drawn vehicles) for those with a BAL above 0.05. This then relates to 50 mg/100 ml alcohol in the bloodstream. The restriction is a result of U.S. statistics which indicate that a person with a BAL of 0.15 is 25 times more likely to have a fatal accident than one with no alcohol. Furthermore, for 41% of Australian men alcohol leads to confrontational behaviour. (For women the figure is 12%.)

Table 2.8.1: Blood alcohol level (BAL in g/(100 ml blood)) and behavioural effects.

BAL	Behavioural effects	
0.05	Lowered alertness, usually good feeling, release of inhibitions, impaired judgement	Dull and dignified
0.10	Slowed reaction times and impaired motor function, less caution	Dangerous and devilish
0.15	Large consistent increases in reaction time	Dizzy
0.20	Marked depression in sensory and motor capability, decidedly intoxicated	Disturbing
0.25	Severe motor disturbance, staggering, sensory perceptions greatly impaired, smashed!	Disgusting and dishevelled
0.30	Stuporous but conscious, no comprehension of what's going on	Delirious and disoriented
0.35	Surgical anaesthesia; minimal level causing death	Dead drunk.
0.40	50 times the minimal level; causing death	Dead!

The alcohol intake into the GI-tract is 'controlled' by the drinker. The amount of alcohol subsequently absorbed into the bloodstream depends on the concentration of alcohol, other liquids and food in the GI-tract, as well as on the weight and sex of the individual. Alcohol is removed from the bloodstream at a constant rate by the liver. This is independent of the body weight, sex of the individual and concentration of alcohol in the bloodstream and assumes that the liver has not been damaged by large doses of alcohol. (As mentioned above *a small percentage leaves through sweat, saliva, breath and urine, however, ignoring this the BAL estimate may be slightly above the true value.)*

Table 2.8.2: Effective alcohol intake during drinking.

Absolute alcohol content	Beverage intake	Number of drinks
14 g	1 oz spirits, or 1 glass wine, or 1 can beer	$n = 1$
28 g	2 oz spirits, or 2 glasses of wine, or 2 cans of beer	$n = 2$
$(n \times 14)$ g	n oz spirits, or n glasses of wine, or n cans of beer	n

Let $x(t)$ be the amount of alcohol (effective BAL) in the GI-tract at time t and let $y(t)$ be the BAL in the bloodstream at time t. Applying the balance law, the rates of change of alcohol in the GI-tract and the bloodstream can be modelled as

$$\frac{dx}{dt} = I - k_1 x,$$
$$\frac{dy}{dt} = k_2 x - \frac{k_3 y}{y + M}, \tag{1}$$

with constants k_1, k_2 and M positive, and I the amount of alcohol consumed per time step (here taken as one hour).

In the case of drinking on an empty stomach $k_1 = k_2$, while if drinking occurs together with a meal (or is diluted) then $k_1 > k_2$. Note that the amount of alcohol leaving the GI-tract, and the amount entering the bloodstream, are directly proportional to the amount present in the GI-tract.

On the other hand the amount of alcohol metabolised through the liver is represented as a more complicated function, a Michaelis–Menten type function (Michaelis and Menten (1913)). As was mentioned above the amount removed by the liver is constant regardless of the amount of alcohol or its concentration.

This has been modelled with the function $-k_3 y/(y + M)$ for the following reason. If y is large compared with M then $y' \simeq -k_3$ and y is decreasing at the constant rate k_3. Alternatively, as y decreases and becomes small compared with M, then $y' \simeq -k_3 y/M$ and y is an exponential function ($y \simeq y_0 e^{-k_3 t/M}$) which ensures that y remains positive. An appropriate value for M in this case is $M = 0.005$ in units of BAL.

What follows are model runs for cases where drinking occurred on an empty stomach or with a meal, and where drinking occurred within a short period or continued over several hours. In each case Maple *was used to solve the equations and display the graphics.*

Drinking on an empty stomach:　*In this case the alcohol is modelled as being very rapidly absorbed into the bloodstream in the first hour, that is $k_1 = k_2 = 6$, and the intent is to model the increasing/decreasing BAL in subsequent hours. In the case where n initial drinks are consumed and no more alcohol is taken we have $x(0) = x_0$, where x_0 is the effective BAL from the initial drink(s) consumed, and $I = 0$ as there is no subsequent drinking in this case. In the bloodstream, the initial amount of alcohol is 0 (that is, there is no alcohol in the bloodstream prior to drinking) and so $y(0) = y_0 = 0$. The system above becomes*

$$
\begin{aligned}
\frac{dx}{dt} &= -k_1 x, \qquad x(0) = x_0 \\
\frac{dy}{dt} &= k_2 x - \frac{k_3 y}{y + M}, \qquad y(0) = 0.
\end{aligned}
\tag{2}
$$

The initial value x_0 is given by $x_0 = Cn$ where n is the number of drinks as listed in Table 2.8.2 and C is the effective BAL produced by a single standard drink. The value of C depends on the sex and weight of the iindividual, since they dictate the total amount of body fluid for that person. One drink produces 14 g effective alcohol. The total volume of body fluids in a woman is approximately $0.67 \times W$ litres, where W is her body weight in kg. (For a man the volume of fluids is approximately $0.82 \times W$ litres.) The concentration of alcohol will be (amount of alcohol/total fluids in grams per litre) and this can be expressed in g/100 ml, that is, BAL.

So for a man of 68 kg who has 3 glasses of wine rapidly and then stops drinking, $x_0 = 3 \times C$. His total body fluids are $\approx 0.82 \times 68 = 55.76$litres. Then the concentration of alcohol per drink is given as $C = 14/55.76$ g/litre which converts to an effective BAL of 0.025. Thus $x_0 = 3C = 0.075$.

Further, k_3 is a measure of the rate at which the liver removes alcohol from the blood, and this value is the same for all persons and is approximately 8g/hr. The associated reduction in BAL depends on the total body fluids of the individual, and for the 68 kg man can be calculated as $8/(55 \times 10) = 0.014$ g/100 ml. Then $k_3 = 0.014$ BAL.

For the case of drinking continuously over time we take $I = nC$ and $x_0 = 0$ in the differential equations (1) where n is the number of drinks consumed each hour. The results are illustrated in Figure 2.8.1 for the 68 kg male. (The constants are as described above.)

Drinking with a substantial meal:　*The rate at which alcohol is absorbed into the bloodstream from the GI-tract is substantially reduced when alcohol*

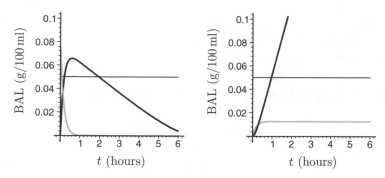

Figure 2.8.1: The BAL of a 68 kg man drinking on an empty stomach, with $n = 3$ drinks per hour. The first diagram is for a single drinking bout, and the second for a continuous drinking binge. The black line illustrates the BAL in the bloodstream and the grey line the level in the GI-tract. The legal limit BAL 0.05 is also included.

is diluted or taken with a meal. The presence of fats in particular reduces the resulting BAL. Returning to equation system (2) above we now have that $k_1 > k_2$, that is, the rate at which alcohol leaves the GI-tract is greater than the rate at which alcohol enters the bloodstream. In fact, after a substantial meal the rate of absorption into the bloodstream is approximately halved so that we can replace k_2 with $k_1/2$. (This is a very simplistic approximation and the model would be improved by including a more realistic mathematical relationship between food intake and alcohol absorption into the bloodstream.) The graphs of Figure 2.8.2 illustrate the effect of a meal on the BAL for the 68 kg man in the case of an initial alcohol intake without further drinking, and in the case of continuous drinking.

Figure 2.8.2: The BAL of a 68 kg man drinking after a meal, with $n = 3$ drinks per hour. The first diagram is for a single drinking bout, and the second for a continuous drinking binge. The black line illustrates the BAL in the bloodstream and the grey line the level in the GI-tract. The legal limit BAL 0.05 is also included.

As for the aftermath of all this? Well the Germans hear the 'wailing of cats', the Italians call it 'out of tune', the French refer to it as 'woody mouth', the

Norwegians say there are 'workmen in my head', the Swedes have 'pain in the roots of my hair' and the English suffer the hangover.

2.9   *Cascades of compartments*

As exemplified in the previous section(s), linear cascades are models with a series of compartments each leading to a new one without any loops as in the case of drug assimilation or a series of lakes.

 linear cascade

In other cases two sources can feed into one, or one into many compartments; such an arrangement is called a branching linear cascade.

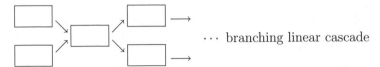 ⋯ branching linear cascade

Usually in these cases the equations can be solved using the top-down approach which we followed in Section 2.7. That is, solving the first equation and using this solution to solve the second, and so on. Such models are also applicable to the decay of radioactive elements such as Uranium-238 which decays to Lead-206 in a decay cascade involving 16 intermediate elements and isotopes (see Borelli and Coleman (1996)):

$$x'_1 = -k_1 x_1$$
$$x'_2 = k_1 x_1 - k_2 x_2$$
$$\vdots$$
$$x'_{17} = k_{18} x_{15} - k_{19} x_{16}$$
$$x'_{18} = k_{19} x_{16} + k_{20} x_{17}$$

half-life   As with the section on exponential decay, these constants k_i are expressed in terms of the half-lives of the radioactive elements using the relation $\ell n\, 2 = \tau_i k_i$ (see equation (9) in Section 2.2). This particular problem is difficult to solve using a computer because of the very different time lengths involved. The half-life of Uranium-238 is approximately 4.5 billion years, whilst that for the next element, Thorium-234, is 24 days. This problem of very different time lengths in two parts of the same problem, a *stiff problem*, is discussed in Chapter 4.

2.10   *First-order linear DEs*

When we examined the solutions to the linear differential equations above   the lake
more closely there were some interesting features. For the lake problems   problem
we had   again

$$\frac{dC}{dt} = \frac{F}{V}c_{\text{in}} - \frac{F}{V}C,$$
$$C = c_0 e^{-Ft/V} + c_{\text{in}}(1 - e^{-Ft/V}).$$

Notice that this solution has two parts: one dependent on the initial value   two parts
of the system c_0 and the other dependent on the input to the system c_{in}.   to the
Further, the effect of c_0 decreases, approaching 0 with time while that of   solution
c_{in} grows with time.

This phenomenon is a general feature of such linear differential equations   generalis-
and there are some elegant general results which follow. (They are in-   ing
cluded here to provide a better 'feel' for expected results and the reader
is referred to other texts for proofs and more details if they are required.)

Consider the *normal form* of a first-order linear ordinary differential equa-   normal
tion (ODE)   form

$$y' + p(t)y = q(t) \tag{1}$$

(normal form just means that the coefficient of y' is one) where the
coefficient $p(t)$ and the driving term (or input) $q(t)$ are continuous over a
finite time interval. If there is no input ($q(t) = 0$) the equation is said to
be *homogeneous*. If $q(t) \neq 0$, then the equation is *inhomogeneous*, that
is, it is *driven* or *forced*.

Such an ODE has infinitely many solutions, each associated with a dis-   the
tinct constant of integration. Given an initial condition for the above   solutions
equation, $y(t_0) = y_0$, the solution will be unique. Further, as we noticed
above the solution to the system can be divided into two parts:

- the response to the initial conditions, and
- the response to the forcing term.

We state this in some theorems: theorems

Theorem 1 (General solution theorem) *Consider the linear ODE*
(1) with $p(t)$ and $q(t)$ continuous on some interval containing t_0. Let
$P(t) = \int_{t_0}^t p(r)\,dr$ and $R(t) = \int_{t_0}^t e^{P(s)}q(s)\,ds$ on that interval. Then
every solution $y(t)$ of the ODE has the form

$$y(t) = e^{-P(t)}K + e^{-P(t)}R(t), \qquad \text{with } K \text{ an arbitrary constant.} \tag{2}$$

Conversely, any function $y(t)$ of the form (2) is a solution of the differential equation on the interval.

existence and uniqueness

Note that one important consequence of this theorem is that every solution of the ODE exists at least in the interval in which $p(t)$ and $q(t)$ are continuous. Another important consequence is that no two solutions corresponding to different values of K can ever intersect. Why? Because suppose they did for $t = T$, then rearranging $y_1(T) = y_2(T)$ gives

$$
\begin{aligned}
0 &= y_1(T) - y_2(T) \\
&= \left[e^{-P(T)}K_1 + e^{-P(T)}R(T)\right] - \left[e^{-P(T)}K_2 + e^{-P(T)}R(T)\right] \\
&= e^{-P(T)}(K_1 - K_2) \\
&\neq 0 \qquad \text{since } K_1 \neq K_2.
\end{aligned}
$$

Since T is arbitrary, no two solutions corresponding to different initial values can intersect.

Theorem 2 (Solution of initial value problem (IVP)) *Let $p(t)$ and $q(t)$ be continuous in some interval containing t_0, and let y_0 be any constant. Then the IVP*

$$
y' + p(t)y = q(t), \qquad y(t_0) = y_0
$$

has exactly one solution.

Writing $P_0(t) = \int_{t_0}^t p(r)\, dr$, this solution is given by

$$
\underbrace{y(t)}_{\substack{\text{total} \\ \text{response}}} = \underbrace{e^{-P_0(t)}y_0}_{\substack{\text{response to} \\ \text{initial data}}} + \underbrace{e^{-P_0(t)}\int_{t_0}^t e^{P_0(s)}q(s)\, ds}_{\substack{\text{response to} \\ \text{input}}}.
$$

There is no need to memorise these theorems as they are just the application of the integrating factor technique.

Rewriting the IVP above as $y' = f(t, y) = q(t) - p(t)y$, $y(t_0) = y_0$, we have

Theorem 3 (Existence theorem) *If f and the partial derivative $\partial f/\partial y$ are continuous on a region R in the (t, y)-plane, and (t_0, y_0) is a point in R, then the IVP above has a unique solution $y(t)$ on an interval containing t_0.*

> **Skills developed here:**
> - Identify a linear differential equation for which there exists a unique solution.
> - Understand the significance of the two separate parts to the solution.

2.11   Exercises for Chapter 2

1. Atmospheric pressure. The earth's atmospheric pressure p is often modelled by assuming that dp/dh (the rate at which pressure p changes with altitude h above sea level) is proportional to p. Suppose that the pressure at sea level is 1013 millibars and that the pressure at an altitude of 20 km is 50 millibars. Answer the following questions using analytical techniques and then check them with `Maple`.

(a) Use an exponential decay model

$$\frac{dp}{dh} = -kp$$

to describe the system, and then by solving the equation find an expression for p in terms of h. Determine k and the constant of integration from the initial conditions.

(b) What is the atmospheric pressure at an altitude of 50 km?

(c) At what altitude is the pressure equal to 900 millibars?

2. The Rule of 72. Continuous compounding for invested money can be described by a simple exponential model, $M'(t) = 0.01rM(t)$, where $M(t)$ is the amount of money at time t and r is the percent interest compounding. Business managers commonly apply the Rule of 72, which says that the number of years it takes for a sum of money invested at $r\%$ interest to double, can be approximated by $72/r$. Show that this rule always overestimates the time required for the investment to double.

3. Dating a sea shell. If an archaeologist uncovers a sea shell which contains 60% of the ^{14}C of a living shell, how old do you estimate that shell, and thus that site, to be? (You may assume the half-life of ^{14}C to be 5568 years.)

4. Olduvai Gorge. *(From Borelli and Coleman (1996).) Olduvai Gorge, in Kenya, cuts through volcanic flows, tuff (volcanic ash), and sedimentary deposits. It is the site of bones and artifacts of early hominids, considered by some to be precursors of man. In 1959, Mary and Louis Leakey uncovered a fossil hominid skull and primitive stone tools of obviously great age, older by far than any hominid remains found up to that time. Carbon-14 dating methods being inappropriate for a specimen of that age and nature, dating had to be based on the ages of the underlying and overlying volcanic strata.*

The method used was that of potassium–argon decay. The potassium–argon clock is an accumulation clock, in contrast to the ^{14}C dating method. The potassium–argon method depends on measuring the accumulation of 'daughter' argon atoms, which are decay products of radioactive potassium atoms. Specifically, potassium-40 (^{40}K) decays to argon (^{40}Ar) and to Calcium-40 (^{40}Ca) by the branching cascade illustrated below. Potassium decays to calcium by emitting a β particle (i.e. an electron). Some of the potassium atoms, however, decay to argon by capturing an extranuclear electron and emitting a γ particle.

The rate equations for this decay process may be written in terms of $K(t)$, $A(t)$ and $C(t)$, the potassium, argon and calcium in the sample of rock:

$$K' = -(k_1 + k_2)K,$$
$$A' = k_1 K,$$
$$C' = k_2 K,$$

where
$$k_1 = 5.76 \times 10^{-11} \text{ year}^{-1}, \qquad k_2 = 4.85 \times 10^{-10} \text{ year}^{-1}.$$

(a) *Solve the system to find $K(t)$, $A(t)$ and $C(t)$ in terms of k_1, k_2, and $k_3 = k_1 + k_2$, using the initial conditions $K(0) = k_0$, $A(0) = C(0) = 0$.*

(b) *Using the above result, show that $K(t) + A(t) + C(t) = k_0$ for all $t \geq 0$. Why would this be the case?*

(c) *Show that $K(t) \to 0$, $A(t) \to k_1 k_0/k_3$ and $C(t) \to k_2 k_0/k_3$ as $t \to \infty$.*

(d) *The age of the volcanic strata is the current value of the time variable t because the potassium–argon clock started when the volcanic material was laid down. This age is estimated by measuring the ratio of argon to potassium in a sample. Show that this ratio is*

$$\frac{A}{K} = \frac{k_1}{k_3} \left(e^{k_3 t} - 1 \right).$$

(e) Now show that the age of the sample in years is

$$\frac{1}{k_3} \, \ell n \left[\left(\frac{k_3 A}{k_1 K} \right) + 1 \right].$$

(f) When the actual measurements were made at the University of California at Berkeley, the age of the volcanic material (and thus the age of the bones) was estimated to be 1.75 million years. What was the value of the measured ratio A/K?

5. Tracers in the body. (Adapted from Borelli and Coleman (1996).) In a biochemical laboratory radioactive phosphorus (^{32}P) was used as a tracer. (A tracer, through its radioactive emission, allows the course followed by a substance through a system to be tracked, which otherwise would not be visible.) ^{32}P decays exponentially with a half-life of 14.5 days and its quantity is measured in curies (Ci). (Although it is not necessary for the calculations, one curie is the quantity of a radioactive isotope undergoing 3.7×10^{-5} disintegrations per second.) After the experiment the biochemists needed to dispose of the contents, but they had to store them until the radioactivity had decreased to the acceptably safe level of 1×10^{-5}Ci. The experiment required 8Ci of ^{32}P. Using a simple model of exponential decay, establish how long they had to store the contents of the experiment before it could be disposed of safely?

6. Lake Burley Griffin. Read the case study on Lake Burley Griffin. The average summer flow rate for the water into and out of the lake is $4 \times 10^6 \, \mathrm{m}^3/\mathrm{month}$.

(a) Using this summer flow, how long will it take to reduce the pollution level to 5% of its current level? How long would it take for the lake with pollution concentration of $10^7 \mathrm{parts}/\mathrm{m}^3$, to drop below the safety threshold? (Assume in both cases that only fresh water enters the lake.)

(b) Use `Maple` to replicate the results in the case study, for both constant and seasonal flow and constant and seasonal pollution concentrations entering the lake. Comment on the solutions.

7. North American lake system. Consider the American system of two lakes: Lake Erie feeding into Lake Ontario. What is of interest is how the pollution concentrations change in the lakes over time. You may assume the volume in each lake to remain constant and that Lake Erie is the only source of pollution for Lake Ontario.

(a) Write down a differential equation describing the concentration of pollution in each of the two lakes, using the variables V for volume, F for flow, $c(t)$ for concentration at time t and subscripts 1 for Lake Erie and 2 for Lake Ontario.

(b) Suppose that only unpolluted water flows into Lake Erie. How does this change the model proposed?

(c) Solve the system of equations to get expressions for the pollution concentrations, $c_1(t)$ and $c_2(t)$.

(d) Set $T_1 = V_1/F_1$ and $T_2 = V_2/F_2$, and then $T_1 = kT_2$ for some constant k as V and F are constants in the model. Substitute this into the equation describing pollution levels in Lake Ontario to eliminate T_1. Then show that, with the initial conditions $c_{1,0}$ and $c_{2,0}$, the solution to the differential equation for Lake Ontario is

$$c_2(t) = \frac{k}{k-1}c_{1,0}\left(e^{-t/(kT_2)} - e^{-t/T_2}\right) + c_{2,0}e^{t/T_2}.$$

(One way of finding the solution would be to use an integrating factor. See Appendix A.4.)

(e) Compare the effects of $c_1(0)$ and $c_2(0)$ on the solution $c_2(t)$ over time.

(f) For Lake Erie $V_1 = 458 \times 10^9\,\mathrm{m}^3$ and $F_1 = 479\,582\,208\,\mathrm{m}^3$. For Lake Ontario $V_2 = 1\,636 \times 10^9\,\mathrm{m}^3$ and $F_2 = 572\,562\,432\,\mathrm{m}^3$. Estimate the time it would take for the pollution level to reach 2% and 5% of its current level.

8. Smoke in the bar. *(Adapted from Fulford et al. (1997).) A public bar opens at 6 p.m. and is rapidly filled with clients of whom the majority are smokers. The bar is equipped with ventilators which exchange the smoke–air mixture with fresh air. Cigarette smoke contains 4% carbon monoxide and a prolonged exposure to a concentration of more than 0.012% can be fatal. The bar has a floor area of 20 m by 15 m, and a height of 4 m. It is estimated that smoke enters the room at a constant rate of 0.006 m³/min, and that the ventilators remove the mixture of smoke and air at 10 times the rate at which smoke is produced. The problem is to establish a wise time to leave the bar. That is, sometime before the concentration of carbon monoxide reaches the lethal limit.*

(a) Starting with a word equation and compartmental diagram, formulate the differential equation for the changing concentration of carbon monoxide in the bar over time.

(b) By solving the equation above, establish at what time the lethal limit will be reached.

9. Detecting art forgeries. Based on methods used in the case study describing the detection of art forgeries (Section 2.3), comment on the authenticity of the following two paintings using the information provided:

(a) 'Washing of Feet', where the disintegration rate for ^{210}Po is 8.2 per minute per gram of white lead, and for ^{226}Ra is 0.26 per minute per gram of white lead.

(b) 'Laughing Girl', where the disintegration rate for ^{210}Po is 5.2 per minute per gram of white lead and for ^{226}Ra is 6 per minute per gram of white lead.

10. Cold pills. In Section 2.7, we developed the model

$$\frac{dx}{dt} = -k_1 x, \qquad x(0) = x_0,$$

$$\frac{dy}{dt} = k_1 x - k_2 y, \qquad y(0) = 0,$$

where $k_1, k_2 > 0$ determine the rate at which a drug, antihistamine or decongestant, moves between two compartments in the body, the GI-tract and the bloodstream, when a patient takes a single pill. Here $x(t)$ is the level of the drug in the GI-tract and $y(t)$ the level in the bloodstream at time t.

(a) Find solution expressions for $x(t)$ and $y(t)$ which satisfy this pair of differential equations, when $k_1 \neq k_2$. Show that this solution is equivalent to that provided in the text.

(b) The solution above is invalid at $k_1 = k_2$. Why is this and what is the solution in this case?

(c) For old and sick people the clearance coefficient (that is, the rate at which the drug is removed from the bloodstream) is often much lower than that for young, healthy individuals. How does an increase or decrease in k_2 change the results of the model. Using `Maple` to generate the time-dependent plots, check your results.

11. Cold pills. In Section 2.7, we also developed a model to describe the levels of antihistamine and decongestant in a patient taking a course of cold *pills:*

$$\frac{dx}{dt} = I - k_1 x, \qquad x(0) = 0,$$

$$\frac{dy}{dt} = k_1 x - k_2 y, \qquad y(0) = 0.$$

Here $k_1, k_2 > 0$ describe rates at which the drugs move between the two sequential compartments (the GI-tract and the bloodstream) and I denotes the amount of drug released into the GI-tract in each time step. The levels of the drug in the GI-tract and bloodstream are x and y respectively. By solving the equations sequentially show that the solution is

$$x(t) = \frac{I}{k_1} \left(1 - e^{-k_1 t} \right), \qquad y(t) = \frac{I}{k_2} \left[1 - \frac{1}{k_2 - k_1} \left(k_2 e^{-k_1 t} - k_1 e^{-k_2 t} \right) \right].$$

12. Antibiotics. *(Adapted from Borelli and Coleman (1996).) Tetracycline is an antibiotic prescribed for a range of problems, from acne to acute infections. A course is taken orally and the drug moves from the GI-tract through the bloodstream, from which it is removed by the kidneys and excreted in the urine.*

(a) *Write down word equations which describe the movement of a drug through the body, using three compartments: the GI-tract, the bloodstream and the urinary tract. Note that the urinary tract can be considered as an absorbing compartment, that is, the drug enters but is not removed from the urinary tract.*

(b) *From the word equations develop the differential equation system which describes this process, defining all variables and parameters as required.*

(c) *The constants of proportionality associated with the rates at which tetracycline (measured in milligrams) diffuses from the GI-tract into the bloodstream, and then is removed, are $0.72 \, \text{hour}^{-1}$ and $0.15 \, \text{hour}^{-1}$, respectively (Borelli and Coleman (1996)). Suppose, initially, the amount of tetracycline in the GI-tract is 0.0001 milligrams, while there is none in the bloodstream or urinary tract. Use* Maple *to solve this system analytically, and thus establish how the levels of tetracycline change with time in each of the compartments. In the case of a single dose, establish the maximum level reached by the drug in the bloodstream and how long it takes to reach this level with the initial conditions as given above.*

(d) *Suppose that, initially, the body is free from the drug and then the patient takes a course of antibiotics: 1 unit per hour. Use* Maple *to examine the level of tetracycline (expressed as units) in each of the compartments over a 24-hour period. Use the constants as given above.*

13. Alcohol consumption. *Use the model from the case study on alcohol consumption (Dull, dizzy or dead, Section 2.8), to establish, for the case of drinking on an empty stomach, the following:*

(a) *Use* `Maple` *to generate graphs to investigate the effects of alcohol on a woman of 55 kg, over a period of time.*

(b) *Compare these results with those for a man of the same weight.*

(c) *Assuming the legal limit to be 0.05 BAL (the Australian limit) establish roughly how much alcohol the man and woman above can consume each hour and remain below this limit.*

(d) *Repeat (a–c) for the case of drinking together with a meal.*

14. Alcohol consumption. *Alcohol is unusual in that it is removed (that is, metabolised through the liver) from the bloodstream by a constant amount each time period, independent of the amount in the bloodstream. This removal can be modelled by a Michaelis–Menten type function $y' = -k_3 y/(y + M)$ where $y(t)$ is the 'amount' (BAL) of alcohol in the bloodstream at time t, k_3 is a positive constant and M a small positive constant.*

(a) *If y is large compared with M then show that $y' \simeq -k_3$. Solve for y in this case.*

(b) *Alternatively, as y decreases and becomes small compared with M, show that then $y' \simeq -k_3 y/M$. Solve for y in this case.*

(c) *Now sketch the solution function for $y' = -k_3 y/(y + M)$ assuming that, initially, y is much greater than M. Indicate clearly how the graph changes in character when y is small compared with M, compared with when y is large compared with M. Show how the solution behaves as $t \to \infty$.*

(d) *When and why would this function be more suitable than simply using $y' = -k_3$ to model the removal rate?*

15. Solving differential equations. *Consider the differential equations*

$$t\frac{dx}{dt} = x, \qquad x(t_0) = x_0,$$

$$y^2\frac{dx}{dy} + xy = 2y^2 + 1, \qquad x(y_0) = x_0.$$

Put each equation into normal form and then use the integrating factor technique to find the solutions. Establish whether these solutions are unique, and which part of each solution is a response to the initial data and which part a response to the input or forcing.

Chapter 3

Models of single populations

In this chapter, we develop models describing the growth and decline of single populations with continuous breeding programs. Initially we model exponential growth. A more realistic model would include the effects on population growth of limited resources, and thus we extend our model to take account of density dependent growth (logistic growth), which describes the population size as stabilising after an initial exponential growth spurt. We examine the effects of harvesting the population and see that there is a critical harvesting rate, above which extinction ensues. Furthermore, we consider single populations with discrete breeding seasons and examine how the concept of 'chaos' can arise in population dynamics. Finally, we briefly introduce the notion of time delayed models through a case study.

3.1   Exponential growth

overview

We begin by developing a very simple mathematical model which describes the growth of a population. This leads to exponential growth.

Background

uses of population models

In many cases mathematical modelling is applied to understand population growth dynamics for animal and human populations. For example, modelling the way fish populations grow, and accounting for the effect of fishing is essential to the fishing industry, as we cannot afford to deplete this resource. Another use of modelling is to understand the manner in which human populations grow: in the world, in individual countries, in towns and in organisations.

world population

The population of the world was estimated in 1990 to be approximately 5.3 billion individuals. Estimates from census data show that this has grown from approximately 1.6 billion at the turn of the century. What will the population be in 5 years' time or even one hundred years time? By developing simple mathematical models we attempt to predict this, based upon certain assumptions about birth rates and death rates.

continuous versus discrete growth

One important factor in modelling populations is whether the population grows continuously with time or in discrete jumps. Many animal populations grow in discrete times, due to having well-defined breeding seasons, whereas human populations grow continuously in time. Some insect populations also have non-overlapping generations, where the adults all die directly after giving birth. Even if a population grows in discrete time jumps it may still be reasonable to use a continuous time model provided the time interval between jumps is small compared with the overall time in which we are interested.

small and large populations

Another important factor in deciding on how to model population growth is the size of the population. Small populations are subject to random fluctuations in that we cannot predict with certainty when a parent will give birth, or what the size of an animal litter will be. For small populations it makes sense to talk about the probability of giving birth in a certain time interval, and the mean population at a given time. When the population is large, however, random fluctuations between individuals are small compared with the whole population size and so we can model this without the need for probability functions.

size or density

For an entire population it would only be appropriate for $X(t)$, a measure of the population size, to be an integer, but for large populations it is

usual to work with a value of $X(t)$ which is continuous in t and round off
to the integer value at the end. Of course for population densities, that
is, number per unit of area, using fractional numbers poses no problem.

General compartmental model

We can consider this problem as a compartmental model with the compartment being the 'world', 'town', 'organisation', 'ocean', etc.

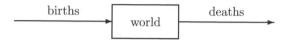

Figure 3.1.1: Input–output compartmental diagram for a population.

This compartmental sketch leads to a word equation describing a chang- word
ing population,   equation

$$\left\{ \begin{array}{c} rate\ of \\ change\ of \\ population\ size \end{array} \right\} = \left\{ \begin{array}{c} rate \\ of \\ births \end{array} \right\} - \left\{ \begin{array}{c} rate \\ of \\ deaths \end{array} \right\}. \qquad (1)$$

We now develop equations by making some assumptions and then, under
these conditions, describe the birth and death processes in symbols.

Model assumptions

When dealing with large populations we can ignore random fluctuations   ignoring
between individuals and treat each individual as being identical. Thus   random
we assume that each individual in the population has an equal chance of   fluctu-
                                                                          ations
giving birth and we also assume that each individual has an equal chance
of dying within a given time interval. It thus makes sense to talk about
a *per-capita birth rate* β per unit time, per member of the population,
and a *per-capita death rate* α.

The per-capita birth rate for the world has been estimated at $\beta = 0.027$ typical
per year per individual and the per-capita death rate is $\alpha = 0.010$ per   per-capita
                                                                                birth and
year per individual (1990 estimate, Microsoft (1995)). This is not ho-   death
mogeneous over all countries in the world. For Australia SBS (1998)   rates
between 1990 and 1995 the per-capita birth rate was $\beta = 0.014$ per year
per individual and the per-capita death rate was $\alpha = 0.007$ per year
per individual. The population of Australia in 1993 was approximately
17.8 million people, with a population density of 2.1 persons per km^2.
Note that some other countries have much higher per-capita birth and
death rates, which is consistent with the higher world per-capita birth
and death rates compared with those for Australia.

summary
of assump-
tions We make the following assumptions and then build the model on them.

- We assume that the populations are sufficiently large so that we can ignore random differences between individuals.
- We assume that births and deaths are continuous in time.
- We assume that per-capita birth and death rates are constant in time.
- In the model development, we ignore immigration and emigration, which can be included later.

Formulating the differential equation

basic steps Let us consider a population whose initial value is x_0, with constant per-capita birth rate given as β, and constant per-capita death rate given as α. Our aim is to find the population size at any time t. The first step is to determine an equation for the population. We assume that the population can only change due to births or deaths, neglecting here any immigration or emigration. Also, we assume that this change in population at any time is proportional to the size of the population at that time. The appropriate word equation was given in (1).

modelling
the
birth/death
rate Since the per-capita birth rate β is assumed constant, the overall birth rate at any time is the per-capita birth rate multiplied by the current population size. Similarly, the overall death rate is the per-capita death rate multiplied by the population size. Thus we write

$$\left\{ \begin{array}{c} rate \\ of \\ births \end{array} \right\} = \beta X(t), \qquad \left\{ \begin{array}{c} rate \\ of \\ deaths \end{array} \right\} = \alpha X(t). \qquad (2)$$

Substituting (2) into (1) we obtain

$$\frac{dX}{dt} = \beta X - \alpha X. \qquad (3)$$

(Note that $X(t)$ can be written as just X, since it is clear that X is evaluated at t.)

initial
condition We have obtained a differential equation for the population size $X(t)$. We need one initial condition to ensure a unique solution since this is a first-order, linear differential equation. For the population of the world, if we define the initial time to be 1990 then we have $X(0) = x_0 = 5.3$ billion people.

Solving the differential equation

We now solve the differential equation (3) for continuous population growth. Let $r = \beta - \alpha$ and then

methods of solution

$$\frac{dX}{dt} = rX.$$

We call r the *growth rate* or the *reproduction rate* for the population. When $r > 0$ this is a model describing exponential growth and when $r < 0$ the process is exponential decay. (Note the similarities here with Section 2.2 on exponential decay.)

This differential equation can be solved using the separation of variables technique covered in Appendix A.3. (The Appendices provide a summary of techniques for solving differential equations.) The general solution is

solution

$$X = Ae^{rt}$$

and applying the initial condition $X(0) = x_0$ to obtain the value of the arbitrary constant A, the solution to the differential equation is

$$X(t) = x_0 e^{rt}. \tag{4}$$

Clearly this describes exponential growth or decay depending on the sign of r.

Interpretation of parameters

Returning to the formulation of the differential equation we can provide an approximate interpretation of the per-capita death rate α. From the rate of deaths we can approximate the number of deaths by multiplying the rate of deaths by the length of the time interval. This approximation would be better if the time interval were short. Thus we can write

approximate number of deaths over a lifetime

$$\left\{ \begin{array}{c} number \\ of\ deaths \\ in\ time\ interval\ \Delta t \end{array} \right\} \simeq \alpha X(t)\Delta t.$$

Let us now suppose that x_0 people will die in time t_1, that is, t_1 is the *average life expectancy*. Then, let $X(t) = x_0$ and $\Delta t = t_1$ so that (from above) we have

life expectancy

$$x_0 \simeq \alpha x_0 t_1.$$

Hence we obtain

$$\alpha \simeq \frac{1}{t_1}$$

giving an estimate for α as the reciprocal of the average life expectancy.

typical values

For human populations $\alpha = 0.7$. This approximates an average life expectancy of 140 years. This is too high, but nevertheless is of the correct order of magnitude compared with the average life expectancy of humans which is 70–80 years. For many animal populations with shorter lifespans the reciprocal of the per-capita death rate is a better approximation to the average life expectancy.

doubling time

Using the model we can predict the time taken for the population size to double. Note the similarity between this concept of doubling time and the half-life of radioactive substances discussed in Chapter 2.

Example 1: *Find an expression for the time for the population to double in size.*

Solution: *We solve*

$$X(t+T) = 2X(t)$$

where T is the time taken to double the current size. So, using the solution (4),

$$\frac{X(t+T)}{X(t)} = 2 = \frac{x_0 e^{r(t+T)}}{x_0 e^{rt}}$$

$$\text{whence} \quad T = \frac{\ln 2}{r}.$$

So the time taken for a population to double in size is $T = \ln 2/r$.

Model validation

model predictions

Let us now see what this model predicts for known values of the parameters. Taking the 1990 world population values $r = 0.017$ and $x_0 = 5.3$ billion, we apply equation (4) to predict the population in 1995 as $X(5) = 5.77$ billion. In 10 years we have $X(10) = 6.28$ billion and in 100 years $X(100) = 29.01$ billion. Thus, in just 100 years, this predicts that the world population will have increased by a 2500%!

limited validity

We should also compare our predictions with population sizes at previous times for which data is available. Looking back to 1960 the world's population was 3.005 billion and in 1900 it was 1.608 billion. Our model predicts in 1960, $X(-30) = 3.18$ billion, which is not too far out and in 1900, $X(-90) = 1.15$ billion. The predictions appear to get worse the further back we go. One possible contributing factor towards this discrepancy is that the per-capita birth and death rates have changed gradually due to improvements in technology and changing attitudes. (In 1970, the reproduction rate was somewhat higher, at $r = 2.0$.)

Another way to test the model is to measure the population of some very simple organism which breeds fast and is sufficiently small, for example, a yeast culture. A comparison with data from an experiment growing a yeast culture is given in Renshaw (1991), and the results illustrated in the diagram of Figure 3.1.2. Note that the model appears to be in good agreement in the earlier stages of the growth, but for later times the population levels out rather than continuing to grow exponentially.

yeast data

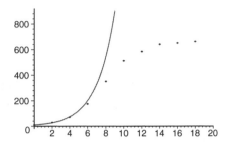

Figure 3.1.2: Illustration of the difference between the trend of experimental data for a yeast culture, and the continuous growth model predictions, over a period of hours (t). $X(t)$ is a measure of the yeast population size. (Adapted from Renshaw (1991).)

The cause of this is that the per-capita death rate increases due to crowding and competition for limited resources, and thus the overall population size growth rate decreases. In certain animal populations the per-capita birth rate (number of births per unit of time) decreases when the death rate increases. For example, rabbits reabsorb their embryos when their population density is high. However, in the model we develop to describe this observed growth pattern in Section 3.2, we consider only a change in the per-capita death rate dependent on the population density.

model limitations

There are several other ways in which this basic population model could be extended. It is easy to incorporate immigration and emigration, which is appropriate for models of populations in specific countries, rather than the world. This is done by adding terms to the word equation for the change in population, and making assumptions about the rates of immigration and emigration. A similar concept is the inclusion of the harvesting of a population which we cover in Section 3.3. Some other important extensions (which are not addressed in this book) include discrete growth for animals with distinct breeding seasons, stochastic (random) growth and age dependent growth.

model extensions

> **Skills developed here:**
>
> - Formulate differential equations for single populations, including immigration, emigration or harvesting.
> - Obtain exact solutions by solving the differential equations.
> - For simple solutions, draw general sketches.

3.2 Limited growth

overview

Realistically, populations cannot continue growing exponentially over time due to limited resources and/or competition for these with other species. If populations are observed over long periods they often appear to reach a limit, or to *stabilise*. We modify the exponential growth model of the previous section to account for competition or limited resources and to include the stabilising effect observed in populations.

Background

carrying capacity

As a population grows, individuals eventually will compete for the limited resources available. In principle this competition means that a given environment can support only a limited number of individuals. This number is called the *carrying capacity* for the population and is usually denoted by the symbol K in biological literature. Technically, we define it as the population size (or density) for which the per-capita birth rate is equal to the per-capita death rate, excluding external factors such as harvesting or interaction with another population. We need to extend the model to include an additional death rate due to the resource limitations, and thus curb the exponential growth and allow the population to stabilise.

Formulating the differential equation

word equation

The population is described by the same word equation as before,

$$
\left\{ \begin{array}{c} rate\ of \\ change\ in \\ population \end{array} \right\} = \left\{ \begin{array}{c} rate \\ of \\ births \end{array} \right\} - \left\{ \begin{array}{c} rate \\ of \\ deaths \end{array} \right\} \tag{1}
$$

and, as in the previous section, we assume a constant per-capita birth rate of β. Thus

$$\left\{\begin{array}{c} rate \\ of \\ births \end{array}\right\} = \beta X(t).$$

Instead of assuming a constant per-capita death rate, we allow the per-capita death rate to increase as the population increases, as can be observed in some populations. We can model this behaviour by assuming a linear dependence of the per-capita death rate on the population size,

$$\left\{\begin{array}{c} per\text{-}capita \\ death \\ rate \end{array}\right\} = \alpha + \gamma X(t)$$

population dependent per-capita death rate

where α (positive) is the per-capita death rate due to natural attrition, and γ (positive) is the per-capita dependence of deaths on the population size. Note that as $X \to 0$ the per-capita death rate tends to α, while for increasing population size the per-capita death rate increases. This linear form is the simplest for a population dependent per-capita death rate which increases with increasing population size. The overall death rate is thus given by multiplying the per-capita death rate by the population size, so that

$$\left\{\begin{array}{c} rate \\ of \\ deaths \end{array}\right\} = \alpha X(t) + \gamma X^2(t). \qquad (2)$$

The word equation (1) then translates to

$$\frac{dX}{dt} = \beta X - \alpha X - \gamma X^2.$$

new differential equation

Writing $r = \beta - \alpha$, which is the reproduction rate, we obtain the model for density dependent growth

$$\frac{dX}{dt} = rX - \gamma X^2. \qquad (3)$$

An alternative formulation of this equation comes from splitting the death rate into a normal death rate and an extra death rate due to members of the population competing with each other for limited resources. Thus we can write

alternative formulation

$$\left\{\begin{array}{c} rate\ of \\ change\ in \\ population \end{array}\right\} = \left\{\begin{array}{c} rate \\ of \\ births \end{array}\right\} - \left\{\begin{array}{c} normal \\ rate\ of \\ deaths \end{array}\right\} - \left\{\begin{array}{c} rate\ of \\ deaths\ by \\ crowding \end{array}\right\}.$$

For the birth rate and normal death rate we assume constant per-capita rates β and α respectively.

<small>extra deaths by crowding</small>

For the extra deaths by crowding we assume that the per-capita death rate increases with population size. If we assume that it is proportional to the population size and thus given by γX (where γ is a positive constant), then the extra overall death rate is the extra per-capita death rate multiplied by the current population size. Thus we have

$$\left\{ \begin{array}{c} \text{extra rate} \\ \text{of deaths} \\ \text{by crowding} \end{array} \right\} = \gamma X^2$$

with γ a positive constant. The result is again (3).

Equilibrium populations

<small>equilib-rium</small>

If we observe the levelling of a population over time, this implies that the rate of change of the population approaches 0, that is, $X' \to 0$. Any value of X which gives a zero rate of change is called an *equilibrium point*.

<small>constant solution</small>

We can deduce this equilibrium value from the original differential equation by looking for a constant solution to the differential equation $X' = 0$. Thus we solve

$$rX - \gamma X^2 = 0.$$

To find all of the solutions for X we factorise this equation, to obtain

$$X(r - \gamma X) = 0.$$

There are two possible solutions, $X = 0$ and $X = r/\gamma$, and they are called *equilibrium solutions*. The former, $X = 0$, is not very informative, but the latter we consider in detail.

<small>equilib-rium stability</small>

Returning to the differential equation in factored form,

$$\frac{dX}{dt} = X(r - \gamma X)$$

we see that if $X < r/\gamma$ then the rate of change of X is positive. This implies the population is increasing with time and will approach $X = r/\gamma$. Similarly, if $X > r/\gamma$ then $dX/dt < 0$ which implies that the population is decreasing towards $X = r/\gamma$. When the population always approaches the equilibrium population we say the equilibrium is *stable*. Conversely, if it is repelled from the equilibrium, we say the equilibrium is *unstable*. (Note that for $0 < X < r/\gamma$, $dX/dt > 0$ and thus the equilibrium point at $X = 0$ is unstable.)

Thus the model predicts that all populations approach the equilibrium
value K which we have already defined as the carrying capacity for the
population. The stable equilibrium point and carrying capacity coincide
in this case.

equilib-
rium and
the
carrying
capacity

Without actually solving the differential equation, we can infer how the
solution would behave. The population has an initial value, x_0. The
derivative dX/dt is always positive for $X < K$ and thus the population
is increasing. The rate of growth increases initially and then slows down
as the population approaches the carrying capacity K. The sketch is
illustrated in Figure 3.2.1.

a sketch of
the
solution

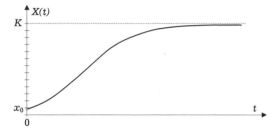

Figure 3.2.1: Sketch of the general solution to the continuous logistic equation, provided $x_0 < K$.

The logistic equation

With $K = r/\gamma$ the differential equation (3) becomes

rewriting
the
differential
equation

$$\frac{dX}{dt} = rX - \frac{r}{K}X^2$$

which can be written as

$$\frac{dX}{dt} = rX\left(1 - \frac{X}{K}\right). \qquad (4)$$

This model leads to a nonlinear differential equation. It is the *logistic
equation* and is also referred to as the limited growth model or the density
dependent model. We consider only $r > 0$ and $K > 0$ to ensure positive
population values.

Interpretation of the parameters

population
dependent
per-capita
growth
rate

Recall the differential equation for unrestricted population growth (see Section 3.1)

$$\frac{dX}{dt} = rX.$$

We can write down a general differential equation for population growth as

$$\frac{dX}{dt} = R(X)X$$

where $R(X)$ represents a population dependent per-capita growth rate. It is interesting to interpret the logistic equation in terms of this population dependent per-capita growth rate, and thus from (4) we identify $R(X)$ as

$$R(X) = r\left(1 - \frac{X}{K}\right).$$

properties
of $R(X)$

Note that $R(X)$, a linear function of X, tends to zero as the population approaches its carrying capacity K, while as the population size tends to zero $R(X)$ approaches r. This form corresponds to a straight line, which passes through the points $R = 0$ when $X = K$ and $R = r$ when $X = 0$ as illustrated in Figure 3.2.2. If $R < 0$ then we have that $X > K$ and the population (X) is decreasing as it approaches the carrying capacity.

Figure 3.2.2: The simplest assumption for a population dependent per-capita growth rate is a straight line.

Solving the logistic equation

two ap-
proaches

We can solve the logistic equation in two ways. First, using `Maple` to obtain a numerical solution, and second, analytically using the separation of variables technique.

using
Maple

`Maple` is a symbolic language which means that it can use analytic methods (where possible) to solve such equations. Often analytic solutions cannot be found and numerical schemes are required. (We have a closer look at some of these numerical methods in Chapter 4.) We can use `Maple`

to solve the equation and draw the family of graphs in Figure 3.2.3 with
the following commands:

```
> restart:with(plots):
> r:=1;K:=1000;
> de1:=diff(x(t),t)=r*x(t)*(1-x(t)/K);
> soln:=x0->dsolve({de1,x(0)=x0},x(t),numeric):
> plot1:=x0->(odeplot(soln(x0),[t,x(t)],0..8)):
> list1:=seq(plot1(i*50),i=1..24):
> line1:=plot([[0,K],[8,K]],color=gray):
> plot11:=display({list1,line1},view=[0..8,0..1200]):display({plot11});
```

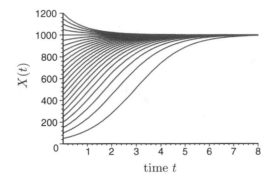

Figure 3.2.3: Logistic growth model with $r = 1$ and $K = 1000$, and for a number of
different initial conditions.

The next example illustrates how to obtain the analytic solution. (For
details on partial fractions and separable techniques see the Appendices.)

analytic
solution

Example 1:   *Solve the logistic differential equation* (4) *given the initial
condition* $X(0) = x_0$.

Solution: *The logistic equation is not linear, but it is separable,*

$$\frac{dX}{dt} = rX\left(1 - \frac{X}{K}\right) = \frac{rX(K - X)}{K} \tag{5}$$

and then separating

$$\frac{K}{X(K - X)}\frac{dX}{dt} = r$$

assuming that $X \neq 0$. *Integrating gives*

$$\int \frac{K}{X(K - X)}\, dX = \int r\, dt.$$

*For the integral on the left hand side (LHS) we need to use partial fractions
(see Appendix B.5) and thus*

$$\frac{K}{X(K - X)} = \frac{a}{X} + \frac{b}{K - X} = \frac{a(K - X) + bX}{X(K - X)}.$$

Solving for the constants a and b gives

$$aK = K,$$
$$(-a + b)X = 0,$$

which implies that $a = b = 1$, and then

$$\frac{K}{X(K - X)} = \frac{1}{X} + \frac{1}{K - X}.$$

Now

$$\int \frac{K}{X(K - X)}\, dX = \int \frac{1}{X}\, dX + \int \frac{1}{K - X}\, dX = \int r\, dt$$

and we can integrate to get

$$\ell n\,|X| - \ell n\,|K - X| = rt + c \quad \text{(c an arbitrary constant)},$$
$$\left| \frac{X}{K - X} \right| = c_1 e^{rt}$$

where $c_1 = e^c$. Assuming $0 < X < K$ then

$$X = c_1 e^{rt}(K - X). \tag{6}$$

Using the initial condition $X(0) = x_0$ we deduce that $c_1 = x_0/(K - x_0)$ and then solving (6) for X we obtain

$$X = \frac{K}{1 + me^{-rt}} \qquad \text{where } m = \frac{K}{x_0} - 1. \tag{7}$$

Alternatively, in the case where $0 < K < X$, then

$$X = c_1 e^{rt}(X - K).$$

With the initial condition $X(0) = x_0$ we have that $x_0 = c_1(x_0 - K)$ and hence $c_1 = x_0/(x_0 - K)$. The solution is as in (7).

It is also possible to use **Maple** to find the analytic solution. The commands are as follows:

```
> restart;
> de := diff(X(t),t) = r*X(t)*(1-X(t)/K);
> dsolve( {de, X(0)=x0}, X(t));
```

Solutions over time

As usual we are interested in the long-term behaviour of the population   long-term
as predicted by our model. The following example shows how to sketch   behaviour
the graph and draw conclusions about the long-term behaviour forecast
by this model.

Example 2:   *Using the differential equation and considering the second
derivative, sketch the graph of the solution.*

Solution: *Since $r > 0$, as t increases we have $\lim\limits_{t \to \infty} X(t) = K$ which is the
carrying capacity of the population. If $0 < x_0 < K$ then $X(t)$ is strictly
increasing (a monotonically increasing function). If $x_0 > K$, then $X(t)$ is
strictly decreasing (a monotonically decreasing function). What happens if
$x_0 = K$?*

Considering the second derivative,

$$X'' = rX' - \frac{2XrX'}{K} = X'r\left(1 - \frac{2X}{K}\right) = rX\left(1 - \frac{X}{K}\right)r\left(1 - \frac{2X}{K}\right)$$

*which changes sign at $X = 0$, $K/2$, and K. These are inflection points of
the curve, that is, where the curve changes from concave to convex or vice
versa. This, together with the limit above and nature of the slope provides
the information required to sketch the graph which was illustrated earlier in
Figure 3.2.1 and generated with* `Maple` *in Figure 3.2.3.*

A limitation of the model

With the logistic model developed above we have incorporated an instan-   time-
taneous reaction to the environment. That is, increased pressure on the   delayed
resources produces an immediate response from the system in terms of,   regulation
for example, more deaths. This is often not realistic in that the response
usually takes effect after some time delay, or time lag. Vegetation needs
time to recover and changed environmental conditions, which may lead
to increased birth rates, will take time to appear in the numbers of an
adult population. This leads us to a model which includes a time delay resulting from a multitude of sources such as maturation times, food
supply, resources, or crowding: each a measurable quantity. If the time
lag is small compared with the natural response time $(1/r)$ then there is
a tendency to overcompensate which may produce oscillatory behaviour.
This is exemplified in Section 3.6 and Section 3.7.

Skills developed here:

- Understand how to modify the differential equation to account for a change in reproduction rate due to crowding.
- Be able to find the equilibrium populations for a given differential equation.
- Determine if the population approaches or is repelled from the equilibrium.

3.3   *Limited growth with harvesting*

overview

The effect of harvesting a population on a regular or constant basis is extremely important to many industries. One example is the fishing industry. Will a high harvesting rate destroy the population? Will a low harvesting rate destroy the viability of the industry?

Formulating the equation

word
equation

Including a constant harvesting rate in our logistic model gives

$$
\left\{ \begin{array}{c} rate\ of \\ change\ in \\ population \end{array} \right\} = \left\{ \begin{array}{c} rate \\ of \\ births \end{array} \right\} - \left\{ \begin{array}{c} normal \\ rate\ of \\ deaths \end{array} \right\}
$$
$$
- \left\{ \begin{array}{c} rate\ of \\ deaths\ by \\ crowding \end{array} \right\} - \left\{ \begin{array}{c} rate\ of \\ deaths\ by \\ harvesting \end{array} \right\}. \tag{1}
$$

Assuming the harvesting rate to be constant, equation (1) translates to the differential equation

$$
\frac{dX}{dt} = rX \left(1 - \frac{X}{K} \right) - h. \tag{2}
$$

Here h is included as the constant rate of harvesting (total number caught per unit time, or deaths due to harvesting per unit time) and it is independent of the population size and thus could be interpreted as a quota.

Solving the differential equation

We can infer much useful information about the solution from the differ- *two ap-*
ential equation, as we see below and in the exercises, or we can obtain *proaches*
an explicit solution for specific parameter values which we do first, using
`Maple`.

The code associated with Figure 3.2.3 was adapted to produce the figures *example*
below (Figure 3.3.1 and Figure 3.3.2), taking a reproduction rate of $r = 1$, *numerical*
a carrying capacity of $K = 1000$ and harvesting rates of $h = 100$ and *solution*
$h = 500$.

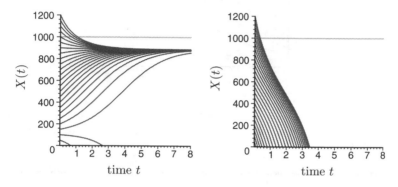

Figure 3.3.1: Logistic growth including harvesting with $h = 100$ in the first diagram
and with $h = 500$ in the second. The grey line is the carrying capacity K. (The
parameter values are $K = 1000$ and $r = 1$.)

The following example illustrates how to gain information about the dy- *analytical*
namics of the model from the differential equation for certain numerical *solution*
values of the parameters. (The purely theoretical case, which covers all
possible parameter values, is examined in the Exercises, Question 5.)

First, we can write (2) in the factored form

$$\frac{dX}{dt} = -\frac{r}{K}\left(X^2 - KX + \frac{Kh}{r}\right)$$

with r, K and h positive constants.

Example 1:   Let $r = 1$, $K = 10$, $h = 9/10$ and $X(0) = x_0$. Use the
*differential equation to investigate the behaviour of the solution and compare
this with the* `Maple` *solution.*

Solution: *The differential equation becomes*

$$\frac{dX}{dt} = -\frac{1}{10}(X^2 - 10X + 9) = -\frac{1}{10}(X - 1)(X - 9).$$

We have the following cases to consider when sketching a graph. If $x_0 < 1$ then $X' < 0$ and the population declines. If $1 < x_0 < 9$ then $X' > 0$ and the population increases. If $x_0 > 9$ then $X' < 0$ and the population declines. If $x_0 = 1$ or $x_0 = 9$ then the population does not change $(X' = 0)$. The `Maple` solution is illustrated in Figure 3.3.2 and clearly satisfies these predictions.

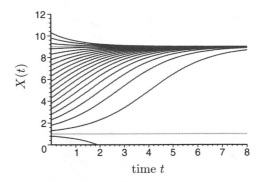

Figure 3.3.2: The logistic equation with harvesting (2) with $r = 1$, $K = 10$ and $h = 9/10$. The grey lines are at $x = 9$ and $x = 1$ across which there is a change in the sign of X'.

solving the differential equation   Using partial fractions we can solve the differential equation for X explicitly, as shown in the following example.

Example 2: *Solve the differential equation.*

Solution: *Separating the variables gives*

$$\int \frac{1}{(X-1)(X-9)}\, dX = -\int \frac{1}{10} dt$$

and using partial fractions,

$$\frac{1}{8} \int \left(\frac{1}{X-9} - \frac{1}{X-1} \right) dX = -\frac{1}{10} \int dt.$$

Carrying out the integration gives

$$\ln \left| \frac{X-9}{X-1} \right|^{1/8} = -\frac{1}{10} t + c$$

where c is some arbitrary constant. Then rearranging with constant $b = e^{8c}$

$$\left| \frac{X-9}{X-1} \right| = b e^{-4t/5}.$$

Now substituting for the initial condition

$$b = \left| \frac{x_0 - 9}{x_0 - 1} \right|$$

and the explicit solution is

$$X(t) = \frac{9 - be^{-4t/5}}{1 - be^{-4t/5}}.$$

From this solution (illustrated in Figure 3.3.2) we can conclude that the interpret-
harvesting rate causes the population to stabilise to a value less than ation
the carrying capacity. In the case of harvesting, or some other disaster,
causing the population to decrease below a critical level, the threshold
level (which in this example is 1), our model predicts that the population
will become extinct.

This concept of a threshold level is critical to many industries and thus threshold
harvesting rates should be set with extreme caution as the following case level
studies indicate.

Skills developed here:
- Be able to model a population following exponential growth or logistic growth and include harvesting.
- Understand the concept of a threshold harvesting level.
- Sketch graphs of the changes in a population over time and ratify them with `Maple` generated solutions.
- Modify the model developed to include immigration and/or emigration, or include the impact of a toxin which kills a percentage of the population.

3.4   Case Study: Anchovy wipe-out

In the following case study we cite a case in which a population was fished annually. Although measures were taken to ensure that over fishing did not occur, so that there would always be sufficient stock for recovery, a natural event combined with the fishing resulted in the total destruction of the industry associated with that species.

The case study is adapted from correspondence with Hearn (1998), and the book, May (1981). A discrete model of the system can be found in Caulkins et al. (1985).

Some decades ago anchovies were a very cheap fish dominating the supermarket shelves: little guys in little tins and very salty. They were a major source of high protein food for people as well as animals. It was noted that the world price of anchovies showed a correlation with the El Niño cycle, that is, El Niño events increased prices due to shortages in the resource. Ironically it was to be a combination of El Niño and fishing practices which destroyed the industry.

In the decades after World War II and the introduction of the nylon fishing nets the anchovy fishery expanded exponentially to become, by the 1960s, the largest fishery in the world and a major part of the Peruvian economy. The anchovies inhabited a narrow strip on the west coast of Peru, where nutrients are available in abundance from strong upwelling in the Pacific Ocean. The upwelling brings nutrients to the surface (into the photic zone) and starts a food chain in that it allows plants to grow on which fish may then feed. Off the coast of Peru, the upwelling is caused by the trade winds, and during the El Niño events these weaken, hence reducing the fishery. Most of the world's major fisheries are in regions of coastal upwelling and this region is particularly remarkable in that it accounts for 22% of the worlds' production of fish. Furthermore, the anchovies themselves were within a food chain and in turn provided 80–95% of the diet of guano birds. (Each bird consuming approximately 35 500 fish a year, or close to 100 per day!) Needless to say the population of these birds was extremely sensitive to that of the anchovies.

Over the years the anchovy catch had increased steadily from an annual 2 million tons in 1959 to 12.3 million tons in 1970. A small El Niño event in 1965, together with heavy fishing, was sufficient to wipe out the guano birds, and thus the associated guano industry collapsed. Fishery 'experts' advised the Peruvian government, now concerned about over fishing, that a maximum sustainable yield was about 10 million tons annually. This was immediately adopted as the quota. No regulation was placed on the number of boats involved or the number of fish taken per boat: it was a free for all and the anchovies declined alarmingly. Still no action was taken to conserve the resource. *In 1971, with hugely expanded fishing and processing, the quota was reached in just 3 months and the fishery had to be closed for the remainder of the year. The catch fell dramatically to 4.5 million tons and then along came a particularly severe El Niño. The anchovy population slumped still further and has never recovered. The little anchovies have gone.*

To be fair, this was before we understood much about El Niño but the case emphasises the concept of population thresholds for the sustainability of both industry and environment. With discrete breeding seasons the population lends itself to discrete modelling, and one such model which includes the effect of El Niño as well as predators such as the guano birds and the efficiency of fishermen with a variety of net types can be found in Caulkins et al. (1985). Unfortunately it is with hindsight that we see that the adoption of the recommendations from these improved models may have changed anchovy history. Still, one is left wondering as fishing in the region continues, with authorities too afraid of retribution to improve restrictions.

3.5 Case Study: How can 2×10^6 birds mean rare?

Perhaps it is not always easy to judge just what level of harvesting is safe, however, the following case study emphasises the importance of biological research into understanding the species' behaviour, before such levels can be set legitimately.

The following is adapted from Quammen (1997).

When it comes to the extinction of species, there is no magic number below which a population ends with extinction. In fact rarity may not mean extinction, just as abundance may not imply survival.

For example, the passenger pigeon, Ectopistes migratorius, was probably once the least rare bird on earth. Inhabiting the eastern half of North America the rookeries covered as much as 500 square kilometres and the birds literally darkened the sky when they took off. Around 1810 a flock was seen in Kentucky taking three days to pass overhead! Biologically this species was a huge success. Then, within the very short space of 36 years, its population went from roughly 3 billion to zero. It was people who reduced its number to a few million, enough to tip the balance, and extinction ensued. The decline happened suddenly during the 1880s and the unbelievably numerous pigeons simply ceased to exist.

The massacre/harvest peaked after the civil war. However, during the last years of the pigeons' existence, their population decline was too steep to attribute to hunting alone. Habitat destruction would have been a further cause for population decline, but even so this was not enough to cause the crash. The explanation proposed by biologist T.R. Halliday in 1980 was in terms of a critical colony size. He argued the benefits to the population of crowding, as protection from predators, cooperative food source location, the maintenance of mating and nesting rhythms, and hence also its breeding rate. He suggested that a remnant population of a couple of million occurring in small flocks would be inadequate to sustain the population. That is, abundance was essential to the species and human interference had pushed the numbers below its particular threshold. Although two or so million sounds like abundance, for Ectopistes migratorius this was rare.

At the opposite extreme we have the Kestrels of Mauritius, Falco punctatus. In that case the population declined slowly. Originally the population would have been roughly 700–800 birds but by the 1950s it had decreased to well below 50 and by 1971 only 4 surviving birds were known to exist. With reduced and fragmented habitat, DDT, introduced exotic predators and plants, human hunters and then a cyclone, the population appeared doomed. It was the most rare bird in the world. Carl Jones, a Welshman and a 'bird person' arrived in 1979 having been sent by the International Council for Bird Preservation (which later withheld funding). He incubated eggs he collected from precarious ledges (replacing them with dummy eggs), used artificial insemination to improve the breeding rates and successfully reared birds in captivity. Furthermore he supplemented the feed of those in the wild. Single handed, and learning on the run, he eventually turned the population decline around.

By 1988 the population had climbed to 80 and although, due to the fragmentation of habitat and exotic predators and plants, the numbers will never reach their original estimates, the population today is relatively healthy.

3.6   Discrete population growth and chaos

overview   Many animal populations have distinct breeding seasons. It is often more appropriate to model such population growth using *difference equations* rather than differential equations. In this section we see how discrete growth models with crowding can predict oscillatory or chaotic growth in populations.

Background

Some observations of the growth of two beetle populations are shown in   observa-
Figure 3.6.1. There appears to be quite a large variation in the type   tions
of growth. In the first example the effect of overcrowding appears to
be levelling out the population. However the population fluctuations in
the second example are more complex. The challenge is to find a single
model that is capable of predicting both types of population growth in
Figure 3.6.1.

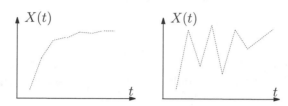

Figure 3.6.1: Schematic diagrams of some experimental data for growth of two different populations of beetle, where $X(t)$ is the number of beetles over a number of
generations t. From May (1981).

Formulating a difference equation

In discrete growth we assume that the population does not change ex-   discrete
cept at discrete intervals, corresponding to breeding seasons. Instead of   breeding
                                                                          interval
thinking about the rate of change of the population at an arbitrary time,
we consider the time interval as the time from one breeding season to the
next. Let us choose to measure time in units of this period. Furthermore,
we assume here that the population changes occur from births or deaths
alone.

We assume that the deaths are due to normal causes and also to crowding   word
effects. Over a single breeding interval we thus write,   equation

$$\left\{ \begin{array}{c} change\ in \\ population \\ size \end{array} \right\} = \left\{ \begin{array}{c} no.\ of \\ births \end{array} \right\} - \left\{ \begin{array}{c} no.\ of \\ deaths \end{array} \right\}. \qquad (1)$$

Let us define X_n to be the population size in the n-th breeding interval.   notation
Here n is an integer, representing time, but taking only discrete values.
Consequently, X_{n+1} is the population in the next breeding interval.

Assuming a constant per-capita birth rate β as before, the overall birth   modelling
rate is βX_n. Since the birth rate is constant over the current breeding   births and
                                                                          deaths
interval, the number of births for that interval is obtained by multiplying

the birth rate by the length of the time interval (which is 1). Thus, for a single breeding interval

$$\left\{\begin{matrix} no.\ of \\ births \end{matrix}\right\} = \beta X_n. \tag{2}$$

For the per-capita death rate we use $\alpha + \gamma X_n$ where the extra term represents an increased per-capita death rate due to competition for limited resources. The number of deaths is obtained by multiplying by the current population size by the time interval. Thus for a single breeding interval

$$\left\{\begin{matrix} no.\ of \\ deaths \end{matrix}\right\} = \alpha X_n + \gamma X_n^2. \tag{3}$$

putting it together
From one breeding interval to the next the change in population is $X_{n+1} - X_n$. Substituting (2) and (3) into the word equation (1) we obtain

$$X_{n+1} - X_n = \beta X_n - \alpha X_n - \gamma X_n^2, \qquad i = 0, 1, 2, 3, \ldots. \tag{4}$$

This type of equation is known as a difference equation. It gives the population at one time in terms of the population at the previous time.

the discrete logistic equation
Suppose we write $r = \beta - \alpha$ and $\gamma = r/K$, as before, then (4) becomes

$$X_{n+1} = X_n + rX_n\left(1 - \frac{X_n}{K}\right), \qquad n = 0, 1, 2, 3, \ldots. \tag{5}$$

This equation is sometimes called the *discrete logistic equation*[1]. Here K is the carrying capacity and r is the per-capita reproduction rate.

Solving the equation

solving by iteration
We solve (5) iteratively, using a computer or calculator, but we first need to choose values for r, K and x_0. One way of gaining an understanding of the population growth is to slowly vary one of the parameters r, K and x_0, observing any changes in the behaviour, or dynamics of the population. In this case we examine the effect on the population growth of varying only r, the intrinsic reproduction rate. We fix the carrying capacity K at

---

[1] Another form of this equation

$$X_{n+1} = rX_n\left(1 - \frac{X_n}{K}\right).$$

is also called the discrete logistic equation. This form arises for populations with non-overlapping generations—where all the adults die after they have given birth.

1000 and the initial population x_0 at 100. Using `Maple` we can generate
the time-dependent values and plot them. A variety of growth patterns
emerge as we increase the per-capita growth rate r: from logistic growth
through oscillatory growth to chaotic growth. These are described below
together with the diagrams generated using the following code:

```
> restart: with(plots):
> X[0]:=100; r:=0.2; K:= 1000;
> for n from 0 to 50 do
    X[n+1] := X[n]+r*X[n]*(1-X[n]/K):
  od:
> points := [seq(X[n], n=0..50)]:
> plot1 := listplot(points,style=point,symbol=circle):
> plot2 := listplot(points, style=line, color=red):
> display(plot1,plot2);
```

Logistic growth

Figure 3.6.2 illustrates the population growth for the small value of r,   results for
$r = 0.2$. Note that the growth is very similar to that for the continuous   small r
case, with the population growing exponentially at first and then levelling
out as the reproduction rate declines due to crowding effects. We see that
the population appears to approach its carrying capacity $K = 1000$.

As we increase the reproduction rate from $r = 0.2$ to $r = 0.8$ we observe   increasing
that the population curve is initially much steeper and then levels off r
more quickly. This is illustrated in Figure 3.6.2.

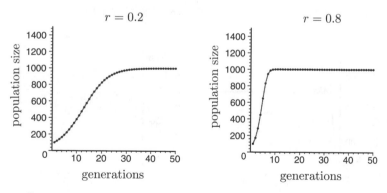

Figure 3.6.2: Computer generated solutions for the discrete logistic equation with
$K = 1000$, $x_0 = 100$. Values of r used were $r = 0.2$ and $r = 0.8$. The growth follows
logistic curves.

Oscillatory growth

When the intrinsic reproduction rate is increased further so that $r > 1$ damped
it becomes possible for the population to increase above the carrying   oscillations

capacity, but in the next iteration it then falls below K. A *damped* oscillation results, where the oscillations become smaller with time, as seen in Figure 3.6.3. As r increases the amplitude of these damped oscillations increases.

interpret-
ation of
oscillations

Once the population is above the carrying capacity K the reproduction rate $r(X_n)$ now is negative (corresponding to the death rate being higher than the birth rate). Thus in the next breeding season the population drops below the carrying capacity. In the continuous model the reproduction rate changed instantaneously; with the discrete model, however, there is a delay of one breeding season before the reproduction rate can adjust to the change in population. This is why it is possible for the population to jump above the carrying capacity in the discrete model, which is not the case in the continuous model.

2-cycles

As we increase r further (see Figure 3.6.3), for $r = 2.2$ the population is again oscillating about the carrying capacity, but the amplitude of the oscillations appears to be constant. We call this oscillation a 2-cycle since the population size is repeated every second breeding cycle once the initial transients have died out. As r increases further, the amplitude of the oscillation increases. It is possible to prove, using the difference equation (5), that stable 2-cycles persist if $2 \leq r < 2.4$. This illustrates an increasingly common approach in mathematics, where computer experiments suggest results which are then proved using analysis. Somewhere between $r = 2.4$ and $r = 2.5$ the 2-cycles become unstable .

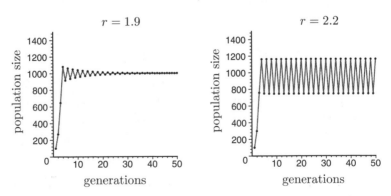

Figure 3.6.3: Computer generated solutions of the discrete logistic equation with $K = 1\,000$, $x_0 = 100$ and with $r = 1.9$ and $r = 2.2$. The graph for $r = 1.9$ shows a damped oscillation. The graph for $r = 2.2$ exhibits a 2-cycle.

Period doubling and chaotic growth

Again, the trend here is for the amplitude of the 2-cycles to increase
but they do not increase indefinitely. We can see from Figure 3.6.4 that
the 2-cycle has become a 4-cycle by $r = 2.5$, where the values repeat
themselves every 4 breeding cycles. As r increases further this becomes
an 8-cycle and then a 16-cycle and so on.

4-cycles
and period
doubling

When we try $r \geq 2.6$ some entirely new behaviour is observed. The
population does not grow in ordered cycles but seems to change randomly
in each breeding period. We call this type of growth chaotic, and it is
illustrated in Figure 3.6.4 for $r = 2.7$.

onset of
chaotic
growth

 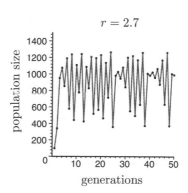

Figure 3.6.4: Computer generated solutions of the discrete logistic equation with
$K = 1\,000$, $x_0 = 100$ and $r = 2.5$ and $r = 2.7$. For $r = 2.5$ we get a 4-cycle. For
$r = 2.7$ we get chaotic growth where there is seemingly a random pattern to the
growth.

Another feature of chaotic growth is that it is very sensitive to a change
in the initial population. This feature of chaotic growth is illustrated in
Figure 3.6.5 with two initial population sizes of 100 and 101.

sensitivity
to initial
condition

Although the difference between the populations is only 1 initially, after
some time they become very different. Both populations are increasing
and decreasing in a random-like pattern but there are significant differences in the graphs.

If we try to plot a similar graph, using initial conditions $x_0 = 100$ and
$x_0 = 101$ for any value of r which gave non-chaotic growth ($r < 2.6$) the
graphs of the two populations become indistinguishable after only a few
breeding seasons. (See Exercises, Question 10.)

chaotic
versus
non-
chaotic
growth

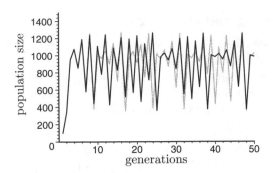

Figure 3.6.5: Sensitivity to initial conditions. Computer generated solutions of the discrete logistic equation for $r = 2.7$ and $K = 1\,000$, and for two different initial populations: $x_0 = 100$ (black line) and $x_0 = 101$ (grey line). As time increases the solutions move further apart.

implica-
tions of
chaotic
growth

The discovery that simple population models like this can give rise to chaotic growth had an extremely profound effect. Previously, when presented with records of population growth which seemed to have the population growing in a random manner, it was thought that this was due to some external factor such as climate, environment, etc. Now, however, it is known that the random-like behaviour may be a natural feature of the way that the population grows, that is, a nonlinear response to a time delay in compensating the reproduction rate to account for crowding. (It should be noted here that $r = \beta - \alpha$ and thus that large values of r are only valid for certain populations.)

Bifurcation diagram

plotting
equilib-
rium
solutions
against r

To summarise the different types of growth as the parameter r varies, it is useful to plot the possible equilibrium solutions against r. This is done in Figure 3.6.6 below. As r increases, we see that there is a single equilibrium population which increases gradually. At $r = 2$ the solution curve splits into two and we have two solutions. This is called a *bifurcation* and hence the diagram is called a bifurcation diagram. As r continues to increase the solution undergoes further bifurcations, corresponding to 4-cycles, 8-cycles and so on. Each bifurcation occurs for a relatively smaller increase in r. The `Maple` code used to produce the bifurcation diagram is as follows:

```
> restart:with(plots):
> n:=125: nrep:=32: K:=1000:
> r := array(1..n):    #values of bifurcation parameter
> X := array(0..n):    #population sizes
> y := array(1..nrep): #population size(s) for large n
> X0 := 0.1*K:
```

```
> rstart:=1.5: rend := 3:
> for i from 1 to n do
>    r[i] := evalf(rstart + i*(rend-rstart)/n);
>    for j from 0 to n-1 do
>       X[0] := X0:
>       X[j+1] := evalf(X[j]+r[i]*X[j]*(1-X[j]/K));
>    od:
>    for k from 1 to nrep do
>       y[k] := X[n-nrep+k]:
>    od:
>    pp := [[r[i],y[jj]] $jj=1..nrep];
>    bif[i] := plot(pp, r=rstart..rend, y=0..1.5*K,
         style=point,symbol=point):
>    col[i] := display([seq(bif[j],j=1..i)]);
> od:
> plot11:=display(col[n]):display({plot11});
```

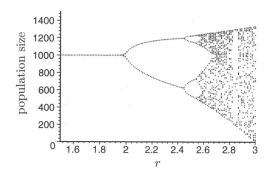

Figure 3.6.6: Bifurcation diagram for the discrete logistic model. This is a plot of the solutions, after a period of time, as the parameter r increases. They are the equilibrium solutions.

The continual bifurcation behavior eventually results in chaos. It might be useful to think of chaos as an infinite number of bifurcations. As the parameter r increases further some unexpected things happen. We appear to get small regions, or windows, of regular behaviour within the chaos. This includes the appearance of 3-cycles.

chaos

Even a simple equation, such as the logistic equation, exhibits a rich variety of behaviour. This is a relatively recent area of research, and for the interested reader there is an excellent series of articles in New Scientist which provide a good introduction to the subject, in particular Vivaldi (1989) and Stewart (1989). For an example of chaos in real populations the articles Costantino et al. (1995) and Costantino et al. (1997) describe experiments on flour beetle populations, where manipulating the adult mortality rate can cause the population to exhibit cycles

studying chaos

or become chaotic. They used three difference equations to describe the larvae, pupae and adult populations.

Limitations of the model

extinction

When $r > 3$ it is possible for the population to become negative. We would interpret this as the population becoming extinct. Of course this may not happen in practice as the stronger members of the population might survive into the next breeding season. Thus this model has a serious weakness if $r > 3$; however, it is useful in demonstrating how extremely complex patterns of growth can evolve from simple assumptions.

an improved model

A more robust model, which does not have negative populations for any values of r, is the model

$$X_{n+1} = X_n e^{a(1 - X_n/K)}, \qquad \text{where} \qquad a = \ell n(r + 1).$$

This model is derived from simple probabilistic arguments[2] and exhibits similar behaviour to the logistic model, with damped oscillations, 2-cycles, 4-cycles and chaotic growth behaviour (although for different values of r). It is superior to the logistic model for discrete growth since it incorporates a constant probability of death for each individual. This model is discussed in May (1981).

Skills developed here:
- Explore a difference equation model on a computer and describe the range of phenomena.
- Understand the different modelling approaches of discrete and continuous growth.
- Be able to discuss the properties of chaotic growth.

3.7   Time-delayed regulation

overview

We discuss, briefly, how to introduce a time lag response into the logistic model of population growth.

---

[2]The model assumes interactions between two different individuals are randomly distributed (as a Possion distribution) with a rate of contact proportional to $(1 - X_n/K)$.

Background

With the logistic model developed in Section 3.2, we incorporated an *instant-* *aneous* *response* instantaneous reaction to the environment. That is, increased pressure on the resources produces an *immediate response* from the system (in terms of, for example, more deaths).

Often this is not realistic in that the response usually takes effect after *time lag* some time delay, or time lag. Vegetation needs time to recover and changed environmental conditions which may lead to increased birth rates will take time to show up in the numbers of an adult population. This leads us to a model which includes a time delay which could result from a multitude of measurable sources (maturation times, food supply, resources, crowding, etc.). If the time lag is small compared with the natural response time $(1/r)$ then there is a tendency to overcompensate, which may produce oscillatory behaviour. This is what we observe in the following case study (Section 3.8).

Logistic equation with time lag

From Section 3.2, the logistic equation for population growth was *modified* *logistic* *equation*

$$\frac{dX}{dt} = rX(t)\left(1 - \frac{X(t)}{K}\right),$$

where $X(t)$ is the population size, r the intrinsic reproduction rate and K the carrying capacity for the environment. The term $r(1 - X(t)/K)$ represents the density dependent reproduction rate. This states that the reproduction rate decreases with population size owing to increased deaths from overcrowding. If we now suppose that this is applied at the earlier time $t - T$, where T represents the time delay between increased deaths and the resulting decrease in population reproduction, then we obtain

$$\frac{dX}{dt} = rX(t)\left(1 - \frac{X(t-T)}{K}\right). \tag{1}$$

This type of equation is known as a *differential-difference* equation.

We cannot use `Maple` directly to solve this, however, we can approximate *approxi-* *mating* *with a* *difference* *equation* it by writing

$$\frac{dX}{dt} \simeq \frac{X(t + \Delta t) - X(t)}{\Delta t}.$$

Let τ be defined as T measured as an integer multiple of Δt, $T = \tau \Delta t$. Defining our time scale so that $\Delta t = 1$ and writing $X(t) = X_n$ and $X(t + \Delta t) = X_{n+1}$, the differential-difference equation (1) becomes the difference equation

$$X_{n+1} = X_n + rX_n \left(1 - \frac{X_{n-\tau}}{K}\right). \tag{2}$$

This can be iterated from an initial condition using `Maple` by adapting the code provided in Section 3.6.

further
reading
We do not examine the model further here, but the interested reader will find details in May (1981). What follows is a case study adopting one such model.

3.8   Case Study: Australian blowflies

We now present an example of a time-delayed model, and, although we do not examine it in any depth, it serves to illustrate how the inclusion of a time lag produces oscillatory behaviour. These model results closely replicate Nicholson's data, which recorded the size of Australian blowfly populations. The following is adapted from May (1981).

Blowflies reproduce by laying their eggs in carcasses and meat or in the open wounds or sores of live animals. In the latter case the maggots hatch, and after eating the rotten flesh burrow into the animal, literally eating it alive. The open wound becomes worse, attracting more flies and so the process can snowball within a single day. Over the years, blowflies have caused a particular problem in the Australian sheep industry where farms may comprise many thousands of hectares, and the fly population is huge.

In the early part of this century A.J. Nicholson conducted single-species experiments on this Australian sheep-blowfly population (Lucilia cuprina). His results are well approximated by the model described by a differential-difference equation, which can itself be approximated by the difference equation

$$X_{n+1} = X_n + rX_n \left(1 - \frac{X_{n-\tau}}{K}\right).$$

Here X is the fly population size, r its reproduction rate and K its carrying capacity. Using `Maple` *this can be solved iteratively with the following relevant parameter values $r = 0.106\,\text{days}^{-1}$, $K = 2.8 \times 10^3$ flies and $\tau = 17$ days.*

The match between the data collected and the time-delayed model predictions can be seen in Figure 3.8.1 where the model output is imposed on Nicholson's data (obtained from May (1981)). The model performs well in spite of being somewhat simplistic in not including accurate details of reproduction, age classes or other details associated with the flies.

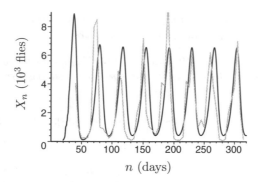

Figure 3.8.1: A plot of Nicholson's measurements (grey) with the model output (black). The model parameter values used are $r = 0.106\,\text{days}^{-1}$, $K = 2.8 \times 10^3$ flies and $\tau = 17$ days.

3.9   Exercises for Chapter 3

1. Farming fish. In a fish farm, fish are harvested at a constant rate of 2100 fish per week. The per-capita death rate for the fish is 0.2 fish per day per fish, and the per-capita birth rate is 0.7 fish per day per fish.

(a) Write down a word equation describing the rate of change of the fish population. Hence obtain a differential equation for the number of fish. (Define any symbols you introduce.)

(b) If the fish population at a given time is 240 000, give an estimate of the number of fish born in one week.

(c) Determine if there are any values for which the fish population is in equilibrium. (That is, look for values of the fish population for which there is no change over time.)

2. Modelling the spread of technology. Models for the spread of technology are very similar to the logistic model for population growth. Let $N(t)$ be the number of ranchers who have adopted an improved pasture technology in Uruguay. Then $N(t)$ satisfies the differential equation

$$\frac{dN}{dt} = aN\left(1 - \frac{N}{N^*}\right)$$

where N^* is the total population of ranchers. It is assumed that the rate of adoption is proportional to both the number who have adopted the technology and the fraction of the population of ranchers who have not adopted the technology.

(a) Which terms correspond to the fraction of the population who have not yet adopted the improved pasture technology?

(b) According to Banks (1994), $N^* = 17\,015$, $a = 0.490$ and $N_0 = 141$. Determine how long it takes for the improved pasture technology to spread to 80% of the population.

Note: This same model can be used to describe the spread of a rumour within an organisation or population.

3. Density dependent births. Many animal populations have decreasing per-capita birth rates when the population density increases, as well as increasing per-capita death rates. Suppose the density dependent per-capita birth rate $B(X)$ and density dependent death rate $A(X)$ is given by

$$B(X) = \beta - (\beta - \alpha)\delta\frac{X}{K}, \qquad A(X) = \alpha + (\beta - \alpha)(1 - \delta)\frac{X}{K}$$

where K is the population carrying capacity, β is the intrinsic per-capita birth rate, α is the intrinsic per-capita death rate and δ, where $0 \le \delta \le 1$, is a parameter describing the extent that density dependence is expressed in births or deaths.

Show that this still gives rise to the standard logistic differential equation

$$\frac{dX}{dt} = rX\left(1 - \frac{X}{K}\right).$$

4. Mouse population model. A population, initially consisting of 1000 mice, has a per-capita birth rate of 8 mice per month (per mouse) and a per-capita death rate of 2 mice per month (per mouse). Also, 20 mouse traps are set each week and they are always filled.

Write down a word equation describing the rate of change in the number of mice and hence write down a differential equation for the population.

5. Harvesting model. Consider the harvesting model from Section 3.3,

$$\frac{dX}{dt} = rX\left(1 - \frac{X}{K}\right) - h.$$

(a) Show there can be two non-zero equilibrium populations, with the larger value given by

$$X_e = \frac{K}{2}\left(1 + \sqrt{1 - \frac{4h}{rK}}\right).$$

For the parameter values used in Figure 3.3.2, calculate the value of this equilibrium population.

(b) If the harvesting rate h is greater than some critical value h_c, the non-zero equilibrium values do not exist and the population tends to extinction. What is this critical value h_c?

(c) If the harvesting rate is $h < h_c$, the population may still become extinct if the initial population x_0 is below some critical level, perhaps due to an ecological disaster. Show that this critical initial value is

$$x_c = \frac{K}{2}\left(1 - \sqrt{1 - \frac{4h}{rK}}\right).$$

(Hint: Consider where $X' < 0$.)

6. Fishing with quotas. In view of the potentially disastrous effects of over fishing causing a population to become extinct, some governments impose quotas which vary depending on estimates of the population at the current time. One harvesting model that takes this into account is

$$\frac{dX}{dt} = rX\left(1 - \frac{X}{K}\right) - h_0 X.$$

(a) Show that the only non-zero equilibrium population is

$$X_e = K\left(1 - \frac{h}{r}\right).$$

(b) At what critical harvesting rate can extinction occur?

Although extinction can occur with this model, as the harvesting parameter h increases towards the critical value the equilibrium population tends to zero. This contrasts with the constant harvesting model in Section 3.3 and Question 5, where a sudden population crash (from a large population to extinction) can occur as the harvesting rate increases beyond a critical value.

7. Predicting with a model. In a population the initial population is $x_0 = 100$. Suppose a population can be modelled using the differential equation

$$\frac{dX}{dt} = 0.2X - 0.001X^2,$$

with an initial population size of $x_0 = 100$ and a time step of 1 month. Find the predicted population after two months. (Use either an analytical solution, or a numerical solution from Maple.)

8. Plant biomass. Let the dry weight of some plant (that is, its biomass) at time t be denoted by $x(t)$. And suppose this plant feeds off a fixed amount of some single substrate, or a nutrient medium, for which the amount remaining at time t is denoted by $S(t)$. Assume that the growth rate of the plant is proportional to its dry weight as well as to the amount of nutrient available, and that no material is lost in the conversion of S into x.

(a) Starting with a word equation, model the rate of plant growth dx/dt with an initial plant biomass of x_0 and with x_f the amount of plant material associated with $S = 0$. (Use the fact that S can be written as a function of x and x_f: $S(t) = x_f - x(t)$.)

(b) Solve the equation using analytical techniques or numerically with `Maple`. (Separable techniques together with partial fractions are one way to solve the equation.)

(c) Using this model why can the plant biomass of x_f never be attained? (In spite of this, and the fact that plants do reach a maximum biomass with finite time, this model does give a reasonable prediction of annual plant growth.)

9. Modelling the population of a country. Consider the population of a country. Assume constant per-capita birth and death rates, and that the population follows an exponential growth (or decay) process. Assume there to be significant immigration and emigration of people into and out of the country.

(a) Assuming the overall immigration and emigration rates are constant, formulate a single differential equation to describe the population size over time.

(b) Suppose instead that all immigration and emigration occurs with a neighbouring country, such that the net movement from one country to the other is proportional to the population difference between the two countries and such that people move to the country with the larger population. Formulate a coupled system of equations as a model for this situation.

In both (a) and (b) start with appropriate word equations and ensure all variables are defined. Give clear explanations of how the differential equations are obtained from the word equations.

10. Sensitivity to initial conditions. Referring to the results generated in Figure 3.6.5 for two separate initial conditions, $x_0 = 100$ and $x_0 = 101$, generate the results with these initial conditions for $r < 2.7$, $r = 2.5$ and $r = 1.9$. What do you notice about the distance between the two graphs at each time step? (The relevant code is given in Section 3.6.)

11. Investigating parameter change. Using `Maple` examine the effect of increasing the parameter r on the solution to the equation

$$X_{n+1} = X_n e^{a(1-X_n/K)}, \qquad \text{where} \qquad a = \ln(r+1).$$

Establish (roughly) for what values of r the system undergoes its first two bifurcations. (Code can be adapted from that in Section 3.6.)

12. Modelling insects with a difference equation. Many insect populations breed only at specific times of the year and all the adults die after breeding. These may be best modelled by a difference equation, such as

$$X_{n+1} = r(X_n - 0.001X_n^2).$$

Using `Maple` investigate what happens as the parameter r (the growth rate) is varied from $r = 0$ to $r = 3$. Sketch all the different types of growth patterns observed, together with the corresponding value of r.

13. Aquatic environments. A chemostat is used by microbiologists and ecologists to model aquatic environments, or waste treatment plants. It consists of a tank filled with a mixture of some medium and nutrients, which microorganisms require to grow and multiply. A fresh nutrient–medium mixture is pumped into the tank at a constant rate F and the tank mixture is pumped from the tank at the same rate. In this way the volume of liquid in the tank remains constant. Let $S(t)$ denote the concentration of the nutrient in the tank at time t, and assume the mixture in the tank is well stirred. Let $x(t)$ denote the concentration of the microorganism in the tank at time t.

(a) Draw a compartmental diagram for the system.

(b) In the absence of the organism, suggest a model for the rate of change of $S(t)$.

(c) If the microorganisms' per-capita uptake of the nutrient is dependent on the amount of nutrient present and is given by $p(S)$, and the per-capita reproduction rate of the microorganism is directly proportional to $p(S)$, extend the model equation above to include the effect of the organism. (The per-capita uptake function measures the rate at which the organism is able to absorb the nutrient when the nutrient's concentration level is S.)

(d) Now develop an equation describing the rate of change of the concentration of the live organism (x') in the tank to derive the second equation for the system.

(e) The nutrient uptake function $p(S)$ can be shown experimentally to be a monotonically increasing function bounded above. Show that a Michaelis–Menten type function

$$p(S) = \frac{mS}{a + S},$$

with m and a positive, non-zero constants, satisfies these requirements. What is the maximum absorption rate? And why is a called the half-saturation constant? (Hint: The maximum absorption rate is the maximum reached by $p(S)$. For the second part consider $p(a)$.)

This system of equations is known as the Monod Model for single species growth and was developed by Jaques Monod in 1950.

14. Stability of 2-cycles. Consider the discrete logistic equation (with $K = 1$)

$$X_{n+1} = X_k + rX_k(1 - X_k).$$

(a) Show that every second term in the sequence $X_0, X_1, X_2, \ldots$ satisfies the difference equation

$$
\begin{aligned}
X_{k+2} &= (1 + 2r + r^2)X_k - (2r + 3r^2 + r^3)X_k^2 \\
&\quad + (2r^2 + 2r^3)X_k^3 - r^3 X_k^4.
\end{aligned}
$$

(b) For equilibrium solutions, let $S = X_{k+2} = X_k$ and obtain a quartic equation (that is, an equation with the unknown raised to the fourth power, at most). Explain why $S = 0$ and $S = 1$ must be solutions of this quartic equation. Hence show that the other two solutions are

$$S = \frac{(2 + r) \pm \sqrt{r^2 - 4}}{2r}.$$

(c) Explain why these values, when they are real numbers, represent the values between which the population oscillates. (Compare this solution to Figure 3.6.3 for $r = 2.2$.)

Chapter 4

Numerical solution of differential equations

This chapter provides a brief overview of numerical procedures on which we rely when employing software as a tool in the solution and analysis of mathematical models. While developments have ensured that in most cases the methods are robust, it is important to understand the trade-offs we must accept when using them, and the possible errors which could accumulate, making interpretation inaccurate. Perhaps the most important part of this chapter is the final discussion.

4.1   *Introduction*

overview   In finding solutions to differential equations, plotting trajectories and dis-
playing time-dependent graphs, we have come to rely on the performance
of computers and software packages. Some packages use analytical solutions where possible, a symbolic approach such as in `Maple`, but many
problems cannot be solved analytically and there is a need to employ
numerical schemes to find a solution. In this case the numerical solution,
whilst possibly extremely accurate, is an approximation to the exact analytic solution.

sources of   There are some major drawbacks in accepting whatever is produced by
error   a computer, particularly when numerical solvers are used, as errors may
accrue for the following reasons:

- there are *round-off errors* and these increase with an increase in
  the number of calculations performed,
- there are *discretisation errors* resulting from the estimation of a
  solution for a discrete set of points; these may decrease with an
  increase in the number of points used,
- there are errors due to the *estimating procedure*, or method of approximation, used by the numerical solver.

drawbacks   Many different numerical procedures are available for finding derivatives,
integrals, sums etc. First we provide a very brief outline of a few used
in finding solutions to differential equations (the main use of solvers in
this book), and we then consider some of the drawbacks in using these
numerical approximations.

4.2   *Basic numerical schemes*

overview   To examine some numerical schemes used in solving differential equations, we consider the general problem $y' = f(t, y)$ with $y(t_0) = y_0$,
and apply the schemes which follow to calculate (estimate) $y(t)$ with
$t_0 < t < T$. We begin with a very simple method, Euler's method, to
gain a basic understanding of the process any method needs to adopt, and
then look briefly at a well-known and widely used method, the Runge–
Kutta method.

Euler's Method

Clearly a computer cannot calculate every point on a curve as there are infinitely many, so an approximation is made for a discrete set of points. Euler's method provides a very simple way of approximating the solution of a differential equation at a discrete set of points.

approximate solution of a DE

The steps in Euler's method are as follows:

the basic idea

- Divide the interval into N equal sections, then $t_n = t_0 + nh$ for $n = 0, 1, \ldots, N-1$ and $h = (T - t_0)/N$ is the step size.
- We know that (t_0, y_0) is on the curve and we approximate $y_n = y(t_n)$. To approximate y_1 we follow the tangent of the curve through (t_0, y_0) (which is known) extending it to t_1. Then

$$y_1 \approx y_0 + hf(t_0, y_0)$$

or

$$y' = f(t_0, y_0) \approx \frac{y_1 - y_0}{h} = \frac{y_1 - y_0}{t_1 - t_0}.$$

- We follow this procedure repeatedly: $y_2 \approx y_1 + hf(t_1, y_1)$ and so on for $y_3, \ldots, y_n$.

Figure 4.2.1 illustrates this procedure.

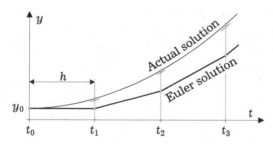

Figure 4.2.1: Diagram of Euler's method for solving differential equations.

In general Euler's method is the recursive scheme

the algorithm for Euler's method

$$y_{n+1} = y_n + hf(t_n, y_n)$$

with

$$t_{n+1} = t_n + h \quad \text{and} \quad 0 \le n \le N - 1.$$

The following example shows how to apply Euler's method.

Example 1:   Use Euler's method, with step size $h = 0.1$, to find $y(0.2)$ for the differential equation

$$\frac{dy}{dt} = y^2 + t, \qquad y(0) = y_0 = 1.$$

Solution: For this equation $f(t, y) = y^2 + t$. Using Euler's method, with $h = 0.1$, and letting $n = 0$, we obtain

$$y_1 = y(0.1) = y_0 + hf(t_0, y_0) = 1 + 0.1 \times (1^2 + 0) = 1.1.$$

Now, with $n = 1$, we obtain

$$y_2 = y(0.2) = y_1 + hf(t_1, y_1) = 1.1 + 0.1 \times (1.1^2 + 0.1) = 1.231.$$

---

link with
Taylor's
Theorem

Taylor's Theorem states that

$$y_{n+1} = y_n + h\frac{dy_n}{dt} + \frac{h^2}{2!}\frac{d^2 y_n}{dt^2} + \dots .$$

(Taylor's Theorem provides an accurate polynomial approximation to a large group of functions. For details see Appendix A.3.)

order of
approxi-
mation

Comparing this approximation with Euler's method, it is clear that the latter consists of the first two terms in this expansion. We say that Euler's method is a *first-order approximation*. (Including further terms in this series would produce second-order, third-order etc., approximations.)

error
analysis

With any approximation, and thus with each numerical scheme, comes an associated error term. Since the approximation in Euler's method consists of only the first two terms of the Taylor series, it would be useful to be able to calculate the error term and then perhaps control the size of the error by choice of the step size. From Euler's method we have the estimated value

$$y_{n+1} = y_n + hf(t_n, y_n) \quad n = 0, \dots, N-1.$$

From Taylor's Theorem we have the actual value

$$y(t_{n+1}) = y(t_n) + hf\left(t_n, y(t_n)\right) + \frac{h^2}{2}f'(\xi_n, y(\xi_n))$$

where ξ_n lies between t_n and t_{n+1}.

Subtracting the estimate from the true value we get

$$y(t_{n+1}) - y_{n+1} = y(t_n) - y_n$$

$$+ h\left[f(t_n, y(t_n)) - f(t_n, y_n)\right] + \frac{h^2}{2}f'(\xi_n, y(\xi_n))$$

which is the error term E_{n+1}. So for each step of the method an error E_{n+1} is incurred. We can find an upper bound for this error term and it can be shown that

$$E_{n+1} \leq \frac{Dh}{2L}\left[e^{T-t_0} - 1\right]$$

where L is an upper bound for f and D is an upper bound for f' on the interval $[t_0, T]$ where $f(t, y)$ is continuously differentiable and $T = t_0 + (N-1)h$.

Clearly $\lim_{h \to 0} E_{n+1} = 0$ indicating increasing accuracy of the estimation, with decreasing step size. (For a full analysis see, for example, Kincaid and Cheney (1991).)

However, it should be remembered that every computational operation (for example, addition, division, etc.) comes at a cost, as each may incur an error. This is a result of the fact that a computer can only store a fixed number of digits and so a number may be rounded off at each step. Thus, while decreasing the step size h (or equivalently increasing the number of points in the interval N), we are simultaneously increasing the number of operations or calculations and are thus increasing the impact of round-off errors. For best performance then, we should aim at finding some optimum h such that the combined effect of these two errors is minimised.

Runge–Kutta Methods

One-step algorithms that use averages of the slope function $f(t, y)$ at two or more points over the interval $[t_{n-1}, t_n]$ in order to calculate y_n are called *Runge–Kutta methods*. They are also examples of *predictor–corrector methods* as they make predictions for 'next' values, and then with a series of weights, correct them.

The fourth-order Runge–Kutta method (RK4) is one of the most widely used methods of any step algorithms. It involves weighted averages of slopes at the midpoint and end points of the subinterval. For the IVP $y' = f(t, y)$, $y(t_0) = y_0$, the fourth order RK method is a one step method (that is, y_n depends only on y_{n-1}) with the constant step size h and is given by

$$y_{n+1} = y_n + \frac{h}{6}(k_1 + 2k_2 + 2k_3 + k_4)$$

where

$$k_1 = f(t_n, y_n)$$
$$k_2 = f(t_n + \frac{h}{2}, y_n + \frac{h}{2}k_1)$$
$$k_3 = f(t_n + \frac{h}{2}, y_n + \frac{h}{2}k_2)$$
$$k_4 = f(t_n + h, y_n + hk_3).$$

Heun's method   Another Runge–Kutta method using a simpler scheme of averaging than the RK4 above, is Heun's method. This is illustrated in Figure 4.2.2.

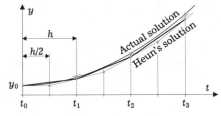

Figure 4.2.2: Diagram of Heun's method for solving differential equations.

accuracy   Like Euler's scheme this method is a one-step method, but it is more accurate. With a constant step size h the general form of Heun's method is given by

$$y_{n+1} = y_n + \frac{h}{2}(k_1 + k_2)$$

where

$$k_1 = f(t_n, y_n),$$
$$k_2 = f(t_n + h, y_n + hk_1).$$

4.3   Computer implementation

Most ODE solver packages use sophisticated methods of controlling er-   general
rors, and being efficient most of the time they have widespread accep-   software
tance. Many schemes choose their own step sizes in order to optimise the
results, and some vary this step size on an interval in order to provide
an optimal result.

Note, once again, that there is a trade-off between decreasing the step   trade-offs
size and increasing the round-off error, due to an increased number of
computations. Another problem is that of instability over long time
periods: sometimes, whilst providing an accurate approximation over an
initial period of time, the solution may 'become' very inaccurate at later
times.

Using `Maple` the method of integration can be stipulated and the results   Maple
compared. `Maple` also allows you to change and choose the step size   implemen-
of the default numerical scheme which, as we see below, is required for   tation
adequate resolution.

In Chapter 2, in the discussion on finding a computer solution to a ra-   stiff
dioactive element decay cascade, we mentioned that it would be difficult   problems
to find a suitable numerical scheme because of the vastly different half-
lives. (For Uranium-238 the half-life is billions of years whereas for the
next element in its decay series, Thorium-234, the half-life is in days.)
This problem requires large time steps to resolve the Uranium decay process and (comparatively) very small time steps to resolve the Thorium
decay: such a problem is known as a *stiff problem*. This discrepancy
between rates can be seen clearly in the following example. Suppose a
solution to some differential equation $X' = f(x)$ is

$$X(t) = e^{-t} + e^{-1000t}, \qquad t \geq 0.$$

Clearly, the second component decays at a much faster rate than the   decay
first. When t is small, the value of $X(t)$ is dominated by e^{-1000t}, and   rates
small step sizes are required for resolution of this behaviour. Alternatively, for t away from 0, the solution is dominated by e^{-t} and large step
sizes (in comparison) can be used for high accuracy. As this is a problem for many numerical methods, and as the solution to the differential
equation problem is unknown before a numerical scheme is chosen, this
group of problems often prompts an error message in the more commonly
used schemes. Other procedures have been developed to deal specifically
with stiff problems, but typically they increase the number of computations incurred substantially when the problem is not stiff. Thus, they are

usually employed only once a problem is known to be stiff. `Maple` provides a stiff solver which can be used for the solution of such problems by stipulating the numerical method (always after `numeric`) in `dsolve`. Applying this to the example of lake pollution (Section 2.5), which is not a stiff problem, it is easy to establish a substantial increase in computation time.

```
>  restart:with(plots):
>  cin:=3;V:=28;F:=4*12;threshold:=4;init_c:=10;
>  de1:=diff(c(t),t)=(F/V)*(cin-c(t));
>  soln:=c0->dsolve({de1,c(0)=c0},c(t),numeric,method=lsode):
>  plot1:=c0->odeplot(soln(c0),[t,c(t)],0..8):
>  list1:=seq(plot1(i/2),i=1..12):
>  line1:=plot([[0,threshold],[8,threshold]]):
>  display({list1,line1});
```

stability of numerical schemes A problem which frequently arises when using `Maple` to sketch solutions in a plane is a very odd shaped or angular scribble. By now, if you have experimented with `Maple`, you will no doubt have encountered this problem in some implementation. Consider a pair of differential equations in time, $de1 = dX/dt$ and $de2 = dY/dt$. Then the following code fragment without the step size stipulated produces the mess in the first diagram of Figure 4.3.1, while with the step size included we obtain the second diagram:

```
>  ...
>  plot1:=DEplot([de1,de2],[X,Y],t=0..80,{inits},X=0..Xlimit,Y=0..Ylimit,
   scene=[P,H],linecolor=black,stepsize=0.5,arrows=none):
>  display(plot1);
```

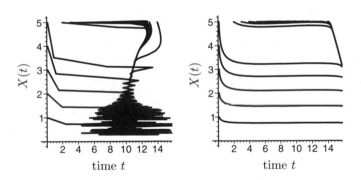

Figure 4.3.1: Solution of a differential equation for different initial conditions. In the diagram on the left a large step size was used, and instability of the numerical scheme prevents convergence. In the diagram on the right the step size of the solver has been reduced.

instability When the step size is specified the problem 'rights' itself as displayed in the second diagram of Figure 4.3.1. This problem is that of instability of the numerical scheme. Typically this can be improved with a decrease in the step size, but sometimes the values of a step size for which convergence to the solution ensues can be extremely small and depend on the differential equation itself. (We see an example of this in Section 4.4.) Furthermore, as the step size decreases, there is an increasing error due to a larger number of operations which once again becomes an issue.

4.4 Instability

Recall that Euler's method is a recursive application of

Euler's method

$$y_{n+1} = y_n + hf(t_n, y_n)$$

where $t_{n+1} = t_n + h$ and $0 \leq N - 1$.

Note also that the logistic equation is given by

logistic equation

$$y' = ry\left(1 - \frac{y}{K}\right) = ry(1-y) \quad \text{when} \quad K = 1$$

with $y(0) = y_0$ and $r > 0$, $K > 0$. We know that the resulting solution is a smooth monotonically increasing, or decreasing, function approaching $K = 1$ (see Chapter 3).

What we now consider is how well Euler's method predicts or approximates the logistic equation solution as h (the step size) increases. Applying Euler's method to the logistic equation (with $K = 1$) gives successive approximations for $y_1, y_2, \ldots$ with

applying Euler's method to the logistic equation

$$y_{n+1} = y_n + hry_n(1 - y_n) \quad n = 0, 1, 2, \ldots \tag{1}$$

Note here that Euler's equation is a difference equation estimating a differential equation. Furthermore, this equation is identical to that considered in Section 3.6 (to describe discrete population growth) if rh of (1) is set equal to parameter r of that section, and the carrying capacity is taken as $K = 1$.

equivalent to discrete growth

Applying the results of that section to this case, for small rh the estimation approaches the analytic solution smoothly, however, for a small increase in rh, the estimation will oscillate about the analytic solution. And for some further increase, the numerical solution predicts chaos.

using previous results

This means that when solving the logistic equation with parameter r using Euler's method, if the parameter rh is sufficiently small then the

relation to numerical methods

method will converge to the analytic solution. If not, then the method will never converge and the sequence of generated values may oscillate or behave chaotically. For any fixed r, there is not one single value for h (the step size of the method) that will ensure convergence to the analytic solution in all cases. Choice of the step size would need to depend on the parameter r, as otherwise the solution would not necessarily be stable, regardless of how small the step size is taken.

4.5   Discussion

choice of method

Thus numerical methods need to be chosen carefully for accurate results. Euler's method, for example, is not a good choice for situations in which high accuracy is desired. Likewise, it brings us to the realisation that it would be very valuable to have some prior knowledge of expected behaviour/results in order to interpret and evaluate numerical solutions correctly.

need for theoretical under-standing

To this end we have already developed some tools, and will develop others, both analytical and graphical, to establish aspects concerning the behaviour of the solution. These can then be compared with the numerical results. When using computational methods, it is commonplace to run experiments determining values which are already known exactly (through some theoretical or experimental means) in order to establish consistency with expected results.

under-standing the dynamics

Another essential aspect of generating numerical solutions to be borne in mind is that a single solution to a single set of parameter values, and for a single set of initial values, may impart very little information about the general behaviour of a system. Analytical tools are essential in providing a framework for choosing which numerical solutions to generate in order to gain an understanding of the system dynamics. In Chapter 6, in particular, we see how well analytical and numerical methods can combine to provide this information.

> **Skills developed here:**
> - Understand the general principles of applying numerical methods, and how the process contributes towards generating errors in computed solutions.
> - Understand the notion of a stable or unstable numerical scheme.
> - Understand the difference between an exact and an estimated solution.
> - Know how to choose a numerical scheme offered by `Maple`.

4.6 Exercises for Chapter 4

1. Round-off errors. *A simple example can illustrate how the round-off errors accumulate to provide inaccuracy.*

Using `Maple` *set*

$$z = (x+1)^{12} - (4x+3)^6.$$

(a) *Set $x = 1 + \sqrt{3}$ and evaluate z.*

(b) *Simplify z and compare this solution with that in (a).*

Their difference arises through round-off errors which accumulate from the calculations required for the evaluation in (a), compared with the analytic approach of (b) in which no numerical calculations occur.

Some useful `Maple` *commands are:*

```
> ...
> x:=(1+sqrt(3));z;
> evalf(z);
> simplify(z);
```

2. Numerical schemes. *A variety of numerical methods can be used to solve the same equation and thus give an indication of their differences, in terms of estimation, compared with the analytical solution. Consider the simple IVP*

$$\frac{dy}{dt} = y, \qquad y_0 = y(0) = 1.$$

(a) *Using* `Maple` *(which uses a symbolic approach) find and plot the analytical solution.*

(b) *Now create further plots choosing particular numerical methods as follows:*

```
> ...
> plot2:=DEplot(eqn1,[y(t)],t=0..4,y=0..maxy,{[y(0)=1]},
    iterations=5, stepsize=0.1, arrows=NONE, linecolor=blue,
    method=classical[foreuler]):
> plot3:=DEplot(eqn1,[y(t)],t=0..4,y=0..maxy,{[y(0)=1]},
    stepsize=0.1, arrows=NONE, linecolor=red,
    method=classical[heunform]):
> plot4:=DEplot(eqn1,[y(t)],t=0..4,y=0..maxy,{[y(0)=1]},
    stepsize=0.1, arrows=NONE, linecolor=green, method=classical[rk4]):
> display(....);
```

Display these on the same axes to see how they differ from the analytical solution.

(c) *Try changing the stepsize in the methods above and compare the results.*

3. Numerical and analytical solutions. *One can get an idea of how the error between the analytical and numerical solutions grows by plotting that error. To this end, for the IVP in Question 2, use the following code to examine these errors, comparing a variety of time steps and methods. Notice the rate of growth in the error functions as t increases.*

```
> restart:with(plots):
> lasttime:=10:
> timelist:=seq(i,i=1..lasttime):
> deq:=diff(y(x),x)=y(x):
> init:=y(0)=1:
> Digits:=20:
Three methods of solution
> ans1:=dsolve({deq,init},y(x),
    type=numeric, method=classical[foreuler],
    value=array([timelist]), stepsize=0.1):
> ans2:=dsolve({deq,init},y(x),
    type=numeric, method=classical[heunform],
    value=array([timelist]), stepsize=0.1):
> ans3:=dsolve({deq,init},y(x),
    type=numeric, method=classical[rk4],
    value=array([timelist]), stepsize=0.1):
Compare the solutions
> plot13:=plot(
    [seq([ans1[2,1][i,1],ans3[2,1][i,2]-ans1[2,1][i,2]],i=1..lasttime)],
    style=line):
> plot23:=plot(
    [seq([ans2[2,1][i,1],ans3[2,1][i,2]-ans2[2,1][i,2]],i=1..lasttime)],
    style=line, colour=blue):
Compare the numerical solutions with the analytical solution
> plot10:=plot(
    [seq([ans1[2,1][i,1],exp(i)-ans1[2,1][i,2]],i=1..lasttime)],
    style=line, colour=red):
> plot20:=plot(
    [seq([ans2[2,1][i,1],exp(i)-ans2[2,1][i,2]],i=1..lasttime)],
    style=line, colour=blue):
```

```
> plot30:=plot(
  [seq([ans3[2,1][i,1],exp(i)-ans3[2,1][i,2]],i=1..lasttime)],
  style=line, colour=green):
Display results
> display({plot13,plot23}, view=[0..lasttime,0..10]);
> display({plot10,plot20,plot30}, view=[0..lasttime,0..10]);
```

4. Discretisation and round-off errors. *The number of steps chosen for a numerical method will also have an impact on the accuracy of the solution, with an increase in the number of steps reducing the discretisation error. However, when the step size approaches the accuracy of the machine (the smallest number representable by the machine) the error produced by the increased number of round-off errors becomes dominant, and the combined error increases reducing the accuracy of the solution. To illustrate this we can use the same IVP as above (Question 2) and set the machine accuracy through the variable* Digits.

(a) *Using the* Maple *code as given below, compare the errors in the RK4 method (fourth-order Runge–Kutta method) as the step size decreases. Use step sizes of 0.1, 0.01 and 0.001 and set the machine accuracy* (Digits) *to 10.*

(b) *What happens when the step size is close to the maximum accuracy of the machine? To see this set the machine accuracy (through* Digits*) to 5 and rerun the code.*

```
> restart:with(plots):
> lasttime:=5:Digits:=10:
> timelist:=seq(i,i=1..lasttime):
> deq:=diff(y(x), x)=y(x):
> init:=y(0) = 1:
Analytic solution
> ans0:=dsolve({deq,init},y(x));
Numerical solutions with different stepsizes
> ans1:=dsolve({deq,init},y(x),type=numeric,method=classical[rk4],
  value=array([timelist]),stepsize=0.1):
> ans2:=dsolve({deq,init},y(x),type=numeric,method=classical[rk4],
  value=array([timelist]),stepsize=0.001):
> ans3:=dsolve({deq,init},y(x),type=numeric,method=classical[rk4],
  value=array([timelist]),stepsize=0.0005):
Compare the solutions
> plot10:=plot(
  [seq([ans1[2,1][i,1],abs(exp(i)-ans1[2,1][i,2])],i=1..lasttime)],
  style=line,colour=red):
> plot20:=plot(
  [seq([ans2[2,1][i,1],abs(exp(i)-ans2[2,1][i,2])],i=1..lasttime)],
  style=line,colour=blue):
> plot30:=plot(
  [seq([ans3[2,1][i,1],abs(exp(i)-ans3[2,1][i,2])],i=1..lasttime)],
  style=line,colour=green):
Display the results
> display({plot10,plot20,plot30});
```

Part II

Models of Interacting Populations

Chapter 5

Formulating interacting population models

Interacting population models are relevant where two or more populations depend on each other. We study, in detail, four examples of interacting populations: (i) a predator–prey interaction, (ii) a competing species interaction, (iii) an epidemic model, and (iv) a model of a battle between two opposing groups. In this chapter, we concentrate on formulating the differential equation model which is governed by two simultaneous first-order differential equations. We solve the differential equations numerically using `Maple`. *In Chapter 6, we return to the models and develop analytic techniques which give more general insights into the models developed here.*

5.1   Introduction

overview

Ecological systems may be extraordinarily complex—an inter-related system of plants and animals, predators, prey, flowering plants, insects, parasites, pollinators, seed dispersing animals etc. In such systems there is a constant stream of arrivals and departures involving time periods of millions of years. New species evolve or arrive, others decline to extinction or migrate. Human interference has impacted hugely on most of the world's ecosystems, particularly over the last 200 years, with one seemingly insignificant species extinction able to spark a cascade of effects throughout the trophic levels.

Interacting populations

interaction and stability

Large numbers of one species may be unaffected by others, however, in some instances removing, introducing or modifying one resource or species through (for example) harvesting or poisoning, may have wide ranging ramifications for the system. Evidence suggests that, typically, communities with many interacting species have greater stability than those comprising much simpler systems. That is, while rainforests are stable, cultivated land and orchards are relatively unstable and the populations of species in laboratory controlled predator–prey systems undergo large oscillations.

mathematical models

The relationships between the species within a system are often highly nonlinear, so that it is extremely difficult to establish, with certainty, a precise mathematical model describing the processes involved. However, there is clearly a need to understand these systems, or aspects of them. To this end we develop simple mathematical models where the independent variable is time and the dependent variables are the densities (numbers per area) of the various different interacting populations. We include case studies of interactions where these simple models, or extensions of them, appear to reflect the main processes involved.

some examples

The most obvious examples come from nature, where different species interact with each other. Examples of populations which interact include different age groups within a population, predator–prey interactions, the spread of a disease between population groups and battles between opposing groups.

predator–prey

There are many different types of interactions between populations in different trophic levels. One such interaction, where some species use other species as their food supply, is the predator–prey interaction. We study the simplest example of it in this chapter.

Another interaction process between populations on the same trophic level is that of competing species, where two or more populations compete for a limited resource, such as food or territory. We model this system in a manner very similar to that developed for the predator–prey system.

competing species

A different type of interaction, which occurs within a single species, is the interaction between those who are infected with a disease and those who are not. This effectively divides the population into two separate sub-populations, where the growth or decline of one sub-population is affected by the size of the other sub-population. We study, in detail, a simple model for the growth and then decline of the number of individuals in a population infected with a disease, and compare the model predictions with data recorded during a flu (influenza) epidemic in a boarding school.

epidemics

The fourth example we study is that of a battle between two opposing groups, such as may occur between two insect or human populations. The rate at which 'soldiers' are wounded, or killed, in a battle depends largely on the number of enemy 'soldiers'. We focus on a very simple human war model for aimed fire, where a soldier aims at a target with a given probability of hitting it. The model is compared with an actual battle in World War II for which daily records of the number of casualties were kept.

battles

Modelling assumptions and approach

Different populations are expected to have different birth and death rates. For some populations the size of another population will affect these parameters, and the populations are then said to interact with each other.

different rates of birth and death

In each of the examples which follow, we make some simplifying assumptions and then build the mathematical model on these. We assume that the populations are sufficiently large, so that we can neglect random differences between individuals. Furthermore, we also assume the growth to be continuous, rather than discrete, over time. The model we derive for the rate of change of each population over time, is based upon the input–output principle of the balance law: the compartmental model technique. This approach leads to two, or more, simultaneous differential equations, or a system of equations.

basic modelling assumptions

Models can describe the number of individuals (population size), as will be the case for the epidemic and battle models we develop. Alternatively, the population density, number per unit area, can be modelled as in the predator–prey and competing species models which follow. For both, the approach is the same.

Systems of differential equations

system of equations

Many processes are described by more than one differential equation. When these equations need to be satisfied simultaneously the set of equations is known as a *system of differential equations*. Although systems can comprise many differential equations and many unknowns, in this book we consider mainly systems of first-order equations where the number of unknowns is the same as the number of equations in the system. This ensures a unique solution to the system when the initial conditions are specified.

dynamical system

The system of equations is known as a *dynamical system* if it allows prediction of future states given present and/or past states. In most of the models we develop, such as in the growth and decay processes examined in previous sections and the population models we develop here, this is the case.

nonlinear equations

We also develop systems in which the equations are *nonlinear*: that is, they include products of the dependent variables or their derivatives. While appearing simple, they do not all have analytic solutions and we rely on the approximate solution to these systems obtained by `Maple`, which employs numerical solvers to find them. Later, in Chapter 7, we develop some theory about linearisation which allows us to predict the behaviour of the system by considering linear approximations to the nonlinear equations.

coupled equations

The systems of equations may be *coupled*, which implies that their solutions are interdependent. For example, in the case of two equations $x'(t) = F(x, y, t)$ and $y'(t) = G(x, y, t)$, if the solution of x' depends on the value of y and the solution of y' depends on x, then the equations are coupled. Alternatively, as was the case in Section 2.7 in the example of drug assimilation, the first equation could be solved independently of the second, and this system is said to be *uncoupled*.

5.2   *Case Study: Nile Perch catastrophe*

We cite here a case where the ramifications of a disturbance within an interacting population system were wide reaching and caused an ecological disaster for the communities living in and around Lake Victoria. It emphasises the responsibility which should accompany any unsuitable modification of environmental processes, and the need for good modelling and forecasting, incorporating a multidisciplinary approach.

Adapted from Murray (1990) and Quammen (1997).

Lake Victoria, which feeds into the Nile River, is the largest and most northern of a series of lakes punctuating the Rift Valley in Africa. Round as a pond and nearly as big as Ireland it is bordered by Kenya, Tanzania and Uganda, and until 1960 it supported many communities along its shores providing them with a large and diverse population of fish as well as fresh water. In particular, it supported an abundance of cichlids (tropical fresh water fish of the family Cichlidae) in both quantity and species. The sort of fish known for their garish appearance and collected for aquaria.

Because of its scalloped shoreline, its irregular patterns of depth and shallows, and its subjection to periods of intense drought and low water levels during the centuries long history of the lake, pockets of cichlids were cut off from others for long periods. In this way they diverged and speciated. Later reconnection allowed the species to mix spatially, but they had become incompatible reproductively and were now in competition. In order to survive they radiated: that is, they settled in different niches within the environment. Thus the lake represented a kind of underwater Galapagos archipelago with species for every niche: rock-scrapers, sand-digging-insect eaters, plant-scrapers, scale-eaters, nibbler-of-other-fishes'-fins, fish-eaters and the famed Haplochromis compressiceps known for biting out eyeballs! Lake Victoria supported about 200 such cichlid species, all of which are thought to have descended from a single ancestral species.

In 1960, supported by the United Nations Food and Agriculture Organisation, the Nile Perch was introduced into the lake to provide an additional source of protein. (The fish can grow to 100 kg or more.) Objections were voiced, however, they were ignored, the introduction went ahead and the perch thrived. Being large and carnivorous the perch all but wiped out the smaller fish species; hundreds of varieties unknown elsewhere. Many of these fish also provided the staple foods and economies for the local fishing communities. The markets became flooded with perch and the overall productivity of the lake was reduced by 80% (from the 1960 levels) in 25 years.

The ramifications spread far wider than the local ecological and economic disaster. The Nile Perch are oily and cannot be sun dried, the traditional means by which fish had been preserved. Instead they had to be smoked which resulted in major felling operations in the local environs to provide the fuel.

Furthermore, the cichlids had controlled the level of a particular snail which lives in and around the lake. The snails are an essential link in the spread of the disease bilharzia (liver fluke disease), which can be fatal to humans. Essentially, bilharzia is a disease produced by the larvae of a flatworm. The larvae first infect certain freshwater snails in streams and rivers, but particularly in lakes. They develop in the digestive gland of the snail and are discharged into the lake as larvae. From there they penetrate the skin of animals, such as humans, with a muscular boring action. Once inside the body they mature to an adult fluke and attach to the intestine or bladder and begin producing eggs. The effect on humans is a range of unpleasant problems of the kidney, liver, lung, intestine or central nervous system and can be seriously debilitating or fatal. Without the cichlids there was a marked increase in the number of snails and hence also the disease.

All in all, the perch introduction was catastrophic. Many of the ramifications could have, and should have, been foreseen and avoided. However, in spite of this, further introductions were being planned in the late 1980s for Nile Perch into other African lakes such as Lake Malawi.

5.3   *Predators and prey*

overview
In the following sections, we develop a predator–prey model for carnivores using the growth of a population of small insect pests which interact with another population of beetle predators. An example of a model for herbivores is examined in Section 8.6, while models for parasitic interactions or cannibalism will be simple to derive from these given examples.

Background

different types of predator–prey interactions
There are several types of predator–prey interactions: that of herbivores which eat plant species, that of carnivores which eat animal species, that of parasites which lives on or in another species (the host), and that of cannibals which eat their own species, often an interaction between the adults and young.

pests in the citrus industry
One interesting example of a predator–prey interaction occurred in the late nineteenth century, when the American citrus industry was almost destroyed by the accidental introduction from Australia of the cottony cushion scale insect. To combat this pest its natural predator, the Australian ladybird beetle, was also imported, but this did not solve the problem and finally DDT was used to kill both predator and prey in a bid to eradicate the pest. Surprisingly, application of DDT to the orchards led to an increase in the scale insects, the original pest, suggesting that the use of pesticide is advantageous to the pest!

Model assumptions

We make a few preliminary assumptions on which to build the model.   some assumptions

- Initially we assume the populations are large, sufficiently large to neglect random differences between individuals.
- We ignore the effect of DDT initially, but modify the model later to incorporate its impact on the system.
- We also assume there are only two populations, the predator and the prey, which affect the ecosystem.
- We assume that the prey population grows exponentially in the absence of a predator.

General compartmental model

There are two separate quantities which vary with time: the number   word
of prey and the number of predators. For populations of animals it is   equations
common to consider the population density, or number per unit area,
rather than population size, as we do here. We need to develop two word
equations, one for the rate of change of prey density and one for the rate
of change of predator density.

Example 1:   *Determine a compartmental diagram and appropriate word equation for each of the two populations, the predator and the prey.*

Solution: *The only inputs for each population are births and the only outputs are deaths. However, the prey deaths occur due to the predators capturing and eating them. This is illustrated in the input–output compartmental diagram of Figure 5.3.1. The appropriate word equations are*

$$\left\{ \begin{array}{c} rate\ of \\ change\ of \\ prey \end{array} \right\} = \left\{ \begin{array}{c} rate\ of \\ prey \\ births \end{array} \right\} - \left\{ \begin{array}{c} rate\ of \\ prey\ killed \\ by\ predators \end{array} \right\},$$

$$\left\{ \begin{array}{c} rate\ of \\ change\ of \\ predators \end{array} \right\} = \left\{ \begin{array}{c} rate\ of \\ predator \\ births \end{array} \right\} - \left\{ \begin{array}{c} rate\ of \\ predator \\ deaths \end{array} \right\}. \tag{1}$$

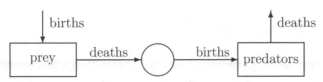

Figure 5.3.1: Input–output diagram for the 2-species predator–prey model. Prey deaths get 'converted' into predator births.

further as-
sumptions

In order to develop mathematical expressions for each of the terms on the right hand side (RHS) of these word equations, we need to make further assumptions concerning how the various birth and death rates depend on the populations.

Formulating the differential equations

prey births
and deaths

Let us assume that the per-capita birth rate for the prey (the scale insect) is a constant β. Remember that the per-capita birth rates give the rate of births from an individual prey. The rate of births from each individual prey is not dependent on the predator density. Alternatively, the deaths of the prey are due to being eaten by the predators. The greater the density of predators, the more likely it is that an individual prey will be eaten. Thus it would seem reasonable for the per-capita death rate of prey to depend on the density of predators. For simplicity, we assume that the prey per-capita death rate is directly proportional to the predator density.

predator
births and
deaths

We assume a constant per-capita death rate for the predators (the lady-bird beetles) independent of the prey density. The predator per-capita birth rate is more complicated. We assume the prey are an essential requirement for births of the predator. At any time the per-capita birth rate will increase with more food available and depend heavily on the amount of prey available. To model this we assume that the overall birth rate for the predators is directly proportional to the food supply, i.e. the number of prey killed by the predators.

summary
of further
assump-
tions

Summarising these further assumptions we have:

- prey per-capita birth rate is constant,
- predator per-capita death rate is constant,
- prey per-capita death rate is proportional to the predator density,
- predator per-capita birth rate is proportional to the rate at which prey are eaten.

Example 2:   *Using the above assumptions and the word equations (1), formulate differential equations for the prey and predator densities.*

Solution: *Let $X(t)$ denote the number of prey per unit area and $Y(t)$ the number of predators per unit area.*

First look at the constant per-capita terms, the prey births and predator deaths. (Subscripts 1 and 2 will be used for the parameters associated with X the prey, and Y the predator, respectively.) Let the prey per-capita birth rate be β_1 and the predator per-capita death rate be α_2. Since the overall rates are the per-capita rates multiplied by the respective population densities, we *can write,*

$$\left\{\begin{array}{c} rate \; of \\ prey \\ births \end{array}\right\} = \beta_1 X(t), \qquad \left\{\begin{array}{c} rate \; of \\ predator \\ deaths \end{array}\right\} = \alpha_2 Y(t). \qquad (2)$$

For the prey deaths we denote the per-capita death rate as $c_1 Y(t)$, since it is proportional to the predator density, with c_1 as the positive constant of proportionality. Thus the rate at which prey are eaten is given by $c_1 Y(t) X(t)$. The predator birth rate is proportional to this rate of prey eaten, so we write

$$\left\{\begin{array}{c} rate \; of \\ prey \; killed \\ by \; predators \end{array}\right\} = c_1 Y(t) X(t), \qquad \left\{\begin{array}{c} rate \; of \\ predator \\ births \end{array}\right\} = f c_1 Y(t) X(t), \qquad (3)$$

where f is also a positive constant of proportionality. Another interpretation of the product $X(t)Y(t)$ is that it is proportional to the rate of encounters between prey and predator.

Now substitute equations (2–3) into the word equations (1). We obtain the pair of differential equations

$$\frac{dX}{dt} = \beta_1 X - c_1 XY, \qquad \frac{dY}{dt} = c_2 XY - \alpha_2 Y \qquad (4)$$

where we have set $c_2 = f c_1$.

This system of equations is called the *Lotka–Volterra predator–prey system* after the two mathematicians who first worked with them. The parameters c_1 and c_2 are known as *interaction parameters* as they describe the manner in which the populations interact. Since there are positive and negative terms on the RHS of each differential equation, we might anticipate that the populations could either increase or decrease. Further, these differential equations are coupled since each differential equation depends on the solution of the other. The differential equations are also nonlinear since they involve the product XY.

Lotka–Volterra system

We now have a model to which we can apply some simple checks to ascertain whether the equations behave as we might expect. For a two-species model we would expect that, in the absence of any predators, the prey would grow without bound (since we have not included any growth limiting effects other than the predators). Also, in the absence of prey, we would expect the predators to die out. The following example indicates that the model incorporates this behaviour.

checking the model

Example 3: *Check the Lotka–Volterra model in the limiting cases of prey with no predators, or predators with no prey.*

Solution: *Suppose there are no predators so that* $Y = 0$. *The equations then reduce to*

$$\frac{dX}{dt} = \beta_1 X,$$

which is the equation for exponential growth. The prey grows exponentially. If there are no prey then $X = 0$ *and the equations reduce to*

$$\frac{dY}{dt} = -\alpha_2 Y,$$

that is, exponential decay, which means that the predator population decreases exponentially and dies out.

Numerical solution

example numerical solution

Although the equations may appear simple, they are not simple to solve. In fact we cannot find an analytic solution. However, to solve (4) numerically we can use Maple. We have chosen the values of the parameters β_1, α_2, c_1 and c_2 arbitrarily, in order to get a feel for how the model behaves. A sample numerical solution is illustrated in Figure 5.3.2 with the code provided below:

```
> restart; with(plots): with(DEtools):
> beta[1]:=1.0: alpha[2]:=0.5: c[1]:=0.01: c[2]:=0.005:
> de1 := diff(X(t),t) = beta[1]*X(t)-c[1]*X(t)*Y(t);
> de2 := diff(Y(t),t) = -alpha[2]*Y(t)+c[2]*X(t)*Y(t);
> inits := [X(0)=200, Y(0)=80];
> myopts := stepsize=0.1, arrows=none:
> plot1 := DEplot([de1,de2], [X,Y], t=0..20,
    [inits],scene=[t,X],linecolor=red,myopts):
> plot2 := DEplot([de1,de2], [X,Y], t=0..20,
    [inits],scene=[t,Y],linecolor=blue,myopts):
> display(plot1,plot2);
```

observations

The prey population oscillates out of phase with the predator population. Further exploration with other values of the parameters appears to indicate that this model always leads to oscillating populations. In Chapter 6 (to come) we use some mathematical analysis to prove this.

lag in oscillations

We also note that the predator population oscillation lags behind the prey cycles. We show (Chapter 6) that the Lotka–Volterra equations (4) always predict oscillations and that the predator oscillations always lag behind those of the prey.

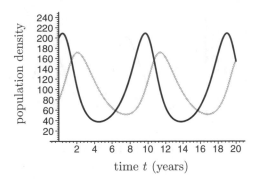

Figure 5.3.2: Sample numerical solution of the predator–prey equations using Maple. The black curve represents the prey and the grey curve the predator. Parameters here are chosen to illustrate the dynamics, and are not based on data from real populations: $\beta_1 = 1$, $\alpha_2 = 0.5$, $c_1 = 0.01$, $c_2 = 0.005$, with initial populations $x_0 = 200$ and $y_0 = 80$.

If DDT is sprayed on the crops we need to modify the equations since this represents a different mechanism for prey and predator deaths. We shall assume the per-capita death rates due to DDT are constant, with different per-capita rates p_1 and p_2 for the prey and predator, respectively. This means the DDT has an equal effect on each individual of a given species. The modified differential equations (4) including the additional terms $p_1 X(t)$ for the prey and $p_2 Y(t)$ for the predator are

include DDT

$$\frac{dX}{dt} = \beta_1 X - c_1 XY - p_1 X, \qquad \frac{dY}{dt} = c_2 XY - \alpha_2 Y - p_2 Y. \qquad (5)$$

Note that the differential equations (5) are of the same form as the original system (4), with the positive constant β_1 replaced with $(\beta_1 - p_1)$ and the positive constant α_2 replaced with $(\alpha_2 + p_2)$. (Note, there is an extra provision here that $\beta_1 - p_1 > 0$.)

To investigate the effect of DDT on the two populations we run the numerical solution again with $p_1 = p_2 = 0.1$. The results are presented in Figure 5.3.3. We see that, compared with Figure 5.3.2, the effect of the pesticide on the predator has been to decrease its overall numbers. It appears, however, that the mean prey population has increased. This is quite the opposite from the desired effect of the DDT and comes as a result of dealing with a nonlinear system. The DDT also reduces the average predator population so that there will be a reduced number of prey deaths due to the prey being eaten by the predators.

numerical results

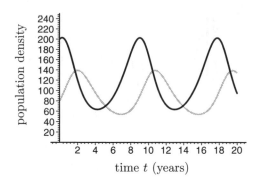

Figure 5.3.3: The scale insect or prey (black) and beetle population or predator (grey) where DDT has been used. Here we have $\beta_1 = 1$, $\alpha_2 = 0.5$, $c_1 = 0.01$, $c_2 = 0.005$, $p_1 = 0.1$ and $p_2 = 0.1$. The initial conditions are $x_0 = 200$, $y_0 = 80$. (These parameter values are chosen to illustrate the dynamics, and are not based on data from real populations.)

Interpretation of parameters

the pa-
rameters

The differential equations (4) have 4 parameters: the prey per-capita birth rate β_1, the predator per-capita death rate α_2 and the two constants of proportionality c_1 and c_2. Making some further assumptions allows us to express c_1 and c_2 in terms of biologically relevant parameters.

searching
rate

Choose an arbitrary time interval Δt. We assume that Δt is sufficiently small, so that the populations $X(t)$ and $Y(t)$ do not change significantly over the time interval. We assume that each individual predator can search an area $a'\Delta t$, where a is the area searched per unit time. We assume the areas covered by each individual predator do not overlap and we neglect the time it takes for the predator to eat the prey. The positive constant a' is called the *searching rate*. For a single predator the number of prey encountered is the prey density (number of prey per unit area) multiplied by the area covered by the predator, which is $X(t)a'\Delta t$. If we assume each encounter results in a kill for the predator, then the rate of prey deaths by a single predator is given by dividing by Δt. So

$$\left\{ \begin{array}{c} \text{rate of} \\ \text{prey killed by} \\ \text{one predator} \end{array} \right\} = a'X(t),$$

and hence the total rate of prey deaths (per unit area) is obtained by multiplying by $Y(t)$,

$$\left\{ \begin{array}{c} \text{rate of} \\ \text{prey killed} \\ \text{by predators} \end{array} \right\} = a'X(t)Y(t).$$

Comparing with equation (3) this gives $c_1 = a'$, for constant a'. This provides a biological interpretation for the parameter c_1 as the searching rate, or the rate of area covered by a predator per unit time.

Recall that $c_1 X(t) Y(t)$ represents the rate of prey deaths and that $c_2 X(t) Y(t) = f c_1 X(t) Y(t)$ the rate of predator births, so the constant ratio $f = c_2/c_1$ represents the amount of prey required for a unit predator birth. This parameter f is interpreted as the *predator efficiency parameter*.

> predator efficiency parameter

Limitations and extensions

A number of attempts have been made to validate the Lotka–Volterra equations. These include experiments by the Russian microbiologist Gause with two different types of protozoa, *Paramecium caudatum* and *Dididium nasutum* (see Kormondy (1976)). Gause found that the prey protozoa died out within only a few oscillations. In Renshaw (1991) a modified version of the Lotka–Volterra equations, which incorporate random births and deaths is investigated. This model exhibits behaviour similar to the Gause experiments. In general, however, the Lotka–Volterra equations are not widely used in practice, but rather form a sound basis for more complicated models, which we investigate in Chapter 8.

> validation

Many mathematics texts point to a well-known data set regarding the number of pelts of lynx and snow-shoe hare as a validation of the Lotka–Volterra equations. This data set has 10-year oscillations. However, Murray (1990) observes that this data is not a validation of the Lotka–Volterra equations at all, since the hare oscillations lag one quarter of a period behind those of the lynx, contradicting what we might expect. It would mean that the hare was eating the lynx, which would be a ridiculous conclusion. In Renshaw (1991), it is pointed out that the hares and the grass they eat show the correct predator–prey lag according to the Lotka–Volterra equations. This simple example demonstrates how easily the over-simplification of an ecosystem, into only two interacting species in this case, can lead to misinterpretation.

> lynx and snow-shoe hare data

An obvious limitation of the model is the inclusion of exponential growth, and in the following section we improve on this by including a density dependent growth rate. Then in Chapter 8 (Section 8.3) we extend this predator–prey model further to improve its performance in predicting the observed behaviour of populations. We consider more realistic functions describing the prey death rate as well as the overall predator growth rate.

> limited growth

Density dependent growth

exponential growth of prey in absence of predator

The oscillatory nature of the predator–prey model above is what one might expect intuitively, and together with the phase lag of the predator oscillation behind that of the prey it is qualitatively what has been observed in some natural systems. However, our model is based on the assumption that the prey grows exponentially in the absence of the predator and this is clearly unrealistic. No matter how abundant the food supply, disease and/or food shortage will eventually curb the growth.

new model

Thus, returning to the case without DDT, we replace the growth term for the prey population in (4) with density dependent growth and a carrying capacity of K. (For details on density dependent growth refer to Section 3.2.) The system becomes

$$\frac{dX}{dt} = \beta_1 X \left(1 - \frac{X}{K}\right) - c_1 XY, \qquad \frac{dY}{dt} = c_2 XY - \alpha_2 Y. \qquad (6)$$

numerical solution

Modifying the `Maple` code used to obtain Figure 5.3.2, and assuming a carrying capacity of $K = 1000$, we get the time-dependent plot of Figure 5.3.4. Note that the amplitude of the oscillations now decreases with an increase in time, for both populations. In fact, as time increases further, the figure suggests that each population might settle to a fixed population density, that is, stabilise. We look more closely at these results in the next chapter where we develop some analytical tools leading to a better understanding of the processes.

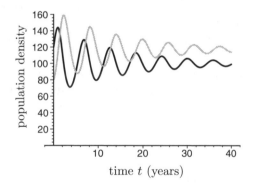

Figure 5.3.4: The scale insect (black line) and beetle population (grey line) where logistic growth for the scale insect has been included. Here we have $\beta_1 = 1$, $\alpha_2 = 0.5$, $c_1 = 0.01$, $c_2 = 0.005$ and $K = 1000$. (These parameter values are chosen to illustrate the dynamics, and are not based on data from real populations.)

Discussion

The control of pests (plant or animal) through the introduction of a nat-   controlling
ural enemy has had some remarkable successes and also some disastrous   pests
failures. The reasons why some programs succeed and others fail are not
always well understood, and thus with current knowledge the outcome is
often difficult to predict. One research paper which discusses this issue
is Beddington et al. (1978), in which the authors attempt to identify features of the interaction process which have a significant impact on the
outcome.

The scale insect/ladybird beetle system discussed in this chapter is just   success or
one example of failure. One example of an extremely successful case   failure
occurred in Australia in 1925. Prior to this date the prickly-pear cactus had been introduced into Australia with disastrous consequences, as
it spread rapidly, thriving in the conditions and rendering thousands of
hectares of farmland useless. In 1925, a natural predator moth was introduced to halt the spread of the cacti, and was spectacularly successful,
nearly wiping out the weed in just two years. The case study is presented in Chapter 8 and provides a mathematical model which forecasts
this success.

Some discussion of more realistic predator–prey models is given in Chap-   further
ter 8 and also in Murray (1990) and Edelstein-Keshet (1988). For a   reading
biological perspective of interactions between a variety of species see (Begon et al., 1990, Chapters 7–12) and also May (1981). Further details
of the effect of DDT on scale insects and ladybird beetles is given in
MacArthur and Connell (1966). A brief discussion of further examples
concerning the use of chemical pesticides on predator–prey systems is
given in Keeton (1972).

Skills developed here:

- Formulate differential equations for two predators and one
  prey species, or two prey and one predator.
- Distinguish between models describing different types of
  interactions, such as models for symbiosis, parasite–host interactions and competition for the same resource.
- Generate numerical solutions to systems of equations.

5.4   Competing species

overview

Another simple ecosystem to model is that of competing species, where two (or more) populations compete for limited resources such as food or territory. There are two aspects of competition: *exploitation*, when the competitor uses the resource itself and *interference*, where the population behaves in such a manner as to prevent the competitor from utilising the resource. This system is very similar to the predator–prey model of the previous section, however, the terms describing the interaction between the species differ.

Model assumptions

some as-
sumptions

We start with a list of assumptions on which to build our first model.

- We assume the populations to be sufficiently large so that random fluctuations can be ignored without consequence.
- We assume that the two species model reflects the ecosystem sufficiently accurately.
- We assume each population grows exponentially in the absence of the other competitor(s), although we later incorporate density dependent growth for each.

General compartmental model

births and
deaths

Let $X(t)$ and $Y(t)$ be the two population densities (number per unit area) where t is again time. As before, we have birth and death rates associated with each population, however, in this case an increase in the number of deaths in one population causes a decrease in the number of deaths in the other population. The compartmental diagram of Figure 5.4.1 illustrates the process.

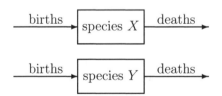

Figure 5.4.1: Input–output diagram for the 2-species competition model.

word
equations

In words, we have for each population in the system

$$\left\{ \begin{array}{c} rate\ of \\ change\ of \\ population \end{array} \right\} = \left\{ \begin{array}{c} rate\ of \\ population \\ births \end{array} \right\} - \left\{ \begin{array}{c} rate\ of \\ population \\ deaths \end{array} \right\}. \qquad (1)$$

Formulating the differential equations

Since neither population is dependent on the other as far as growth rates   birth rates
are concerned (unlike in the predator–prey example of Section 5.3), we
let the positive constants β_1 and β_2 describe the per-capita birth rates
for species X and Y, respectively.

Since the two populations are competing for the same resource, the   death
density of each population has a restraining effect, proportional to this   rates
density, on the other. So the per-capita death rate for Y is proportional
to X, and that for X is proportional to Y, which in symbols is

$$\left\{\begin{array}{c} \text{rate of} \\ \text{species-}X \\ \text{deaths} \end{array}\right\} = (c_1 Y)X, \qquad \left\{\begin{array}{c} \text{rate of} \\ \text{species-}Y \\ \text{deaths} \end{array}\right\} = (c_2 X)Y.$$

Here c_1 and c_2 are the constants of proportionality for this restraining
effect.

Our model becomes   Gause's
equations

$$\frac{dX}{dt} = \beta_1 X - c_1 XY, \qquad \frac{dY}{dt} = \beta_2 Y - c_2 XY. \qquad (2)$$

While β_1 and β_2 were per-capita birth rates, we may also consider them
as overall per-capita growth rates which incorporate deaths (independent
of the other species) as well as births. Thus they are per-capita growth
rates, or per-capita reproduction rates, while the parameters c_1 and c_2
are the interaction parameters. These equations are known as *Gause's
equations* and are a coupled pair of first-order, nonlinear differential equations.

Numerical solution

Again, although they appear reasonably simple, the equations cannot   example
be solved analytically, and we use Maple to draw the time-dependent   numerical
solution
graphs, obtained using approximate numerical techniques as discussed in
Chapter 4. The Maple code is given below and the results are graphed
in Figure 5.4.2.

```
> restart; with(plots): with(DEtools):
> beta[1]:=3.0: beta[2]:=2.5: c[1]:=2: c[2]:=1:
> de1 := diff(X(t),t) = beta[1]*X(t)-c[1]*X(t)*Y(t);
> de2 := diff(Y(t),t) = beta[2]*Y(t)-c[2]*X(t)*Y(t);
> inits := [X(0)=3, Y(0)=1.5];
> mydeopts := arrows=none, method=rkf45:
> plot1 := DEplot([de1,de2], [X,Y], t=0..3,
> [inits],scene=[t,X],linecolor=black,mydeopts):
```

```
> plot2 := DEplot([de1,de2], [X,Y], t=0..3,
> [inits],scene=[t,Y],linecolor=gray,mydeopts):
> display(plot1,plot2,view=[0..3,0..20]);
```

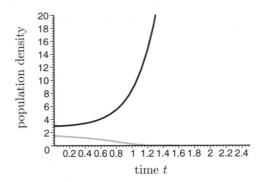

Figure 5.4.2: A sample numerical simulation of the competing species model with each species having unrestricted growth in the absence of the other species. The parameter values are $\beta_1 = 3$, $\beta_2 = 2.5$, $c_1 = 2$ and $c_2 = 1$, with the initial condition $x_0 = 3$, $y_0 = 1.5$. Species X (black), species Y (grey).

long-term behaviour

As usual, we are interested in what happens in the long run. From the time-dependent diagram of Figure 5.4.2 it appears that one of the species dies out over time. Varying the initial conditions illustrates that it is possible to choose values such that the other species dies out instead (e.g. with $X(0) = 2.6$, $Y(0) = 1.6$).

Gause's Principle

The results suggest that in the case of two species competing for the same resource, in the long run one species will survive and the other become extinct. This is known as *Gause's Principle of Competitive Exclusion*, from a series of experiments conducted by the microbiologist on competing species of yeast cells. Furthermore, unlike in the predator–prey model, the populations do not appear to oscillate with time.

Interpretation of parameters

the parameters

The system of equations developed so far for the model of competitive species has four parameters: the two per-capita growth rates β_1 and β_2 which are independent of any other species, and the two constants of proportionality c_1 and c_2 which describe the interaction between the species.

interpreting the interaction

The competition interaction can be interpreted as stating that the more deaths there are in species Y, the more resources will be available to species X, and thus the fewer deaths there will be for X. To explain the interaction we consider a small time interval Δt, sufficiently small so that

X and Y do not change significantly. Let a' be the rate of area 'used' by one individual in species Y (for food), hence making it unavailable to X (for food). Assume that no such areas can overlap. Then one individual 'uses' an area of $a'\Delta t$, and Y individuals will 'use' $a'Y(t)\Delta t$. Assuming that each area of resource 'used' by Y implies it is unavailable to X, the rate at which area is made unavailable to X by a single Y individual is the density (number per unit area) of X in an area divided by Δt, which is $a'X(t)$. And hence the total effect of the removal of resources from X by Y becomes $a'X(t)Y(t)$. So a', or c_1 in our model, can be interpreted as the rate at which X dies off due to the removal of resources by Y.

If we let $f = c_2/c_1$, then f is a measure of the efficiency of the competitive interaction. It can be considered as the number of units of species Y required to reduce species X by one unit.

efficiency measure

Limitations of the model

One immediate and obvious limitation of this model is that each population grows exponentially in the absence of the other. In the next section, we improve on this by including density dependent growth.

exponential growth

In the time-dependent diagram, Figure 5.4.2, it appears that only one of the species will survive and the other will become extinct. (In the next chapter we show this to be the case for the competition model under nearly all conditions.) This idea was emphasised in the results of experimental work carried out by Gause.

exclusion principle

Early this century, research tended to accept and support Gause's principle, however, more detailed research has led to a questioning of this model. How does the degree of competition modify the predictions? What actually constitutes a competitive model when some resources are shared and others are not? By the 1980s, there was a lively debate underway amongst the leaders in the field, some rejecting outright the credibility of such models, while others argued that there was a value in, and a place for, the years of research already accomplished on these competition models. Two articles Lewin (1983a) and Lewin (1983b) provide an informative insight into this controversy and the positions adopted by the various players.

validity of competition modelling

Density dependent growth

In 1932, the Russian microbiologist, Gause, described an experiment with two strains of yeast, *Saccharomyces cervevisiae* and *Schizosaccharomyces kefir*, hereafter called Species X and Species Y. These are described in Renshaw (1991). Gause found that, grown on their own, each species

experimental results

exhibited a logistic growth curve, but when grown together the growth pattern changed with Species X dying out.

extended model

We now extend the competition model to account for logistic growth in both species, in the absence of the other species. Let K_1 and K_2 be the carrying capacities for species X and Y respectively. Then including density dependent growth in (2) we have

$$\frac{dX}{dt} = \beta_1 X \left(1 - \frac{X}{K_1}\right) - c_1 XY, \qquad \frac{dY}{dt} = \beta_2 Y \left(1 - \frac{Y}{K_2}\right) - c_2 XY. \quad (3)$$

With $d_1 = \beta_1/K_1$ and $d_2 = \beta_2/K_2$ the system (3) becomes

$$\frac{dX}{dt} = \beta_1 X - d_1 X^2 - c_1 XY, \qquad \frac{dY}{dt} = \beta_2 Y - d_2 Y^2 - c_2 XY. \quad (4)$$

Using Gause's data (Renshaw (1991)), it is possible to estimate the parameters in the model as

$$\beta_1 = 0.21827, \qquad \beta_2 = 0.06069,$$
$$d_1 = 0.01679, \qquad d_2 = 0.01046,$$
$$c_1 = 0.05289, \qquad c_2 = 0.00459.$$

Note that these parameters correspond to the carrying capacities (in the absence of the other species) of $K_1 = \beta_1/d_1 = 13.0$ and $K_2 = \beta_2/d_2 = 5.8$.

numerical solutions

Re-running the `Maple` code used to produce Figure 5.4.2 with an adjustment made for the inclusion of the density dependent growth for both X and Y and the parameter values from Gause's experiment, produces Figure 5.4.3.

validation

We find that after a certain time, the figure suggests that only one of the species survives, stabilising to a fixed density, which is the carrying capacity for that species. This is what Gause observed (see Renshaw (1991) for a comparison between data and numerical solutions).

further analysis

To get a better understanding of how this model behaves as we change initial conditions, or the parameter values, we could generate further numerical graphs. However, this is a somewhat hit-and-miss approach. For example, does one population always die out, or can they sometimes coexist? We return to a more detailed analysis of the dynamics of this system in Section 6.3, and also in Chapter 7, as we develop some helpful analytical tools.

Figure 5.4.3: Graph of the two species of yeast populations (species X in black, species Y in grey) as functions of time for the model with density dependence and with parameter values from Gause's experiment.

Renshaw (1991) gives a detailed discussion of the competing species of   further
yeast. See Kormondy (1976) for a discussion of competition between   reading
species of microscopic protozoa (*Paramecium aurelia* and *Paramecium
caudatum*).

Skills developed here:
- Formulate differential equations for variations on the competition model presented here, including models combining three or more competing species.
- Obtain numerical solutions to the competition model and extensions of it.
- Understand what is meant by, and the implications of, the 'random' approach of generating numerical solutions.
- Understand the competitive exclusion principle.

5.5   Case Study: Aggressive protection of lerps and nymphs

The following case study does not incorporate the above theory specifically, but describes the occurrence of a competing species scenario where a colony of birds protects its territory in order to maintain an adequate food supply to support itself. It is included for interest and to provide an example of the value of interspecific territoriality, a system for which the above theory would have relevance.

Adapted from the article by Loyn et al. (1983).

In regions of southeastern Australia infestations of psyllids have inflicted severe damage on the eucalypt forests causing deterioration of the foliage and, in extreme cases, completely destroying certain trees. This has prompted an examination of the conditions under which it may occur. Many species of birds live in the canopy of these eucalypts and compete to feed on these insects: the Bell Miners are one such species.

The Bell Miner (Manorina melanophrys) is a honeyeater living in large colonies which they defend aggressively against other competing birds, often substantially larger than themselves. They feed on the nymphs, sweet secretions and lerps (protective carbohydrate covers) of psyllids, however, where their colonies occur, the canopy foliage appears unhealthy and infested with these insects. The Miners aggressively protect this abundance of prey. In an experiment a colony of Bell Miners was removed to join another colony 45 km away. They settled into this new location and did not return.

Other bird flocks soon moved into the site of the old colony and rapidly the foliage improved. The insect infestation declined, with the numbers of psyllids remaining very small thereafter.

It appears that the insect infestations can be controlled by the removal of the Bell Miner colonies. The interspecific territorial behaviour of the Miners prevents this control and introduces a control of its own. Miners effectively maintain an abundant and exclusive food supply, albeit detrimental to the eucalypts, enabling a colony to remain within the same territory for up to 40 years.

5.6   An epidemic model for influenza

We develop a model to describe the spread of a disease in a population   overview
and use it to describe the spread of influenza in a boarding school. To do
so the population is divided into three groups: those susceptible to catching the disease, those infected with the disease and capable of spreading it
and those who are immune from the disease. Modelling these interacting
groups leads to a system of two coupled differential equations.

Background

Over the centuries, there have been dramatic examples of how epidemics   examples
of various diseases have had a significant effect on the human popula-   of
tion. One of the most well-known is the Black Death in Europe in the   epidemics
fourteenth century. Today epidemics are still prevalent, the most notable
being possibly AIDS and the Ebola virus. If we can understand the nature of how a disease spreads through a population, then we are better
equipped to contain it through vaccination or quarantine. Or in the case
of the biological control of pests we may wish to determine how to increase the spread of the disease (for example, myxomatosis or calicivirus
in rabbits) so as to find an efficient way of reducing the population. Unfortunately, humans themselves have been subjected to this means of
'control'. In colonial times, the spread of European diseases, such as
measles and smallpox, had a disastrous impact on certain indigenous
populations who had no resistance to them.

Many diseases are spread by infected individuals in the population com-   disease
ing into close contact with susceptible individuals. These include in-   character-
fluenza, measles, chicken pox, glandular fever and AIDS. On the other   istics
hand, malaria is transmitted through a host, a mosquito, which car-
ries the disease from individual to individual. Certain diseases are more
contagious than others. Measles and influenza are highly contagious,
whereas glandular fever is much less so. Many diseases, such as mumps
and measles confer a lifelong immunity, however, influenza and typhoid
have short periods of immunity and can be contracted more than once.

New diseases are developing constantly, some within our own bodies as   coping
in the case of new strains of bacteria, which have become resistant to   with new
antibiotics, and others in the bodies of animals where bacteria or viruses and
harmful to humans may mutate to a form transmittable to humans.   spreading
Furthermore, the time taken for these diseases to spread has changed   diseases
drastically. In earlier times, it may have taken many years for a disease
to become widely spread between population groups, however, with air

travel a number of hours is now all that is required for a disease to spread world wide. Clearly we need to understand the general characteristics governing the manner in which these different diseases spread within our community, so that strategies can be developed to cope with possible outbreaks.

incubation period and latent period

There are some natural definitions which we require in order to proceed with our modelling. The *incubation period* of a disease is the time between infection and the appearance of visible symptoms. This should not be confused with the *latent period*, which is the period of time between infection and the ability to infect someone else with the disease. The latent period is shorter than the incubation period so that an individual can be spreading the disease, unaware of having it. For measles the incubation period is approximately 2 weeks and the latent period is approximately 1 week.

influenza epidemic

Influenza pandemics have caused much loss of life. The worst on record occurred in 1918–1919, after World War I, when such an epidemic claimed between 15 and 25 million lives world wide (Garret (1995)). Because of its ability to mutate, the flu virus continues to pose a very real threat to our society. Below we consider a simple mathematical model for a flu (influenza) epidemic at a boarding school over a period of about 30 days. For this period it is reasonable to assume that reinfection does not occur.

Model assumptions

susceptibles and contagious

When considering a disease, the population can be divided into distinct classes: susceptibles $S(t)$ and contagious infectives $C(t)$, where t denotes time. The susceptibles are those liable to catch the disease, while the contagious infectives are those infected with the disease who are capable of communicating it to a susceptible. There are also those who have recovered and who are now immune from further infection of the disease.

some assumptions

Initially, we make some assumptions and then build the model based on them.

- We assume the populations of susceptibles and contagious infectives are large so that random differences between individuals can be neglected.
- We ignore births and deaths in this model and assume the disease is spread by contact.
- We neglect the latent period for the disease, setting it equal to zero.
- We assume all those who recover from the disease are then immune (at least within the time period considered).

- We also assume that, at any time, the population is homogeneously mixed, i.e. we assume that the contagious infectives and susceptibles are always randomly distributed over the area in which the population lives.

General compartmental model

We need to start with an input–output compartmental diagram and then describe the rate of change in the number of susceptibles and contagious infectives with word equations. (For epidemic models it makes more sense to consider the population size rather than the density.) The following example illustrates this process.

inputs and outputs

Example 1: *Construct a compartmental diagram for the model and develop appropriate word equations for the rates of change of susceptibles and contagious infectives.*

Solution: *The only way the number of susceptibles can change is the loss of those who become infected, since there are no births and none of those who become contagious infectives can become susceptibles again. The number of infectives changes due to the susceptibles becoming infected and decreases due to those infectives who die, become immune or are quarantined. The latter cannot become susceptibles again (from the assumptions made). This is illustrated in the compartmental diagram of Figure 5.6.1*

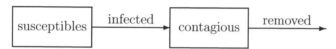

Figure 5.6.1: Input–output diagram for the epidemic model of influenza in a school, where there is no reinfection.

The appropriate word equations are

$$\left\{ \begin{array}{c} \text{rate of} \\ \text{change in no.} \\ \text{susceptibles} \end{array} \right\} = - \left\{ \begin{array}{c} \text{rate} \\ \text{susceptibles} \\ \text{infected} \end{array} \right\}$$

$$\left\{ \begin{array}{c} \text{rate of} \\ \text{change in no.} \\ \text{infectives} \end{array} \right\} = \left\{ \begin{array}{c} \text{rate} \\ \text{susceptibles} \\ \text{infected} \end{array} \right\} - \left\{ \begin{array}{c} \text{rate} \\ \text{infectives} \\ \text{removed} \end{array} \right\} .$$

(1)

Note that the term on the RHS of the equation for the susceptibles is exactly the same term as the first term on the RHS for the infectives.

Formulating the differential equations

modelling transmission of infection

To model the total rate of susceptibles infected, first consider the susceptibles infected by a single infective. It is evident that the greater the number of susceptibles, then the greater the increase in the number of infectives. Thus the rate of susceptibles infected by a single infective will be an increasing function of the number of susceptibles. For simplicity, let us assume that the rate of susceptibles infected by a single infective is directly proportional to the number of susceptibles, with a constant of proportionality b. If we denote the number of susceptibles at time t by $S(t)$ and the number of infectives[1] at time t by $C(t)$, then the rate susceptibles are infected by a single infective is $bS(t)$. The total rate of susceptibles infected by $C(t)$ infectives is obtained by multiplying $bS(t)$ by the number of infectives, and thus

$$\left\{\begin{array}{c} rate \\ suseptibles \\ infected \end{array}\right\} = bS(t)C(t). \qquad (2)$$

The constant b is called the *transmission coefficient* or infection rate.

modelling removal

We must also account for those removed from the system. They consist of fatalities due to the disease, those who become immune to the disease, or those infectives who are quarantined. The number of infectives removed in the time interval should not depend in any way upon the number of susceptibles, but only on the number of infectives. We assume that the number of infectives removed is directly proportional to the number of infectives. We write

$$\left\{\begin{array}{c} rate \\ infectives \\ removed \end{array}\right\} = rC(t) \qquad (3)$$

where r is a positive constant of proportionality, called the *removal rate*.

further assumptions

Thus we have some further assumptions:

- the rate at which susceptibles become infected is proportional to both the numbers of infectives and susceptibles,
- the rate at which infectives recover and are removed is proportional to the number of infectives only.

the differential equations

The rate of change in the number of susceptibles in the time interval is given by dS/dt and the rate of change in the number of infectives is

---

[1]It is conventional to use $I(t)$ to denote the number of infectives, however, we break with convention here and use $C(t)$ to emphasise that infectives transfer the disease to susceptibles. Another reason is that Maple reserves the symbol I for $\sqrt{-1}$.

given by dC/dt. Making use of equations (2) and (3), the population word equations (1) become

$$\frac{dS}{dt} = -bSC, \qquad \frac{dC}{dt} = bSC - rC, \qquad (4)$$

subject to the initial conditions $S(0) = s_0$ and $C(0) = c_0$. Equations (4), a coupled system of nonlinear differential equations, were originally derived by Kermack and McKendrick in 1927 (Kermack and McKendrick (1927)).

Numerical solution

We can solve the differential equations (4) by a numerical method (e.g. Euler's method, Runge–Kutta methods, or `Maple` built-in solvers). Using some data from an influenza epidemic at a British boarding school, the results are given in Figure 5.6.2. The `Maple` code from which the figure was generated is given below: *example numerical solution*

```
> restart: with(plots): with(DEtools):
> b := 2.18*10^(-3): r := 0.44:
> de1 := diff(S(t),t) = -b*S(t)*C(t);
> de2 := diff(C(t),t) = b*S(t)*C(t) - r*C(t);
> inits := [S(0)=762, C(0)=1]:
> myopts := stepsize=0.1:
> plot1 := DEplot([de1,de2], [S(t),C(t)], t=0..30, [inits],
    scene=[t,S],arrows=none, myopts):
> plot2 := DEplot([de1,de2], [S(t),C(t)], t=0..30, [inits],
    scene=[t,C],linecolor=red,
    arrows=none, myopts,linecolour=red):
> display(plot1,plot2);
```

The data seems to agree well with the model's predictions. Notice that the number of infectives starts small and increases substantially over 6 days then decreasing gradually. What is happening is that the number of susceptibles is being 'used up'. Thus, in the latter stages of the epidemic there is a much smaller chance of any given infective coming into contact with someone who has not yet been infected. *model validation*

Interpretation of parameters

Larger values of the transmission coefficient b imply that the disease is more easily spread. While it reflects how contagious the disease is, it also reflects the degree to which the population interacts. So a more crowded population, where an individual experiences a greater number of contacts per unit time, will lead to a higher value of the transmission constant. *transmission coefficient*

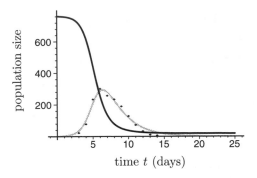

Figure 5.6.2: Numerical solution of the differential equations for an influenza epidemic in an English boarding school in 1978. The epidemic was started by one infective, thus $c_0 = 1$ and the total number of susceptibles was $s_0 = 762$. The values of the infection rate and removal rate were estimated as $b = 2.18 \times 10^{-3}$ susceptibles^{-1} day^{-1} and $r = 0.44$ day^{-1} respectively. The black dots correspond to the original data with the black line, the susceptibles and the grey line, the infectives. (Data from Murray (1990).)

estimating b Let us assume there are $\lambda \Delta t$ contacts made by a single individual in a population with a constant size of N individuals (per given area). (We define an effective contact, such that if it were between a contagious infective and a susceptible, it would result in the susceptible catching the disease.) Here λ is the per capita contact rate. A fraction $S(t)/N$ of these contacts are between the given infective and a susceptible. So the total number of contacts potentially leading to infection is $\lambda \Delta t S(t)/N$. Dividing by Δt and multiplying through by $C(t)$ gives

$$\left\{ \begin{array}{c} rate \\ susceptibles \\ infected \end{array} \right\} = \frac{\lambda}{N} S(t) C(t).$$

Comparing this with equation (2) gives $b = \lambda/N$.

estimating r An estimate of r, the removal rate, can be obtained from equation (3). Let t_{inf} be chosen as the average period of time for which an individual is infectious. Thus the number of infectives removed will be the number of infectives at the beginning of the time interval, which we denote by C_{rem}. From equation (3)

$$C_{\text{rem}} = \left\{ \begin{array}{c} number \\ infectives \\ removed \end{array} \right\} \simeq r C_{\text{rem}} t_{\text{inf}},$$

and an estimate for r is $r \approx 1/t_{\text{inf}}$. For a model such as system (4), t_{inf} is given by the incubation period t_{inc}. However for a model which includes a latent class (see Exercises, Question 12), a better estimate of t_{inf} is given by $(t_{inc} - t_{lat})$.

Table 5.6.1 provides incubation and latent periods for some common diseases. For measles the incubation period is about 8–14 days, and thus as a first estimate we have $r \approx 0.1\,\text{days}^{-1}$. Note that the latent period for measles is 4–10 days, so that a model with a latent class may be more appropriate. Also, as the infectious period may extend for up to 4 days after the appearance of symptoms, a better estimate of t_{inf} (for a measles model with a latent class) is 8 days, suggesting $r \approx 0.125\,\text{days}^{-1}$. Some diseases have an infectious period extending well beyond the disappearance of the symptoms, and in the case of glandular fever it is believed that the patient may remain infectious for up to a year. Thus, in many cases, a better estimate would be obtained using statistical estimation techniques from specific data on the numbers of infectives and susceptibles over time.

r for measles

Table 5.6.1: Incubation and latent periods for some common viral infections from NHMRC (1997).

Disease	t_{inc}	t_{lat}
Influenza	1–3 days	0–1 days
Measles	8–14 days	4–10 days
Chicken pox	13–17 (avg. 14–15) days	11–15 days
Rubella	14–21 (avg. 16) days	7–14 days
Mumps	12–25 (avg. 18) days	6–19 days
Glandular fever	4–6 weeks	unknown

Limitations of the model

There are a number of different variations of the basic model to consider: for example, the effect of births which continually introduce more susceptibles into the population, or diseases without immunity, where infectives become susceptible again after removal.

model variants

Births will provide a source of additional susceptibles and this can lead to oscillations in the number of infectives. Indeed, before the days of vaccinations, for diseases such as measles regular outbreaks occurred, every 2 years in some countries, and every year in others.

effect of births

While some diseases confer permanent immunity, (e.g. measles, chicken-pox) many others do not (e.g. colds, influenza over a longer period than 25 days). This can easily be modelled, by allowing the infectives to become susceptibles again. Thus the compartmental diagram Figure 5.6.1 changes as the output from the contagious compartment becomes the input for the susceptible compartment.

diseases without immunity

Discussion

some
important
questions

One question we might ask is whether a rapid increase in the number of infectives is always followed by a decrease? Also, by adjusting any of the parameters, could we limit the increase or even prevent it? Changing parameters could correspond to, for example, adopting certain vaccination strategies.

obtaining
general
analytic
results

To answer these questions it is useful to gain more qualitative information about this epidemic model. We would like to be able to say what happens for any values of the parameters. An exact solution of the simultaneous equations, however, is not easily obtained because the differential equations are nonlinear. An alternative approach is to use the chain rule to eliminate time and reduce the pair of differential equations to a single first order differential equation, from which some insight can be gained. This analysis is covered in the next chapter. In particular, we prove that according to this model, all of the susceptibles can never be infected.

TB in
possums

In Section 8.7, we include a case study for the spread of bovine tuberculosis in possums, which involves an extension of the model developed above, together with the comparison of a variety strategies to control the epidemic.

modified
rate of
infection

We can also modify the basic model considered by making more realistic assumptions about the rate of infection of susceptibles. Recall that we can write the transmission coefficient b as $b = \lambda/N$, where λ is the per-capita rate of effective contact between two individuals in the population. For a disease where all those infected become immune, there are no inputs into or outputs from the total population, and thus N is a constant. Consequently, the transmission coefficient b is constant. Then the model is the same as that considered in this section. However, if the disease is fatal then N changes with time and so does b.

variable
contact
rate

A further possibility is to allow the per-capita contact rate λ to vary with population size. The contact rate is an increasing function of the total population $N(t)$ (the larger the population, the more contacts occur). Suitable forms of the contact rate function (as discussed in Greenhalgh and Das (1993) and Roberts and Heesterbeek (1993)) include

$$\lambda(N) = \frac{\lambda_m N}{K(1 - \epsilon) + \epsilon N},$$

where ϵ is constant and $0 \le \epsilon \le 1$, K is the carrying capacity of the population and λ_m is the maximum contact rate at carrying capacity. The value of ϵ, depends on the social behaviour of the population. For

$\epsilon \simeq 1$, $\lambda(N)$ is approximately constant, corresponding to individuals which tend to interact in social groups of the same size, no matter what the population size is. For example, animals which group together in herds. For $\epsilon \simeq 0$ the contact rate $\lambda(N)$ is approximately proportional to population N. This corresponds to animals in an enclosure of fixed area with random mixing between all animals, where an increase in population size has a proportional increase in contact rate. For any population, the value of ϵ will lie between these two extremes.

Murray (1990) discusses the same model developed above, and the in- *further* fluenza epidemic used to validate the model. Also considered is the Black *reading* Plague, amongst other examples. Braun (1979) outlines an extension to the model for sexually transmitted diseases. For an introduction to both deterministic and stochastic approaches to modelling human epidemics, see Daley and Gani (1999) and for some further extensions of the basic models, see Grenfell and Dobson (1995).

Skills developed here:

- Formulate differential equations for variations on the model presented here, such as, recurring diseases, births and deaths, continuous vaccination programs, and diseases without immunity.
- Obtain a numerical solution.
- Explain how to estimate the recovery rate parameter in terms of the incubation time of a disease.
- Understand the meaning of the saturation of infectives and contact rates.

5.7   Simple model of a battle

We now consider a novel type of population interaction: a destructive *overview* competition or battle between two opposing groups. These may be battles between two hostile insect groups, athletic teams or human armies. While the models we derive here apply to the last case, the principles can be generalised and would apply to many other examples. The model we develop turns out to be a system of two coupled, linear differential equations.

Background

some
history

Battles between armies have been fought since antiquity. In ancient times battles were primarily hand-to-hand combat. With the development of archery and then gunpowder a crucial feature of battles has been aimed fire. Although many factors can affect the outcome of a battle, experience has shown that numerical superiority and superior military training are critical. The model we present was first developed in the 1920s by F. W. Lanchester who was also well-known for his contributions to the theory of flight.

aims of
the model

Our aim is to develop a simple model which predicts the number of soldiers in each army at any given time, provided we know the initial number of soldiers in each army. (As with epidemics, we consider the number, rather than the density, of individuals.)

Model assumptions

some as-
sumptions

First we make some basic assumptions and then develop the model based on these.

- We assume the number of soldiers to be sufficiently large so that we can neglect random differences between them.
- We also assume that there are no reinforcements and no operational losses (i.e. due to desertion or disease).

These are assumptions that can easily be relaxed at a later stage if the model is inadequate.

General compartmental model

input–
output
principle

The first step is to develop two word equations which describe how the two populations change, based on the input–output principle of the balance law. Suppose the two opposing groups or populations are the red army and the blue army.

Example 1:   *Determine the appropriate input–output diagram and associated word equations for the number of soldiers in both the red and blue armies.*

Solution: *Since there are no reinforcements or operational losses, each population changes by the number of soldiers who are wounded by the other army. Thus we can set up the input–output diagram of Figure 5.7.1.*

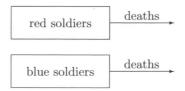

Figure 5.7.1: Compartmental diagram for the simple battle model.

Thus, in any given instant of time,

$$
\left\{ \begin{array}{c} \text{rate of} \\ \text{change of} \\ \text{red soldiers} \end{array} \right\} = - \left\{ \begin{array}{c} \text{rate red soldiers} \\ \text{wounded by} \\ \text{blue army} \end{array} \right\}
$$

$$
\left\{ \begin{array}{c} \text{rate of} \\ \text{change of} \\ \text{blue soldiers} \end{array} \right\} = - \left\{ \begin{array}{c} \text{rate blue soldiers} \\ \text{wounded by} \\ \text{red army} \end{array} \right\}.
$$

(1)

In a real battle there will be a mixture of shots; those fired directly at   firing an enemy soldier and those fired into an area known to be occupied by   shots an enemy, but where the enemy cannot be seen. Some battles may be dominated by one or the other firing method. We consider these two idealisations of shots fired as *aimed fire* and *random fire*. For the model we assume only aimed fire for both armies.

In the aimed fire idealisation, we assume all targets are visible to those   aimed fire firing at them. If the blue army uses aimed fire on the red army, then each time a blue soldier fires he/she takes aim at an individual red soldier. The rate of wounding of the red army depends only on the number of blue soldiers firing at them and not on the number of red soldiers. We see later that this assumption is equivalent to assuming a constant probability of success (on average) for each bullet fired.

For random fire a soldier firing a gun cannot see his/her target, but fires   random randomly into an area where enemy soldiers are known to be. The more   fire enemy soldiers in that given area then the greater the rate of wounding. For random fire we thus assume the rate of enemy soldiers wounded is proportional to both the number firing and the number being fired at.

In summary then we make the following further assumptions: further assumptions
- for aimed fire the rate of soldiers wounded is proportional to the number of enemy soldiers only,
- for random fire the rate at which soldiers are wounded is proportional to both numbers of soldiers.

Formulating the differential equations

modelling the wounding rates

Let $R(t)$ denote the number of soldiers of the red army and $B(t)$ the number of soldiers of the blue army. We assume aimed fire for both armies. This information is expressed mathematically by writing

$$\left\{\begin{array}{c} \text{rate red soldiers} \\ \text{wounded by} \\ \text{blue army} \end{array}\right\} = a_1 B(t), \qquad \left\{\begin{array}{c} \text{rate blue soldiers} \\ \text{wounded by} \\ \text{red army} \end{array}\right\} = a_2 R(t) \quad (2)$$

where a_1 and a_2 are positive constants of proportionality. The constants a_1 and a_2 measure the effectiveness of the blue army and red army respectively, and are called *attrition coefficients*.

another assumption

We thus assume that attrition rates are dependent only on the firing rates and are a measure of the success of each firing.

the differential equations

We now substitute (2) into the basic word equation (1), where the rate of change in the number of red soldiers is dR/dt and for the blue soldiers it is dB/dt. The two simultaneous differential equations are thus

$$\frac{dR}{dt} = -a_1 B, \qquad \frac{dB}{dt} = -a_2 R. \qquad (3)$$

Numerical solution

Battle of Iwo Jima

During the Battle of Iwo Jima in the Pacific Ocean (1945) daily records were kept of all U.S. combat losses. This data is graphed and referenced in Braun (1979). The values of the attrition coefficients a_1 and a_2 have been estimated from the data[2] as $a_1 = 0.0544$ and $a_2 = 0.0106$, and the initial numbers in the red and blue armies respectively were $r_0 = 66\,454$ and $b_0 = 18\,274$.

example numerical solution

Using `Maple` we can obtain accurate numerical solutions to the differential equations (3). The results are shown in Figure 5.7.2 with the red army as the U.S. forces and the blue army as the Japanese. The model shows a remarkably good agreement with the data.

```
> restart: with(plots): with(DEtools):
> a[1] := 0.0544: a[2] := 0.0106:
> de1 := diff(R(t),t) = -a[1]*B(t);
> de2 := diff(B(t),t) = -a[2]*R(t);
```

[2]A program called SBModelMaker has been used to do this. Essentially, it obtains a numerical solution of the differential equations for an initial guess of the parameters. It then varies the parameters to minimise a function which is the sum of the squares of the errors between the numerical solution and the data points.

```
> inits := [R(0)=66, B(0)=18]:
> myoptions := stepsize=0.1, dirgrid=[10,10]:
> plot1 := DEplot([de1,de2], [R,B], t=0..30, [inits],
    scene=[t,R], arrows=none, myoptions):
> plot2 := DEplot([de1,de2], [R,B], t=0..30, [inits],
    scene=[t,B], arrows=none, myoptions):
> display(plot1,plot2);
```

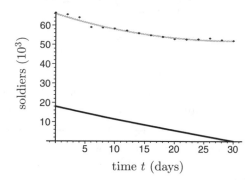

Figure 5.7.2: Numbers of U.S. soldiers in the battle of Iwo Jima measured in thousands. The solid lines (U.S.-grey, Japanese-black) are the model predictions from Maple and the dots are the measured data, from Braun (1979).

Despite the excellent fit to the data there are still some reservations about using this model for the Battle of Iwo Jima. The values of the parameters indicate the Japanese soldiers were approximately five times more effective than the U.S. soldiers, since $a_1 \simeq 5a_2$. In this battle, the Japanese army occupied the island and were well dug in. Perhaps, a better model would be to assume the U.S. soldiers used random fire with the Japanese soldiers using aimed fire. It turns out (see Exercises Question 17) that this model gives an equally good fit to the data.

discussion

Interpretation of parameters

We can further refine the model by trying to express the parameters a_1 and a_2 in terms of possible quantities which could be measured. The rate at which soldiers are wounded depends on both the firing rate and the probability of a shot hitting a target.

further interpretation

We return to equations (2). Consider a single blue soldier firing at the red army. We assume each blue soldier fires at a constant rate f_b. Then

soldiers wounded

$$\left\{ \begin{array}{c} \text{rate red soldiers} \\ \text{wounded by} \\ \text{single blue soldier} \end{array} \right\} = \left\{ \begin{array}{c} \text{rate bullets} \\ \text{fired in time} \\ \text{interval} \end{array} \right\} \times \left\{ \begin{array}{c} \text{probability of} \\ \text{a single bullet} \\ \text{hitting target} \end{array} \right\}$$

$$= f_b p_b,$$

where p_b is the probability (constant) that a single bullet from the blue soldier wounds an red soldier. Hence, for the entire blue army we multiply by the number of blue soldiers, $B(t)$, to obtain the total rate of red soldiers wounded by the blue army (per unit time). This gives

$$\left\{\begin{array}{c} \textit{rate red soldiers} \\ \textit{wounded by} \\ \textit{blue army} \end{array}\right\} = f_b p_b B(t). \tag{4}$$

expres-
sions for
attrition
coeffi-
cients

Equating this to (2) we obtain the attrition rates, or coefficients, a_1 and a_2 as

$$a_1 = f_b p_b, \qquad a_2 = f_r p_r, \tag{5}$$

where f_r is the firing rate by a single red soldier and p_r to be the probability that a single red bullet hits its target. We can think of f_b, f_r and p_b, p_r as factors influenced by morale, training and technology.

random
fire

For random fire we do not assume the probability of a single bullet wounding a soldier to be constant. It will vary depending on the number of target soldiers within a given area. Thus, this probability will depend on both the number of target soldiers and the area into which fire is being directed.

Exact solution

linear DEs

The differential equations (3) are linear, unlike the systems of differential equations in previous sections (Sections 5.3, 5.4 and 5.6) and it is possible to obtain an exact solution. We can use the following `Maple` code to do so:

```
> restart;
> de1 := diff(R(t),t) = -a[1]*B(t);
> de2 := diff(B(t),t) = -a[2]*R(t);
> inits := R(0)=r[0],B(0)=b[0];
> dsolve({de1,de2, inits}, {R(t),B(t)});
```

and the solution can be written as

$$\begin{aligned} R(t) &= r_0 \cosh(\alpha t) + b_0 v \sinh(\alpha t), \\ B(t) &= b_0 \cosh(\alpha t) + r_0 v^{-1} \sinh(\alpha t), \end{aligned} \tag{6}$$

where

$$\alpha = \sqrt{a_1 a_2}, \qquad v = \sqrt{\frac{a_1}{a_2}}.$$

This solution is also obtained in the exercises (see Question 19) by substitution of one differential equation into the other. For details on the hyperbolic functions (sinh and cosh) see Appendix A.2 and Appendix B.4. (Matrices and eigenvalues may also be used to find this solution, as is explained in Section 7.2.)

Limitations and extensions of the model

The model we have developed here was based on the assumption of aimed fire. More generally, battles occur where one army uses aimed fire and the other uses random fire (for example, guerilla warfare) or where both armies use random fire (e.g. long-range artillery). These models lead to the differential equations

aimed or random fire

$$\frac{dR}{dt} = -a_1 B, \qquad \frac{dB}{dt} = -c_2 RB,$$

in the case of guerilla warfare and

$$\frac{dR}{dt} = -c_1 RB, \qquad \frac{dB}{dt} = -c_2 RB,$$

for long-range artillery or trench warfare.

There are several other obvious extensions and variations of the basic model. These include incorporating both random and aimed fire, or modelling regular reinforcement and/or operational losses, such as from disease.

other extensions

In Braun (1979) there is a clear discussion of this model and a similar model for guerilla warfare, where one army is hidden (e.g. jungle warefare). Also discussed is the Battle of Iwo Jima. See Taylor (1980) for a substantial discussion of these types of models, and ways of estimating attrition coefficients. Stochastic (random) effects may also be included into the basic models (see Przemieniecki (1994)). Both Taylor (1980) and Przemieniecki (1994) discuss further extensions of the basic models to include reinforcements, range dependent firing and geometric mean fire.

further reading

Skills developed here:

- Modify the model to account for one, or both, of the armies using random fire.
- Modify the model to account for loss due to disease and/or gains due to reinforcements.
- Obtain numerical solutions to the model developed and some extensions.

5.8   Exercises for Chapter 5

1. Two prey and one predator. Develop a model with three differential equations describing a predator–prey interaction, where there are two different species of prey and one species of predator. What assumptions have been made in this model formulation? (Base these on the assumptions used in Section 5.3.)

2. Effect of DDT. Consider the predator–prey model developed in Section 5.3, which describes the threat to the American citrus industry posed by the accidental introduction of the Australian scale insect, and the later introduction of the ladybird to combat the pest.

(a) Use `Maple` to generate the time-dependent graphs of the populations over time with $\beta_1 = 1$, $\alpha_2 = 0.5$, $c_1 = 0.01$ and $c_2 = 0.005$. Initialise with $(x_0, y_0) = (200, 80)$ and $(x_0, y_0) = (80, 200)$. Comment on any differences in the long-term behaviour.

(b) Verify that $X' = 0$ and $Y' = 0$, if $X = 100$ and $Y = 100$ simultaneously. Plot these values (as horizontal lines) in the time-dependent diagrams.

(At these values there is no change in the populations over time, so they are equilibrium populations. We see how to find such equilibrium solutions in Chapter 6.)

(c) Now include the effect of DDT in the model, with $p_1 = 0.1$ and $p_2 = 0.1$. Generate the time-dependent plots as before. Verify that $X' = 0$ and $Y' = 0$ when $X = 120$, $Y = 80$, and plot these values on the time-dependent diagrams.

(d) Comment on your observations from the above.

3. Predator–prey with density dependence. Starting with the Lotka–Volterra model, we wish to include the effect of logistic growth for the prey and DDT on both species. The differential equations are

$$\frac{dX}{dt} = \beta_1 X \left(1 - \frac{X}{K}\right) - c_1 XY - p_1 X, \qquad \frac{dY}{dt} = c_2 XY - \alpha_2 Y - p_2 Y.$$

(a) With parameter values as in Figure 5.3.3, and also Figure 5.3.4, use `Maple` to generate time-dependent graphs for the populations over time.

(b) Verify (by substitution) that $X' = 0$ and $Y' = 0$, if $X = 50.5$ and $Y = 84.95$. Include them in the time-dependent plots of (a) as horizontal lines.

(c) *Comment on your observations, particularly with respect to Figure 5.3.4.*

4. Competing species with no density dependence. *In Section 5.4, a simple model describing the interaction between two competing species was developed, where each population exhibits exponential growth.*

(a) *Use* `Maple` *to generate time-dependent graphs for the populations over time, with parameter values as given in Figure 5.4.2, but using a variety of initial conditions.*

(b) *Show that $X' = 0$ and $Y' = 0$ simultaneously, when $X = 2.5$ and $Y = 1.5$. Include them in the time-dependent plots as horizontal lines.*

(c) *In this case it is not so easy to make some 'general conclusions' about the population behaviour over long periods of time. Why?*

5. Competing species with density dependence. *Consider the following model for two competing species, with densities, $X(t)$ and $Y(t)$, given by the differential equations*

$$\frac{dX}{dt} = X(\beta_1 - c_1 Y - d_1 X), \qquad \frac{dY}{dt} = Y(\beta_2 - c_2 X - d_2 Y),$$

with parameter values $\beta_1 = 3$, $\beta_2 = 3$, $c_1 = 2$, $c_2 = 1$, $d_1 = 2$ and $d_2 = 2.5$.

(a) *What is the carrying capacity for each of the species, evaluated for the given parameter values? (Hint: Compare with equations (3) in Section 5.4.)*

(b) *With the above parameter values, and the initial values $X = 2$ and $Y = 2$, use* `Maple` *to draw time-dependent plots for these populations. Over a period of time what population densities do you estimate they will approach?*

6. Symbiosis. *Symbiosis is where two species interact with each other, in a mutually beneficial way. Starting with a compartmental diagram, formulate a differential equation model describing this process, based on the following. Assume the per-capita death rate for each species to be constant, but the per-capita birth rate to be proportional to the density of the other species. In other words, the presence of the other species is necessary for continued existence. (Define all parameters and variables of the model.)*

7. Simple age-based model. *Consider a population split into two groups: adults $A(t)$ and juveniles $J(t)$. Starting with word equations formulate a pair of differential equations describing the density of adults and the density of juveniles. Define all variables and parameters used. You may assume constant per-capita birth and death rates for the population, and also assume that the young mature into adults at a constant per-capita rate γ.*

8. Wine fermentation. *In the fermentation of wine, yeast cells digest sugar from the grapes and produce alcohol as a waste product, which is toxic to the yeast cells. Develop a model consisting of three coupled differential equations for the density of yeast cells, the amount of alcohol and the amount of sugar. In the model assume the yeast cells have a per-capita birth rate proportional to the amount of sugar, and a per-capita death rate proportional to the amount of alcohol present. Assume the rate of alcohol produced is proportional to the density of yeast cells, and the rate of sugar consumed is proportional to the density of yeast cells.*

9. Numerical solution of basic epidemic model. *In Section 5.6, a model for an epidemic was developed, which led to the differential equations*

$$\frac{dS}{dt} = -bSC, \quad \frac{dC}{dt} = bSC - rC.$$

(a) *Use parameter values $b = 0.002$ and $r = 0.4$, and assume that initially there is only one infective but there are 500 susceptibles. Use* `Maple` *to generate the time-dependent plot on the interval $t = [0, 20]$.*

(b) *How many susceptibles never get infected?*

(c) *What happens as time progresses if $S(0) = 100$?*

(d) *Suppose the transmission coefficient b is doubled. How does this effect the maximum number of infected individuals? Is this what you expect?*

10. Contagious for life. *Consider a disease where all those infected remain contagious for life. Ignore all births and deaths.*

(a) *Write down suitable word equations for the rate of change of numbers of susceptibles and infectives. Hence develop a pair of differential equations. (Define any notation you introduce.)*

(b) *With a transmission coefficient of 0.002, and initial numbers of susceptibles 500 and infectives 1, use* `Maple` *to sketch time-dependent plots for the sub-populations (susceptibles and infectives) over time.*

11. **Continuous vaccination.** *Consider a model for the spread of a disease where life long immunity is attained after catching the disease. The susceptibles are continuously vaccinated against the disease at a rate proportional to their number. Write down suitable word equations to describe the process, and hence obtain a pair of differential equations.*

12. **Latent period.** *Many diseases have a latent period, which is when there is a period of time between infection and when this infected individual becomes contagious. One example is measles, where the latent period is approximately 5 days.*

Extend the basic epidemic model to one with an additional population class $E(t)$, corresponding to individuals who have been exposed to the disease, so they are no longer susceptibles, but are not yet contagious. You may assume the per-capita rate at which an individual in the exposed class becomes an infective, is constant. (Give a suitable compartmental diagram, or set of word equations and define all parameters.)

13. **Disease with carriers.** *It is possible for some diseases, e.g. typhoid, to have carriers who are infected with a disease, and can pass it on to other susceptibles, but who do not show any symptoms or suffer from the disease themselves. However those who develop symptoms die shortly after exposure. The population can be divided into three sub-populations: susceptibles $S(t)$, sick infectives $C(t)$ and carrier (exposed but healthy) infectives $H(t)$.*

Assume that a susceptible, when contacted by a carrier will become either a carrier or a sick infective. Neglect any contacts between susceptibles and sick infectives. Once infected, all individuals become carriers for the rest of their lives, whereas those who are infectives, and suffer from the disease, may infect others for a time but then die.

Develop a model comprising three differential equations to describe this process.

14. **Cycles of measles epidemics.** *For the spread of measles it is thought that the effect of births, continuously refreshing the number of susceptibles, is an important contributing factor towards the periodic outbreaks of the disease.*

Develop a model for the spread of measles which incorporates this refreshing of the susceptible numbers, assuming a constant per-capita birth rate for susceptibles, and also that all those infected become immune. Start with suitable word equations and define all symbols introduced.

15. **Density dependent contact rate.** *For a fatal disease, if the basic epidemic model of Section 5.6 is modified to include density dependent disease*

transmission, the resulting differential equations are

$$\frac{dS}{dt} = -\frac{\lambda(N)}{N}SC, \qquad \frac{dC}{dt} = \frac{\lambda(N)}{N}SC - rC,$$

where $N = S + C$, and the contact rate function λ is given by

$$\lambda(N) = \frac{\lambda_m N}{K(1 - \epsilon) + \epsilon N},$$

where ϵ is a positive constant between 0 and 1.

Consider $\epsilon = 0.5$, and using `Maple` with $\lambda_m = 1.62$, $r = 0.44$, $K = 1000$, and initial values $c_0 = 1$, $s_0 = 762$ (time is measured in days) graph the number of susceptibles over time and determine when the number of infectives is at a maximum. Compare this model to those with $\epsilon = 1$ and $\epsilon = 0$, and discuss the differences.

16. Battle loss due to disease. Develop a model (a pair of differential equations) for a battle between two armies where both groups use aimed fire. Assume that the red army has a significant loss due to disease, where the associated death rate (from disease) is proportional to the number of soldiers in that army.

17. Jungle warfare. In Section 5.7, we developed a simple model for a battle between two armies. We assumed that the probability of a single bullet hitting its target is constant. This is not a good assumption in jungle warfare where one, or both, of the armies may be hidden from view.

Suppose that soldiers from the red army are visible to the blue army, but soldiers from the blue army are hidden. Thus, all the red army can do is fire randomly into an area and hope they hit something. The blue army uses aimed fire.

(a) Write down appropriate word equations describing the rate of change of the number of soldiers in each of the armies.

(b) By making appropriate assumptions, obtain two coupled differential equations describing this system.

(c) Extend the model to include reinforcements if both of the armies receive reinforcements at constant rates.

18. Jungle warfare (continuation of previous question). This question refers to the model you developed in Question 17. Suppose the blue army defends an area of $10^5 \, \text{m}^2$, with an initial number of 150 soldiers. The red army has 500 soldiers initially, all of whom are exposed to fire from the blue army. Each soldier, in either army, fires at the same rate of 400 bullets per day. Field data has shown that a single bullet from the blue army has the constant probability of one in one hundred of wounding a red soldier.

(a) Write down a formula for the probability of a single bullet fired from a single red soldier wounding a blue soldier.

(b) Hence, estimate the coefficients in your model (i.e. write the rate of wounding in terms of the probability in (a) and the firing rate.)

(c) With your estimates in (b), use Maple to calculate the number of soldiers left in each army after 5 days.

19. **Exact solution for battle model.** *Consider the aimed fire battle model developed in the text*

$$\frac{dR}{dt} = -a_1 B, \qquad \frac{dB}{dt} = -a_2 R.$$

The exact solution can be found using theoretical techniques and then checked using Maple *as follows:*

(a) Take the derivative of the first equation to get a second-order differential equation, and then eliminate dB/dt from this equation by substituting the second equation (given above) into this second-order equation.

(b) Now assume the solution to be an exponential of the form $e^{\lambda t}$. Substitute it into the second-order equation and solve for the two possible values of λ. The general solution for R will be of the form

$$R(t) = c_1 e^{\lambda_1} + c_2 e^{\lambda_2},$$

where c_1 and c_2 are the arbitrary constants of integration. The solution for B is then found using the equation $dR/dt = -a_1 B$.

(c) Now find the arbitrary constants by solving the simultaneous equations for $R(0) = r_0$ and $B(0) = b_0$, when $t = 0$. The final solution is given in the text in Section 5.7.

(d) Using Maple (some code is provided in the text) check the solution above.

(Further details about methods for solving second-order differential equations, in particular for differential equations with constant coefficients, as used here, can be found in Appendix A.5.)

Chapter 6

Phase-plane analysis

For pairs of coupled first-order differential equations it can be useful to use the chain rule to eliminate time and reduce the pair of equations to a single first-order differential equation. The graph of all the solutions to this single equation is called the phase-plane. We apply this method to the basic models from Chapter 5. Consequently, we are able to draw some general conclusions about the models and confirm the observations which were made using only the numerical solutions.

6.1   Introduction

overview

In the previous chapter, we developed models for interacting populations which resulted in pairs of nonlinear, coupled differential equations and we then used `Maple` to solve the equations numerically. However, each solution was for a particular pair of initial conditions and a particular combination of the parameters. Clearly there are infinitely many such combinations and we need some other tools to understand the system behaviour as a whole. What would be valuable is to understand how any changes in the initial conditions or values of the parameters might affect these solutions and the subsequent system dynamics.

general
behaviour
of
solutions

We see how to gain insight into the behaviour of the solutions to these systems by finding the equilibrium points (points corresponding where the derivatives are zero) and eliminating time from the differential equations (by converting the coupled differential equations to a single first-order differential equation). Further, we are interested in the long-term behaviour of the systems. We ask questions such as: does a population die out or settle down to some fixed size, or, can a fatal disease wipe out a population completely or could it recur periodically? The techniques we develop to deal with such questions are designed to investigate the behaviour of solutions to the systems and require an understanding of equilibrium points and the phase-plane. First, we introduce these techniques using a very simple example system and then apply them to the systems developed in the previous chapter.

the
example
problem

Consider the coupled pair of first-order (linear) differential equations

$$\frac{dX}{dt} = Y, \qquad \frac{dY}{dt} = -X. \tag{1}$$

Equilibrium points

equilib-
rium
points

Equilibrium points correspond to solutions of a coupled system of differential equations where the solutions are constant, i.e. where $dX/dt = 0$ and $dY/dt = 0$, simultaneously. Hence, for the differential equations (1) we obtain

$$Y = 0, \qquad X = 0$$

and so $(X, Y) = (0, 0)$ is the only equilibrium solution.

Trajectories and the phase-plane diagram

Let us consider the (X, Y)-plane: this is called the *phase-plane*. Dividing   trajectory directions
the plane into four quadrants in the manner illustrated in Figure 6.1.1, we
have, in the first quadrant where $X > 0$ and $Y > 0$, that $dX/dt = Y > 0$
and $dY/dt = -X < 0$. Thus $X(t)$ is increasing and $Y(t)$ is decreasing
and we obtain a direction vector for any solution in that quadrant, given
by the arrow in Figure 6.1.1. Each quadrant can be considered in the
same manner. We can thus infer that the solutions, that is the *phase-
plane trajectories*, move in a clockwise direction.

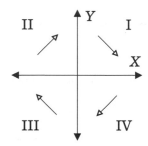

Figure 6.1.1: Direction vectors for the trajectories in the phase-plane of example
equations (1).

Using the chain rule

We note that neither of the differential equations involves the time vari-   independence from time
able t explicitly. We mean by this that t does not appear on the RHS of
either equation. (Of course, the solutions will be time dependent, since
the derivatives on the LHS are with respect to time.) This means that
we should be able to eliminate the time variable and find an expression,
independent of t, which relates X and Y. Alternatively stated, we express Y as a function of X. That is, we are making X the independent
variable where it was previously a variable dependent on t.

An expression for the chain rule is   using the chain rule

$$\frac{dY}{dt} = \frac{dY}{dX}\frac{dX}{dt}$$

which gives the derivative of Y with respect to t in terms of the derivative
of Y with respect to X and the derivative of X with respect to t. Dividing
by the derivative of X with respect to t gives

$$\frac{dY}{dX} = \frac{dY/dt}{dX/dt}. \qquad (2)$$

We substitute from the coupled pair of differential equations (1) into (2) to give

$$\frac{dY}{dX} = -\frac{X}{Y}.$$ (3)

Here we have a first-order differential equation with Y a function of X.

separating variables It is not always a simple matter to solve the differential equation, however, in this case the differential equation (3) is separable since we can write it in the form

$$Y\frac{dY}{dX} = -X.$$ (4)

Example 1: *Solve the differential equation (4) using the separation of variables technique.*

Solution: *Integrate both sides with respect to the independent variable X to obtain*

$$\int Y\frac{dY}{dX}\,dX = \int -X\,dX.$$

By the substitution rule for integration this simplifies to

$$\int Y\,dY = \int -X\,dX.$$

Carrying out both integrations gives

$$\frac{1}{2}Y^2 = -\frac{1}{2}X^2 + C$$

where C is the constant of integration. Multiplying throughout by 2, and rearranging the terms, we have that

$$X^2 + Y^2 = K$$

with $K = 2C$. The value of K will be determined by the initial conditions.

This solution equation is that of a circle. It describes the paths traced out by the (X, Y) pair over time, depending on the initial values or starting conditions. These are the exact solutions to the phase-plane trajectories.

Interpretation of the phase-plane

movement along a trajectory As an interpretation of these trajectories, if a system has initial values x_0 and y_0 then the system starts at the point (x_0, y_0) in the phase-plane and, as time evolves, it traces out the trajectory curve (in this case a circle) in a clockwise direction as was established in Figure 6.1.1. At any subsequent time the values of $X(t)$ and $Y(t)$ will be the coordinates of this trajectory. Since, for this example, the trajectory is a circle the motion is repeated continuously in time. In fact, this would be the case for any closed trajectory.

To see how the system evolves in time we would normally need an exact solution of the original coupled equations giving both Y and X as functions of time. Often this is not possible. Nevertheless, we can still use the chain rule to infer useful information about the system. We see how to do this in the following sections which examine the systems of differential equations for populations, developed in the previous chapter. *determining the time evolution*

In essence, the idea of phase-plane analysis is to draw a phase-plane diagram together with the phase-plane trajectories in order to understand some general features of the system. (The terminology arises from the use of these ideas to calculate planetary orbits.) If the differential equations are sufficiently simple, we can go further and use the chain rule to eliminate time. In this way, we may obtain an exact expression relating the two dependent variables which describes the trajectory path. *basic idea of phase-plane analysis*

The phase-plane diagram is useful for determining the behaviour of solutions for a variety of initial conditions. In the above example, we saw that all solutions of the differential equations have phase-trajectories which are circles. This means that the plots for both variables as functions of time must be oscillations, which follows because X and Y must always return to their original values as we move along the trajectory. Furthermore, as the initial point approaches the equilibrium point the amplitude of the oscillation is reduced, with the equilibrium point itself corresponding to a solution which is constant in time. *use of the phase-plane*

Note, however, that in this procedure we have lost information about time. This was the price for reducing the coupled, first-order differential equations to a single first-order equation. *losing information about time*

> **Skills developed here:**
> - Understand the concept of equilibrium solutions.
> - Establish the directions of trajectories.
> - Use the information on equilibrium points and trajectory directions to draw a phase-plane.
> - Understand how the chain rule can eliminate time and reduce a coupled pair of differential equations to a single differential equation.

6.2   Analysis of a predator–prey model

overview
We explore the Lotka–Volterra predator–prey equations with the aim of gaining a more general understanding of the dynamics using phase-plane analysis. We first obtain the phase-plane diagram for a specific set of parameters, using `Maple` to generate a numerical solution, and thereafter we infer what the phase-plane looks like for a general parameter set.

Review of model

the differential equations
In Section 5.3, we formulated a simple model for predator–prey interactions. The resulting differential equations were

$$\frac{dX}{dt} = \beta_1 X - c_1 XY, \qquad \frac{dY}{dt} = -\alpha_2 Y + c_2 XY. \tag{1}$$

Recall that X denotes the prey population and Y denotes the predator population. For the positive constant parameters we had c_1 and c_2 as the interaction parameters, β_1 as the prey per-capita birth rate and α_2 as the predator per-capita death rate.

general behaviour
We found these equations produced oscillations in time for the various initial conditions and parameter combinations considered. Here we address the question of whether this model always predicts oscillations for any (positive) values of the parameters.

The phase-plane diagram from Maple

We let $\beta_1 = 1$, $\alpha_2 = 0.5$, $c_1 = 0.01$ and $c_2 = 0.005$, which are the same values for the parameters we used in Section 5.3. We can use Maple to plot some phase-plane trajectories for this set of parameter values, and for some different initial conditions.

example numerical solution

```
<   restart: with(plots): with(DEtools):
>   beta[1]:=1.0: alpha[2]:=0.5: c[1]:=0.01: c[2]:=0.005:
>   de1 := diff(X(t),t) = beta[1]*X(t) - c[1]*X(t)*Y(t);
>   de2 := diff(Y(t),t) = -alpha[2]*Y(t) + c[2]*X(t)*Y(t);
>   inits := [0,100,80], [0,50,50], [0,100,170];
>   DEplot([de1,de2], [X,Y], t=0..15, [inits], scene=[X,Y],
>        stepsize=0.1);
```

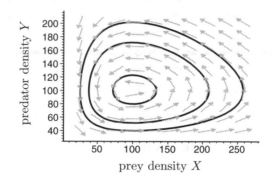

Figure 6.2.1: Maple generated phase-plane plot for the Lotka–Volterra equations. The parameter values $\beta_1 = 1$, $c_1 = 0.01$, $\alpha_2 = 0.5$ and $c_2 = 0.005$ have been used with various initial conditions.

The fact that the curves which appear in this phase-plane diagram of Figure 6.2.1 are closed is very significant. (Closed trajectories do not occur in all phase-plane diagrams.) Recall that as time evolves we travel along the curve. From the arrows in Figure 6.2.1 (the direction field) the direction of travel is in the anticlockwise direction. Thus, if initially the populations are x_0 and y_0 then as we travel along the curve we eventually return to the initial values x_0 and y_0. In fact every point on the curve will be repeated each time we make a complete circuit. These closed curves imply that both populations must follow *periodic cycles*. This is consistent with the time-dependent graphs obtained in Section 5.3.

significance of closed trajectories

We now see what we can deduce about the model for general parameter values. We begin by finding a general expression for the equilibrium points. Then we analyse the directions which the trajectories follow in the phase-plane.

analysis for general parameter values

Equilibrium populations

significance of equilibrium solutions

As we saw from Figure 6.2.1, for different initial values we get different phase-plane trajectories enclosed within one another. We can imagine the trajectories shrinking to a single point. This point would correspond to equilibrium populations, where the populations do not change with time. The equilibrium solutions correspond to those values where the rates of change are zero. These are the constant solutions of the differential equations (1).

finding equilibrium solutions

We can find these values analytically, as the following example shows, by solving a pair of simultaneous equations. When solving simultaneous nonlinear equations it is important to ensure you have found all the possible solutions and to this end it is often useful to factorise, as in the following example.

Example 1: *Find the equilibrium solutions of the differential equations (1).*

Solution: *We set $dX/dt = 0$ and $dY/dt = 0$ in (1) and obtain the equations*

$$\beta_1 X - c_1 XY = 0, \qquad -\alpha_2 Y + c_2 XY = 0.$$

To find solutions we first write the two simultaneous equations in factored form as

$$X(\beta_1 - c_1 Y) = 0, \tag{2}$$
$$Y(-\alpha_2 + c_2 X) = 0. \tag{3}$$

From (2) there are two possible solutions: $X = 0$ or $\beta_1 - c_1 Y = 0$. We need to look at each case. Recall that the parameters β_1, α_2, c_1 and c_2 are positive (non-zero) constants.

If $X = 0$, then substituting this into (3) gives $-\alpha_2 Y = 0$ so that $Y = 0$. This gives one possible solution of both equations at $(X, Y) = (0, 0)$.

Taking the other case, $\beta_1 - c_1 Y = 0$, then $Y = \beta_1/c_1$. Substituting this into (3) gives $-\alpha_2 + c_2 X = 0$, for which the only solution is $X = \alpha_2/c_2$. Hence we obtain a second solution to both equations $(X, Y) = (\alpha_2/c_2, \beta_1/c_1)$.

To ensure we have all the solutions, we should also solve the second equation and substitute the solution into the first, but this yields exactly the same two solutions as we obtained above.

---

geometric interpretation

Summarising, we have obtained two equilibrium solutions

$$(X, Y) = (0, 0) \quad \text{and} \quad (X, Y) = \left(\frac{\alpha_2}{c_2}, \frac{\beta_1}{c_1} \right).$$

Another way of being sure we have all the solutions is to think of the problem of finding the solutions geometrically. Solving the equation $dX/dt = 0$ gave $X = 0$ and $Y = \beta_1/c_1$ which represents the equations of two lines, which we denote as L_0 and L_1, in the (X, Y)-plane. Similarly, $dY/dt = 0$ yields two lines, L_2 and L_3, given by $X = \alpha_2/c_2$ and $Y = 0$, respectively. The equilibrium points occur where the pair L_0 and L_1 intersect with L_2 and L_3 (not L_0 with L_1, or L_2 with L_3). These lines (or curves) are known as *nullclines* and their points of intersection are illustrated in Figure 6.2.2.

We can also use `Maple` to solve the equations for the equilibrium solutions. The code is given below:

using Maple

```
> restart:
> eq1 := beta[1]*X(t) - c[1]*X(t)*Y(t);
> eq2 := -alpha[2]*Y(t) - c[2]*X(t)*Y(t);
> solve({eq1,eq2}, {X(t),Y(t)});
```

Consider the phase-plane trajectories, which are equivalent to the oscillations in the time-dependent diagrams. We may regard the equilibrium populations in this case as the average density of the prey and predator populations, about which they oscillate.

biological interpretation

Direction of trajectories

In order to examine the direction vectors in the phase-plane, we first rewrite the original differential equations (1) in the factored form

increasing or decreasing X and Y

$$\frac{dX}{dt} = X\left(\beta_1 - c_1 Y\right), \qquad \frac{dY}{dt} = Y\left(-\alpha_2 + c_2 X\right). \qquad (4)$$

We draw the nullcline curves corresponding to $dX/dt = 0$ and $dY/dt = 0$. They divide the phase-plane into regions where the trajectories have different directions.

The nullcline curves here are the lines $X = 0$, $Y = \beta_1/c_1$ (for $dX/dt = 0$) and the lines $Y = 0$, $X = \alpha_1/c_2$ (for $dY/dt = 0$). Effectively they divide the phase-plane into the four regions labelled I, II, III and IV in Figure 6.2.2.

the nullclines

From the differential equations (4) we can establish whether $X(t)$ and $Y(t)$ are increasing or decreasing in each of the regions. The following example illustrates how this can be done.

X and Y increasing or decreasing?

Example 2: *Determine the trajectory directions for region II in Figure 6.2.2.*

Solution: *Everywhere in region II, $X < \alpha_2/c_2$ and $Y > \beta_1/c_1$.*

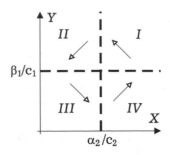

Figure 6.2.2: Regions in the phase-plane illustrating the trajectory directions. The dashed lines are the non-zero nullclines.

From (4), $dX/dt < 0$ everywhere in region II, since $\beta_1 - c_1 Y < 0$ if $c_1 Y > \beta_1$.

Similarly, $dY/dt < 0$ everywhere in region II, since $-\alpha_2 + c_2 X < 0$ when $c_2 X < \alpha_2$.

Thus both X and Y are decreasing in region II. This gives a vector with negative x-component and negative y-component, as shown in Figure 6.2.2.

By a similar argument

- in region I, X is decreasing and Y is increasing,
- in region III, X is increasing and Y is decreasing,
- in region IV, X is increasing and Y is increasing.

interpret-
ation
This tells us that as we move along a phase-trajectory we must move in an anticlockwise direction. However, it does not tell us whether this is a closed curve or a spiral. In Section 6.6, we use the chain rule to prove that all trajectories for this system are closed.

stability
Furthermore, for the equilibrium point $(0,0)$, we note from Figure 6.2.2 that the trajectories move towards it in one direction and away from it in another. It is thus an unstable equilibrium point in the sense that trajectories move away from it with time and we show in Section 6.6 that this is always the case, regardless of the parameter values. It is called a saddle point, which we discuss in detail in Chapter 7.

Density dependent growth

modified
model
In Chapter 5, the basic predator–prey model was extended to include density dependent growth (logistic) for the prey species in order to curb the unrealistic exponential growth. It is left as an exercise (Question 4) to generate the associated phase-plane diagram, and examine how this change in growth alters the equilibrium points and/or the direction vectors of the system.

Skills developed here:
- Find equilibrium solutions.
- Establish and draw the nullclines for a system.
- Find directions of trajectories in the phase-plane.
- Deduce how populations behave in time from their phase-plane diagram.

6.3 Analysis of competing species models

We now apply the phase-plane theory to the model we developed to de- overview
scribe the interaction between competing species. Again, we first look
at the phase-plane obtained from numerical solutions (using `Maple`) for
a specific set of parameter values and then extend the analysis to consider general values of the parameters, finding equilibrium points and
trajectory directions.

Review of model

In Section 5.4, we formulated two models for competing species. The first differential
model assumed exponential growth in the absence of a competitor, and equations
later we introduced an improved model which assumed density-dependent
growth for both species in the absence of a competitor. We consider the
second improved model here, leaving analysis of the first model as an
exercise (Question 8).

From Section 5.4 our model is   the
differential
equations

$$\frac{dX}{dt} = \beta_1 X - d_1 X^2 - c_1 XY, \qquad \frac{dY}{dt} = \beta_2 Y - d_2 Y^2 - c_2 XY \quad (1)$$

where β_1 and β_2 are per-capita birth rates or growth rates, d_1 and d_2 are
density dependent coefficients (independent of the other species) and c_1
and c_2 are interaction coefficients (the effect of species interaction on the
death rates). In the absence of species Y, species X has logistic growth
with carrying capacity $K_1 = \beta_1/d_1$. Similarly, if $X = 0$ then species Y
has logistic growth with carrying capacity $K_2 = \beta_2/d_2$.

The phase-plane diagram from `Maple`

example
numerical
solution

We can use `Maple` to draw the phase-plane, but only for a given set of parameter values. Take, for example, per-capita birth rates $\beta_1 = 0.21827$, $\beta_2 = 0.06069$, interspecies interaction parameters $c_1 = 0.05289$, $c_2 = 0.00459$ and intraspecies interaction parameters $d_1 = 0.01679$, $d_2 = 0.01046$, which are those from Gause's experiments with yeast (see Section 5.4 and Edelstein-Keshet (1988)).

numerical
results

The results in Figure 6.3.1 show clearly that the trajectories approach equilibrium populations on the X and Y axes, for these values of the parameters.

```
> restart: with(plots): with(DEtools):
> tend:= 360;
> beta[1]:=0.21827: beta[2]:=0.06069:
> d[1]:=0.01679: d[2]:=0.01046:
> c[1]:=0.05289: c[2]:=0.00459:
> K[1]  := beta[1]/d[1]; K[2]  := beta[2]/d[2];
> de1:=diff(X(t),t) = beta[1]*X(t) - c[1]*X(t)*Y(t)
       - d[1]*X(t)*X(t);
> de2:=diff(Y(t),t) = beta[2]*Y(t) - c[2]*X(t)*Y(t)
       - d[2]*Y(t)*Y(t);
> inits:= [0,0.5,0.5], [0,20,4], [0,20,8], [0,20,10];
> plot1:=DEplot([de1,de2], [X,Y], t=0..tend, [inits],
     X=0..20,Y=0..10, scene=[X,Y], stepsize=0.5, dirgrid=[10,10],
     arrows=medium):
> display(plot1);
```

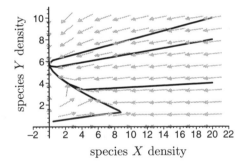

Figure 6.3.1: `Maple` produced phase-plane diagram for the competing species model including density dependent growth, with parameter values as given for Gause's experiment.

observations

Note that the trajectory which starts at $(0.5, 0.5)$ moves to the right before turning around and heading towards an equilibrium point at approximately $(0, 6)$. The other trajectories, from a variety of initial conditions, tend to the same equilibrium point. This is consistent with the time-dependent graph in Section 5.4 where one species died out and the other tended to an equilibrium point.

Equilibrium populations

As in Section 6.1 and Section 6.2, we find the equilibrium points for arbitrary parameter values by setting the rates of change to 0. This is done in the following example.

Example 1: *Find all the equilibrium points for the system of differential equations (1).*

Solution: *We need to solve (simultaneously) $dX/dt = 0$ and $dY/dt = 0$. This implies (after factorising)*

$$X(\beta_1 - c_1 Y - d_1 X) = 0 \tag{2}$$
$$Y(\beta_2 - c_2 X - d_2 Y) = 0. \tag{3}$$

From (2) we have two possibilities: $X = 0$ or $\beta_1 - c_1 Y - d_1 X = 0$. If $X = 0$, then (3) simplifies to $Y(\beta_2 - d_2 Y)$ which has two possible solutions, $Y = 0$ or $Y = \beta_2/d_2$. Thus we obtain two solutions to the equations, $(X, Y) = (0, 0)$ and $(0, \beta_2/d_2)$.

From (3) we have two possibilities: $Y = 0$ or $\beta_2 - c_2 X - d_2 Y = 0$. If $Y = 0$, then (2) simplifies to $X(\beta_1 - d_1 X) = 0$. This has two solutions $X = 0$ and $X = \beta_1/d_1$. Thus, simultaneous solutions (X, Y) of both equations are $(0, 0)$ (already obtained above) and $(\beta_1/d_1, 0)$.

One further possibility not yet considered is where (2) is satisfied by letting $\beta_1 - d_1 X - c_1 Y = 0$ and where (3) is satisfied by letting $\beta_2 - c_2 X - d_2 Y = 0$. By substituting one equation into the other it can be shown that these two equations are satisfied when

$$X = \frac{c_1 \beta_2 - d_2 \beta_1}{c_1 c_2 - d_1 d_2}, \qquad Y = \frac{c_2 \beta_1 - d_1 \beta_2}{c_1 c_2 - d_1 d_2}.$$

Summarising, we have found (at most) four equilibrium solutions

$$(0, 0), \quad \left(0, \frac{\beta_2}{d_2}\right), \quad \left(\frac{\beta_1}{d_1}, 0\right), \quad \left(\frac{c_1 \beta_2 - d_2 \beta_1}{c_1 c_2 - d_1 d_2}, \frac{c_2 \beta_1 - d_1 \beta_2}{c_1 c_2 - d_1 d_2}\right),$$

summary of equilibrium points

where the fourth equilibrium point is relevant only if both components are positive. Two of the equilibrium points suggest the extinction of one species and the survival and stabilisation of the other.

We can also obtain the equilibrium points using `Maple`. The code for this is given below:

using Maple

```
> restart:
> eq1 := beta[1]*X(t) - c[1]*X(t)*Y(t) -d[1]*X(t)*X(t);
> eq2 := beta[2]*Y(t) - c[2]*X(t)*Y(t) -d[2]*Y(t)*Y(t);
> solve({eq1,eq2}, {X(t),Y(t)});
```

geometric interpretation

As before, define the nullcline lines L_0, L_1, L_2, L_3 as

$$L_0: \quad X = 0, \qquad L_1: \quad \beta_1 - d_1 X - c_1 Y = 0,$$
$$L_3: \quad Y = 0. \qquad L_2: \quad \beta_2 - c_2 X - d_2 Y = 0,$$

There is an equilibrium point corresponding to the intersection of the pair of lines L_0 and L_1 with the pair of lines L_2 and L_3 (but not L_0 with L_1 or L_2 with L_3). This is shown in Figure 6.3.2. The diagrams confirm that there are four equilibrium points if the lines L_1 and L_2 intersect in the positive quadrant of the (X, Y)-plane and only three equilibrium points otherwise.

Direction of trajectories

sectioning the phase-plane

As usual, we are interested in the predictions of the model over time. We establish the trajectory behaviour by determining trajectory directions and considering only positive values of X and Y. The linear nullclines, L_0, L_1, L_2 and L_3, divide the phase-plane (with X and Y both positive) into different regions in four different ways. These four cases are illustrated in Figure 6.3.2.

directions

We determine the directions of the trajectories (seen in Figure 6.3.2) by looking at a typical point in each of the regions bounded by the nullclines. The example below illustrates this process for one of the cases in Figure 6.3.2.

Example 2: *Find the directions of the trajectories in each of the regions bounded by the nullclines for Case 3 of Figure 6.3.2.*

Solution: *We examine each of the four regions of Case 3 separately. The differential equations (1) are written as*

$$X' = X(\beta_1 - d_1 X - c_1 Y), \qquad Y' = Y(\beta_2 - d_2 Y - c_2 X)$$

where X' denotes dX/dt and Y' denotes dY/dt.

Consider the region below both L_1 and L_2. From the equations for L_1 and L_2 ($\beta_1 - d_1 X - c_1 Y = 0$ and $\beta_2 - d_2 Y - c_2 X = 0$) a point in the region below L_1 has $Y < (\beta_1 - d_1 X)/c_1$ and a point in the region below L_2 has $Y < (\beta_2 - c_2 X)/d_2$. Thus

$$c_1 Y < \beta_1 - d_1 X \qquad \Rightarrow \qquad 0 < \beta_1 - d_1 X - c_1 Y \qquad \Rightarrow \qquad X' > 0,$$
$$d_2 Y < \beta_2 - c_2 X \qquad \Rightarrow \qquad 0 < \beta_2 - c_2 X - d_2 Y \qquad \Rightarrow \qquad Y' > 0.$$

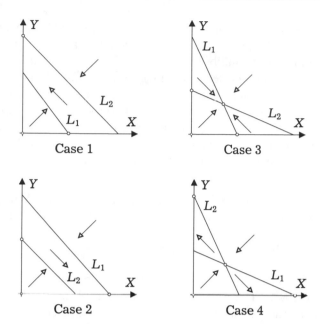

Figure 6.3.2: The four cases of the phase-plane diagram for competing species includ-ing density dependent growth. The direction vectors for the trajectories are illustrated with equilibrium points marked with open circles.

Consider the region below L_1 and above L_2. Here

$$c_1Y < \beta_1 - d_1X \quad \Rightarrow \quad 0 < \beta_1 - d_1X - c_1Y \quad \Rightarrow \quad X' > 0,$$
$$d_2Y > \beta_2 - c_2X \quad \Rightarrow \quad 0 > \beta_2 - c_2X - d_2Y \quad \Rightarrow \quad Y' < 0.$$

Similarly, for the region above L_1 and below L_2

$$c_1Y > \beta_1 - d_1X \quad \Rightarrow \quad 0 > \beta_1 - d_1X - c_1Y \quad \Rightarrow \quad X' < 0,$$
$$d_2Y < \beta_2 - c_2X \quad \Rightarrow \quad 0 < \beta_2 - c_2X - d_2Y \quad \Rightarrow \quad Y' > 0.$$

Finally, for the region above both L_1 and L_2

$$c_1Y > \beta_1 - d_1X \quad \Rightarrow \quad 0 > \beta_1 - d_1X - c_1Y \quad \Rightarrow \quad X' < 0,$$
$$d_2Y > \beta_2 - c_2X \quad \Rightarrow \quad 0 > \beta_2 - c_2X - d_2Y \quad \Rightarrow \quad Y' < 0.$$

This provides all the information required to establish the direction vectors of Case 3.

In Figure 6.3.2 we show the directions for each of the four different ways of sectioning the phase-plane with nullclines. In Case 1, the trajectories all interpret-ing the results move towards the equilibrium point on the Y-axis. This corresponds to extinction of the X-species, in the long term. In Case 2, the trajectories all move towards the equilibrium point on the X-axis. This corresponds to extinction of the Y-species, in the long term. In Case 3, the trajectories move to the equilibrium point where the nullclines L_1 and L_2 intersect. This corresponds to a coexistence of the two species, in the long term. In Case 4, the trajectories may end up at either of the equilibrium points on the axes, but not at the equilibrium point where the nullclines L_1 and L_2 intersect. Which equilibrium point is approached, or which species survives, will depend on the initial conditions.

using the nullcline intercepts

We can use the differences between the intercepts on the X and Y axes to determine which of the above cases provides the solution. Let

$$\text{Int}_X = \frac{\beta_1}{d_1} - \frac{\beta_2}{c_2} \quad \text{and} \quad \text{Int}_Y = \frac{\beta_2}{d_2} - \frac{\beta_1}{c_1},$$

the signs of which establish the equilibrium point towards which the trajectories move. The results are displayed in Table 6.3.1. (In Chapter 7, we develop further theory to establish which of these equilibrium points will attract or repel the trajectories.)

Table 6.3.1: Predicting the equilibrium point from the sign of Int_X and Int_Y.

Int_X	Int_Y	equilibrium point approached
$+$	$-$	$\left(\dfrac{\beta_1}{d_1}, 0\right)$
$-$	$+$	$\left(0, \dfrac{\beta_2}{d_2}\right)$
$-$	$-$	$\left(\dfrac{c_1\beta_2 - d_2\beta_1}{c_1c_2 - d_1d_2}, \dfrac{c_2\beta_1 - d_1\beta_2}{c_1c_2 - d_1d_2}\right)$
$+$	$+$	$\left(\dfrac{\beta_1}{d_1}, 0\right)$ or $\left(0, \dfrac{\beta_2}{d_2}\right)$ depending on starting conditions

Interpretation of parameters

carrying capacities

Above we established that β_1/d_1 was the carrying capacity K_1 for species X and that β_2/d_2 was the carrying capacity K_2 for population Y. Thus we can interpret the trajectory approach to $(\beta_1/d_1, 0) = (K_1, 0)$ as the survival of X and the extinction (or migration) of Y. Likewise, an approach to $(0, \beta_2/d_2) = (0, K_2)$ can be seen as the survival of Y and the extinction of X.

Initially, when our model allowed for only exponential growth (that is, *interference factors* infinite carrying capacity in the absence of the other species) it can be shown there is an equilibrium point at $(\beta_2/c_2, \beta_1/c_1)$, the only point for which both species can possibly exist simultaneously. Now Int_X is defined as the difference between the X coordinate β_2/c_2 and the carrying capacity of the species as was introduced with logistic growth. (Similarly for Int_Y.) Thus Int_X and Int_Y can be considered as *interference factors*, or a measure of how much one species interferes with the other.

Discussion

Our competition model predicts, in almost all cases, the extinction of *model limitation and extension* one species and the survival of the other. However, observations indicate that there is substantial diversity in many ecosystems with competing species coexisting. As we have discussed previously, these ecosystems do not exist in isolation but are integral components of some much larger system.

In Chapter 8, we present a system which uses the competition model *a system with three species* developed here, and embeds it within another system, a predator–prey model, establishing that in this more complicated system the two competing species can coexist. This lends credence to the earlier assertion of larger systems having greater stability than isolated ecosystems.

Skills developed here:

- Reinforcement of methods for finding equilibrium points by solving simultaneous equations.
- Reinforcement of methods for finding direction vectors for trajectories and establishing the general behaviour of a system from the phase-plane.

6.4 Analysis of an epidemic model

In the previous chapter we developed a model for a particular type of *overview* epidemic. We now use the chain rule and some analysis to prove that the disease, described by this model, can never infect the entire population.

Review of the model

the differential equations

The epidemic model, developed in Section 5.6, assumed the population N was divided into susceptibles, denoted by $S(t)$, infectives, denoted by $C(t)$ and removals $N - S(t) - C(t)$. We assumed the disease to confer life-long immunity and we neglected to include natural births and deaths. The model pair of differential equations we obtained was

$$\frac{dS}{dt} = -bSC, \qquad \frac{dC}{dt} = bSC - rC. \tag{1}$$

The parameter b is the transmission coefficient and r is the recovery (or removal) rate.

our aim

For the parameter values we considered in the example of Section 5.6, our results indicated that the number of infectives always tended to zero while the number of susceptibles approached some finite number. It seemed there were always certain individuals who never contracted the disease. It is interesting to speculate whether, according to our model, this is always the case. We use the chain rule below to show this is indeed so.

The phase-plane diagram from Maple

example numerical solution

We use Maple to get a picture of the phase-plane for specific values of the parameters b and r. We set $b = 10^{-4}$ and $r = 0.01$ and generate phase-plane trajectories for different sets of initial conditions. The results are illustrated in Figure 6.4.1.

```
> restart: with(plots): with(DEtools):
> b := 2.18*10^(-3): r := 0.44:
> de1 := diff(S(t),t) = -b*S(t)*C(t);
> de2 := diff(C(t),t) = b*S(t)*C(t) - r*C(t);
> inits := [0,762,1], [0,600, 20], [0,400,50]:
> plot1 := DEplot([de1,de2], [S(t),C(t)], t=0..30, [inits],
      scene=[S,C], stepsize=0.1,dirgrid=[10,10], arrows=MEDIUM):
> display(plot1);
```

Equilibrium populations

equilib-rium points

For general values of the parameters b and r we can determine the equilibrium points analytically.

Example 1: *Find all equilibrium points for the differential equations* (1).

Solution: *The equilibrium points are obtained by setting the derivatives to zero,*

$$-bSC = 0, \qquad bSC - rC = 0.$$

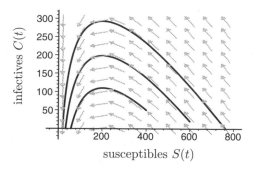

Figure 6.4.1: Maple generated phase-plane diagram, with $b = 10^{-4}$ and $r = 0.01$ with a variety of initial values.

Factorising gives

$$-bSC = 0, \qquad C(bS - r) = 0.$$

Both the first and second equations are satisfied if $C = 0$. Hence there are an infinite number of equilibrium points

$$(S, C) = (S, 0)$$

where S can take on any value. In the phase-plane all points along the S-axis are equilibrium points.

Direction of trajectories

The direction of trajectories in the phase-plane is determined from the differential equations, as below.

Example 2:   Determine the directions of trajectories in the phase-plane.

Solution: From $dS/dt = -bSC$ we see that for $S > 0$ and $C > 0$, dS/dt is always negative. This means that $S(t)$ is always decreasing. Similarly, from the other differential equation, dC/dt is positive provided $C(bS-r) > 0$. Thus $dC/dt > 0$ if $S > r/b$ and negative if $S < r/b$. The phase-plane, for positive S and C is therefore divided into two regions, as shown in Figure 6.4.2.

Applying the chain rule

For the simple differential equations comprising this model it is possible to obtain an analytic expression for the trajectories. We can do this by eliminating time (using the chain rule) which results in a differential equation involving the variables C and S. The solution to the differential   eliminating time

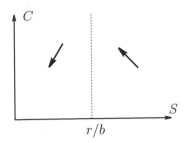

Figure 6.4.2: General diagram showing directions of trajectories in the phase-plane for the epidemic model.

equation establishes a relation between these two variables, as derived in the following example.

Example 3: Use the chain rule to find C in terms of S, given that the initial number of susceptibles is s_0 and the initial number of contagious infectives is c_0.

Solution: Using the chain rule

$$\frac{dC}{dS} = \frac{dC/dt}{dS/dt}.$$

Substituting from the differential equations (1) we eliminate the time variable, obtaining the single differential equation

$$\frac{dC}{dS} = \frac{bSC - rC}{-bSC}$$

which simplifies to

$$\frac{dC}{dS} = -1 + \frac{r}{bS}. \tag{2}$$

This differential equation relates the number of infectives C to the number of susceptibles S but does not retain explicit information about time.

The differential equation (2) is a first-order separable differential equation of a trivial kind and to solve it we have only to integrate both sides with respect to S. The solution is

$$C = -S + \frac{r}{b}\ell\mathrm{n}(S) + K \tag{3}$$

where K is an arbitrary constant of integration.

The initial numbers of susceptibles and infectives are s_0 and c_0, respectively, and hence the initial condition for the differential equation (2) is

$$C(s_0) = c_0.$$

Applying this to the general solution (3) gives an equation for K from which we deduce that

$$K = c_0 + s_0 - \frac{r}{b} \ell n(s_0).$$

Returning to the phase-plane

The solution to the differential equation (2), given by (3), is not very complicated so it is worthwhile to try and obtain a general sketch of C versus S.

sketching the phase-plane

Example 4:   Determine and sketch the family of phase-plane curves given by (3).

Solution: To find turning points we set $dC/dS = 0$. That is

$$-1 + \frac{r}{bS} = 0.$$

Solving for S gives

$$S = \frac{r}{b},$$

corresponding to a potential turning point. Since there is only one potential turning point we do not need to find a second derivative to determine whether it is a maximum, minimum or inflection point. Instead we can use the values of the function at $S = 0$ and $S \to \infty$.

As $S \to 0$ we see that $C \to -\infty$ since $\ell n(S) \to -\infty$. As $S \to \infty$ the limit is harder to work out. Note that $\ell n(S) \to \infty$ and $-S \to -\infty$ as $S \to \infty$. To determine the limit let us consider the derivative dC/dS. Note that from (3), $dC/dS \to -1$ as $S \to \infty$. This implies that $C \to -\infty$ since S grows faster than $\ell n S$ for large S. (A formal method for showing this is to use l'Hôpital's rule.)

Since $C \to -\infty$ as $S \to 0$ and $S \to \infty$ then the point corresponding to $S = r/b$ must be a local maximum. Hence the curve must cross the positive S-axis twice, for sufficiently large K. This is sketched in Figure 6.4.3.

The resulting sketch of Figure 6.4.3 agrees with the earlier theory, which indicated (see Figure 6.4.2) that from any initial condition (s_0, c_0), the trajectory of $S(t)$ decreases continuously.

directions along the trajectory

Interpretations

A useful conclusion from Figure 6.4.3 concerns the number of susceptibles left after the disease has run its course. We see that it is impossible for the disease to infect all of the susceptibles. (Although, the number

always some sus-ceptibles

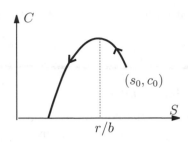

Figure 6.4.3: General sketch of the phase-plane diagram, illustrating a typical phase-plane curve with the direction in which the curve is traced indicated by the arrow.

of susceptibles may become sufficiently small for it to be regarded as effectively zero.)

threshold value
We can interpret some interesting results from Figure 6.4.3. Note that if s_0 is greater than the critical value r/b then the number of infectives must increase for a time before it decreases, whereas if $s_0 < r/b$ then the disease dies out. Thus for an epidemic to occur, in which the number of infectives increases from an initial number c_0, there needs to be at least some threshold number of susceptibles present prior to the outbreak, that is $s_0 > r/b$.

reproduction ratio
The combination rs_0/b is known as the *basic reproduction ratio*. It is a measure of the number of infections caused by a single infective in a fully susceptible population. This is an important quantity in epidemiology: a basic reproduction ratio of less than 1 indicates the propensity of the disease to die out.

vaccination strategies
The analysis of the model can be used to provide insight into effective vaccination strategies. The purpose behind a program of vaccinations is to decrease the number of susceptibles to below the threshold value r/b. This has been successful with smallpox, which has now been eradicated by a successful WHO program. Chapter 8 includes a case study where different strategies are tested in a bid to eradicate bovine tuberculosis in possums in New Zealand. However, the analysis in this case is a little different from models of diseases in humans; for the possums (an introduced and destructive feral pest) we are unconcerned if the population is eradicated in the process!

Skills developed here:

- Use the chain rule to eliminate time from any epidemic model.
- Further practice at graph sketching.
- Use this model (and extensions) to examine vaccination or other disease eradication strategies.

6.5   Analysis of a battle model

We apply the chain rule to the coupled system of differential equations for the Lanchester battle model developed in Section 5.7. This reduces the coupled equations to a single first-order differential equation. By solving this, we obtain an expression which relates the numbers of soldiers in one army to the numbers of those in the other. This in turn provides a more general understanding of the system dynamics.

overview of procedure

Review of the model

In the previous chapter we developed a model describing a battle between two armies. The model resulted in the pair of coupled differential equations, where $R(t)$ and $B(t)$ denote the numbers of soldiers in the red army and blue army, respectively. We assumed both armies to use only aimed fire. We derived the pair of differential equations

the differential equations

$$\frac{dR}{dt} = -a_1 B, \qquad \frac{dB}{dt} = -a_2 R, \tag{1}$$

where a_1 and a_2 are positive constants (attrition coefficients). Recall that these attrition coefficients can also be expressed in terms of firing rates and probabilities of hitting a target.

The phase-plane diagram from `Maple`

example
numerical
solution

We can use `Maple` to plot some typical phase-plane trajectories. We let $a_1 = 0.0544$ and $a_2 = 0.0106$. The phase-plane trajectories corresponding to three different initial conditions are shown in Figure 6.5.1.

```
> restart: with(plots): with(DEtools):
> a[1] := 0.0544: a[2] := 0.0106:
> de1 := diff(R(t),t) = -a[1]*B(t);
> de2 := diff(B(t),t) = -a[2]*R(t);
> inits := [R(0)=66, B(0)=18],
            [R(0)=45, B(0)=18],
            [R(0)=30, B(0)=18]:
> plot1 := DEplot([de1,de2], [R,B], t=0..30, [inits],
      scene=[R,B], method=rkf45):
> display(plot1);
```

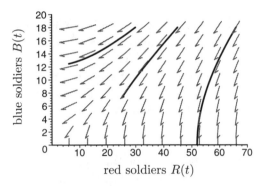

Figure 6.5.1: Phase-plane trajectories generated by `Maple` using parameter values $a_1 = 0.0544$ and $a_2 = 0.0106$ with various initial conditions.

Equilibrium points

finding the
equilib-
rium
points

Typically we set the two rates of change to zero and solve the equations simultaneously. In this case, the exercise is trivial and the only equilibrium solution is $(R, B) = (0, 0)$.

Direction of trajectories

determin-
ing
direction
of trajec-
tories

The direction of the trajectories can be obtained from the original differential equations (1). This is done in the following example.

Example 1: *Determine possible directions of phase-plane trajectories in the phase-plane.*

Solution: *From these equations, since R and B are both positive, and since the constants a_1 and a_2 are positive, then dX/dt is always negative and dY/dt is always negative. Hence $R(t)$ and $B(t)$ are always decreasing with time.*

Because of the directions in the phase-plane, it is clear that the trajectory interpret-ation must head towards one of the axes. Since the only equilibrium point is the origin, the trajectory (unless it passes through the origin) must cross the R or B axis in finite time, as the rate of change is not zero there. This means that the battle would be over in finite time. To determine where the trajectory intersects the axes we need to determine the explicit form of the trajectories, and fortunately we can do this using the chain rule.

Applying the chain rule

We use the chain rule as before to eliminate the time variable t. Substi-tuting from the differential equations (1) we obtain

$$\frac{dB}{dR} = \frac{dB/dt}{dR/dt} = \frac{a_2}{a_1}\frac{R}{B} \qquad (2)$$

using the chain rule to eliminate time

Thus we obtain a single first order differential equation which relates B and R, but does not involve t. The solution is given by the following example.

Example 2: *Solve the differential equation (2).*

Solution: *The solution of the first-order differential equation (2) is obtained by separating the variables and then integrating with respect to the indepen-dent variable R. This gives*

$$\int B\frac{dB}{dR}\,dR = \int \frac{a_2}{a_1}R\,dR.$$

Using the substitution rule for integration, the LHS can be converted to an integral involving the variable B,

$$\int B\,dB = \int \frac{a_2}{a_1}R\,dR.$$

Carrying out the integrations gives the equation

$$\frac{1}{2}B^2 = \frac{a_2}{2a_1}R^2 + C,$$

where C is an arbitrary constant of integration. Multiplying both sides of the equation by 2 we obtain

$$B^2 = \frac{a_2}{a_1}R^2 + K, \qquad (3)$$

where $K = 2C$ is also an arbitrary constant.

Applying the initial conditions $R(0) = r_0$ and $B(0) = b_0$ we obtain

$$b_0^2 = \frac{a_2}{a_1} r_0^2 + K$$

so that

$$K = b_0^2 - \frac{a_2}{a_1} r_0^2. \tag{4}$$

who wins the battle?

If we suppose the battle is fought until one of the sides is wiped out, then we can use (3) to determine who wins the battle. We can also see this graphically by plotting the trajectories described by (3).

Returning to the phase-plane

phase-plane and trajectories

Let us now examine the solution (3) graphically. From (3) we have

$$B = \sqrt{\frac{a_2}{a_1} R^2 + K} \tag{5}$$

where the positive square root is the only valid one (as it only makes sense to have a positive number of soldiers), and K is a constant which is determined by the initial conditions.

graph sketching

To sketch the family of phase-plane trajectories satisfying (5) we use standard graph sketching techniques. First we look for possible turning points, then we examine the asymptotic behaviour for large R and finally, look for intercepts on the axes.

Example 3: *Sketch the phase-plane trajectories.*

Solution: *For turning points we require $dB/dR = 0$. Rather than differentiate (5) directly we can use the differential equation (2). For R and B both positive, dB/dR is positive and so B is always an increasing function of R and there are no turning points.*

We can also check the behaviour of B as R becomes large. For sufficiently large R the term R^2 will always be large compared with K. So for large R the trajectory asymptotes to $B = \sqrt{a_2/a_1}\, R$, which is the equation of a straight line.

We also calculate where the curves cross the axes. For B-intercepts, we set $R = 0$ and then solve for B to obtain the intercept $B = \sqrt{K}$, for $K > 0$. This only makes sense if K is positive. Similarly, for R-intercepts, we set $B = 0$ and then solve for R to obtain the intercept $R = \sqrt{-a_1 K/a_2}$, for $K < 0$. There is a real solution only if K is negative. We thus get a different family of curves, depending on whether K is positive or negative.

Putting together the information that B is an increasing function of R, the fact that the curves all approach a straight line asymptotically, and the known values of the intercepts, allows us to infer the general form of the phase-plane curves. A sketch of these is given below in Figure 6.5.2.

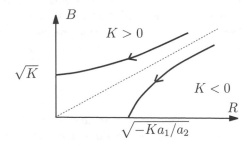

Figure 6.5.2: Typical phase-plane trajectories for the combat model with two armies exposed to each other's fire. The directions along the trajectories are indicated.

For a given initial condition (r_0, b_0) we start at that point on one of the curves. As time increases we move along the curve in a direction towards one of the axes (since the number of soldiers can only decrease in this model). *(travelling along the trajectories)*

From Figure 6.5.2 we see that if $K > 0$ then $R \to 0$ and the blue army wins. However, if $K < 0$ then $B \to 0$ and the red army wins. *(interpretation)*

Discussion

An interesting use of this battle model is for the analysis of tactics. Although the model studied here is probably too simple to represent most real battles, it is still useful in understanding why some battle tactics work. One such example is the 'divide and conquer' strategy. In this strategy you divide your enemy's force, while not dividing your own, so that you fight two battles against two smaller forces. It has been used by many famous generals including Napoleon. Why should the outcome of this be any different from fighting a single battle against the total force? The following example illustrates the reason. *(divide and conquer strategy)*

Example 4:   *For simplicity let us assume that both armies have equal attrition coefficients $a_1 = a_2$. Let us suppose the red army has 10 000 soldiers initially and the blue army has 8000. Determine who wins if (i) there is one battle between the two armies, and (ii) there are two battles: the first with half the red army against the entire blue army and the second with the other half of the red army against the blue army survivors of the first battle.*

Solution: *If there is only one battle the red army wins since $a_1 = a_2$ and $r_0 > b_0$. In other words, $a_1 K = a_1 \times (8 \times 10^3)^2 - a_2 \times (10 \times 10^3)^2 = -36 \times 10^6 a_1$ so $K < 0$ which is the condition for the red army to win.*

Instead, suppose that half the red army meets the entire blue army in a first battle and then the remaining half of the red army meets the remaining blue army soldiers in a second battle. For the first battle, we can calculate the number of blue soldiers left. For this calculation we use equation (3) with $r_0 = 5000$, $b_0 = 8000$ and obtain the surviving number of blue soldiers (as $R \to 0$)

$$B \to (b_0^2 - r_0^2)^{1/2} = (64 \times 10^6 - 25 \times 10^6)^{1/2} \approx 6.25 \times 10^3.$$

In the second battle we now have $r_0 = 5 \times 10^3$ and $b_0 = 6.25 \times 10^3$. Clearly, the blue army now wins the second battle since it has the greater numbers (and $a_1 = a_2$).

import-
ance of
tactics

This example demonstrates how an initially inferior army is able to defeat a superior army by forcing the superior army to engage in two separate battles. It illustrates the important military concept of concentrating forces and provides an indication of the role of this type of theory in the design of battle tactics and strategies.

> **Skills developed here:**
> - Apply the chain rule method to coupled systems.
> - Draw and interpret phase-plane diagrams for systems of differential equations.
> - Extend the results of this and more complicated battle models to give tactical advice.

6.6   The predator–prey model revisited

overview

We return to the analysis of the basic predator–prey model, the Lotka–Volterra system, introduced in Section 6.2. We show by applying the chain rule that, for all parameter values in this model, the phase-plane curves are always closed curves. This allows us to infer that, according to this model, both populations always oscillate over time.

Review of model

The Lotka–Volterra predator–prey equations are

> the differential equations

$$\frac{dX}{dt} = \beta_1 X - c_1 XY, \qquad \frac{dY}{dt} = -\alpha_2 Y + c_2 XY. \qquad (1)$$

where the parameters β_1, α_2, c_1 and c_2 are all positive constants. We also found in Section 6.2 that the equilibrium points were given by

$$(X, Y) = (0, 0) \quad \text{and} \quad (X, Y) = \left(\frac{\alpha_2}{c_2}, \frac{\beta_1}{c_1} \right).$$

Furthermore, we deduced the directions of the phase-plane trajectories from which we could infer that the trajectories were either closed curves or spirals, traversed in an anticlockwise direction. We show, by applying the chain rule to find a relation between X and Y, that we can prove there is never a spiral, regardless of the parameters we may choose.

> directions of trajec-tories

Applying the chain rule

We use the chain rule to eliminate time from the differential equations, thus obtaining a differential equation relating the predator population Y to the prey population X.

> eliminat-ing time

Example 1: *Use the chain rule to find an equation relating X and Y.*

Solution: *By the chain rule*

$$\frac{dY}{dX} = \frac{dY/dt}{dX/dt}.$$

Hence, using (1) we obtain the first-order differential equation

$$\frac{dY}{dX} = \frac{Y(-\alpha_2 + c_2 X)}{X(\beta_1 - c_1 Y)}. \qquad (2)$$

This differential equation is separable. Separating the variables and integrating both sides with respect to X we obtain

$$\int \left(\frac{\beta_1}{Y} - c_1 \right) dY = \int \left(-\frac{\alpha_2}{X} + c_2 \right) dX.$$

Carrying out both integrations yields

$$\beta_1 \ell n(Y) - c_1 Y = -\alpha_2 \ell n(X) + c_2 X + K \qquad (3)$$

where K is an arbitrary constant of integration.

Proving all trajectories are closed

closed
curves or
spirals

From the direction vector diagram of the earlier section (Figure 6.2.2) it is clear that the trajectories must be closed curves or spirals. We now provide an argument that the phase-plane trajectories are closed curves for all positive values of the constant parameters.

the trajec-
tories
cannot be
spirals

To get further information we return to the formula (3). Suppose we hold the variable X constant, $X = X_1$. Then we have the equation

$$\ell n(Y) = \frac{c_1}{\beta_1} Y + K_1 \tag{4}$$

where K_1 is the constant defined by $K_1 = (-\alpha_2 \ell n(X_1) + c_2 X_1 + K)/\beta_1$. If the phase-plane curve were a closed curve then it could have, at most, two intersection points with any vertical line, in particular $X = X_1$. On the other hand, if it were a spiral it would have an infinite number of intersection points. By sketching the curves $Z = \ell n(Y)$ and $Z = (c_1/\beta_1)Y + K_1$ we see there are, at most, two solutions. This is illustrated in Figure 6.6.1.

Figure 6.6.1: A general sketch of the intersection between the two curves $Z = \ell n(Y)$ and $Z = (c_1/\beta_1)Y + K_1$ showing at most two intersection points.

Returning to the phase-plane

general
phase-
plane
trajectory

We have concluded that all the phase-plane curves are closed curves. Using this, together with Figure 6.2.2 and the fact that X and Y are positive, we can sketch the general form of a typical phase-plane trajectory. Recall, from Section 6.2, that the trajectory is traversed in an anticlockwise direction. This is illustrated in Figure 6.6.2.

interpret-
ation

All the phase-plane curves are traversed in the anticlockwise direction along closed trajectories enclosing an equilibrium point, which in this case represents average population densities for the two populations. We note that as we trace out the trajectory, the prey (X) reach their maximum

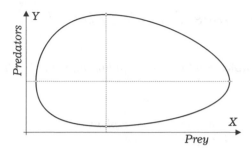

Figure 6.6.2: Phase-plane diagram showing a typical trajectory for the predator–prey problem enclosing the equilibrium point.

population before the predator population (Y) reaches its maximum value. Thus the predator oscillation cycle always lags behind the prey oscillation, no matter what parameter values or initial conditions we use.

A feature of this model is that if you suddenly perturb the populations, limitations they begin to move on a different phase-plane trajectory. This is one undesirable feature of the Lotka–Volterra equations from a biological perspective. Biologists would prefer a more realistic model where the populations tend to return to the original phase-plane curve after they have been perturbed slightly. This can be achieved by using a modified version of the Lotka–Volterra equations, which we consider in Chapter 8.

Skills developed here:

- Use the chain rule to prove that the trajectories of the the Lotka–Volterra equations are always closed.
- Consider the points of intersection between two simple graphs to infer information about solutions to a more complicated equation.

6.7   Exercises for Chapter 6

1. Simple example. *Find the only equilibrium point for the system*

$$x' = x - 5y, \qquad y' = x - y.$$

Then use Maple *to show that the phase-plane trajectories for this system are closed, and enclose this equilibrium point.*

2. Finding equilibrium points. *For the following, find all the equilibrium solutions.*

(a) $\dfrac{dX}{dt} = 3X - 2XY, \qquad \dfrac{dY}{dt} = XY - Y,$

(b) $\dfrac{dX}{dt} = 2X - XY, \qquad \dfrac{dY}{dt} = Y - XY,$

(c) $\dfrac{dX}{dt} = Y - 2XY, \qquad \dfrac{dY}{dt} = XY - Y^2.$

3. Using the chain rule. *For the differential equations*

$$\frac{dX}{dt} = -XY, \qquad \frac{dY}{dt} = -2Y,$$

use the chain rule to find a relation between Y and X.

4. Predator–prey with density dependent growth of prey. *Consider the system*

$$\frac{dX}{dt} = \beta_1 X \left(1 - \frac{X}{K} \right) - c_1 XY, \qquad \frac{dY}{dt} = c_2 XY - \alpha_2 Y,$$

for the dynamics of a predator–prey model, with density dependent growth of the prey, and all parameters positive constants.

(a) *Find all the equilibrium points. How do they differ from those of the Lotka–Volterra system*

(b) *Use* Maple *to explore the phase-plane diagram for different parameter values.*

5. Predator–prey with DDT. *The predator–prey equations with additional deaths by DDT are*

$$\frac{dX}{dt} = \beta_1 X - c_1 XY - p_1 X, \qquad \frac{dY}{dt} = -\alpha_2 Y + c_2 XY - p_2 Y,$$

where all parameters are positive constants.

(a) *Find all the equilibrium points.*

(b) *What effect does the DDT have on the non-zero equilibrium populations compared with the case when there is no DDT. Compare this with Figure 5.3.3 in Section 5.3, which was for a specific set of parameter values. What general conclusions can you draw?*

(c) *Show that the predator fraction of the total average prey population is given by*

$$f = \frac{1}{1 + \left(\frac{c_1}{c_2} \frac{(\alpha_2 + p_2)}{(\beta_1 - p_1)} \right)}.$$

What happens to this proportion f as the DDT kill rates, p_1 and p_2, increase?

6. Predator–prey with density dependence and DDT. *For the predator–prey model with density dependence for the prey and DDT acting on both species,*

$$\frac{dX}{dt} = \beta_1 X \left(1 - \frac{X}{K} \right) - c_1 XY - p_1 X, \qquad \frac{dY}{dt} = c_2 XY - \alpha_2 Y - p_2 Y,$$

show that (X_e, Y_e), where

$$X_e = \frac{\alpha_2 + p_2}{c_1}, \qquad Y_e = \frac{\beta_1 \left(1 - \frac{X_e}{K} \right) - p_1}{c_1},$$

is one equilibrium point, and determine if there are any other equilibrium points.

7. One prey and two predators. *A model of a three species interaction is*

$$\frac{dX}{dt} = a_1 X - b_1 XY - c_1 XZ, \qquad \frac{dY}{dt} = a_2 XY - b_2 Y, \qquad \frac{dZ}{dt} = a_3 XZ - b_3 Z,$$

where a_i, b_i, c_i, for $i = 1, 2, 3$, are all positive constants. Here $X(t)$ is the prey density and $Y(t)$ and $Z(t)$ are the two predator species densities.

(a) *Find all possible equilibrium populations. Is it possible for all three populations to coexist in equilibrium?*

(b) *What does this suggest about introducing an additional predator into an ecosystem?*

8. Competing species without density dependence. *For the competing species model developed in Section 5.4, exhibiting exponential growth in the absence of a competitor, find the equilibrium points and trajectory directions in different parts of the (X, Y)-plane. Hence sketch the phase-plane diagram. Show that this diagram is consistent with the numerical results generated in Section 5.4, and illustrated in Figure 5.4.2.*

9. Competing species with density dependence. *Consider the competition population model with density dependent growth*

$$\frac{dX}{dt} = X(\beta_1 - c_1 Y - d_1 X), \qquad \frac{dY}{dt} = Y(\beta_2 - c_2 X - d_2 Y).$$

(a) *Find all four equilibrium points for the system.*

(b) *Establish all four possible direction vector diagrams for the phase-plane. (This will complete Example 2 in Section 6.3, which considers only Case 3 in detail.)*

(c) *From your results above, choose appropriate parameter values to represent each of the four cases, and then using* Maple *with these parameter values, plot a phase-plane diagram for each case.*

(d) *When the parameter values are $\beta_1 = 3$, $\beta_2 = 3$, $c_1 = 2$, $c_2 = 1$, $d_1 = 2$ and $d_2 = 2.5$, and the initial values are $X = 2$ and $Y = 2$, what population densities do they approach over time? Are your results consistent with those estimated previously in Question (5) of the Exercises from Chapter 5?*

10. Rabbits and foxes. *A population of sterile rabbits $X(t)$ is preyed upon by a population of foxes $Y(t)$. A model for this population interaction is the pair of differential equations*

$$\frac{dX}{dt} = -aXY, \qquad \frac{dY}{dt} = bXY - cY$$

where a, b and c are positive constants.

(a) *Use the chain rule to obtain a relationship between the density of foxes and the density of rabbits.*

(b) *Sketch typical phase-plane trajectories, indicating the direction of movement along the trajectories.*

(c) *According to the model, is it possible for the foxes to completely wipe out the rabbit population? Give reasons.*

11. Microorganisms and toxins. *The pair of differential equations*

$$\frac{dP}{dt} = rP - \gamma PT, \qquad \frac{dT}{dt} = qP,$$

where r, γ and q are positive constants, is a model for a population of microorganisms P, which produces toxins T which kill the microorganisms.

(a) *Given that initially there are no toxins and p_0 microorganisms, obtain an expression relating the population density and the amount of toxins. (Hint: Use the chain rule.)*

(b) *Hence, give a sketch of a typical phase-plane trajectory. Using this, describe what happens to the microorganisms over time.*

12. Contagious for life. *Consider a disease where all those infected remain contagious for life. A model describing this is given by the differential equations*

$$\frac{dS}{dt} = -bSC, \qquad \frac{dC}{dt} = bSC.$$

(a) *Use the chain rule to find a relation between S and C.*

(b) *Obtain and sketch the phase-plane curves. Determine the direction of travel along the trajectories.*

(c) *Using this model, is it possible for all the susceptibles to be infected?*

13. Disease with reinfection. *A model for the spread of a disease, where all infectives recover from the disease and become susceptibles again, is given by the pair of differential equations*

$$\frac{dS}{dt} = -bSC + rC, \qquad \frac{dC}{dt} = bSC - rC,$$

where b and r are positive constants, $S(t)$ denotes the number of susceptibles and $C(t)$ denotes the number of infectives at time t.

(a) *Use the chain rule to find a relationship between the number of susceptibles and the number of infectives given the initial number of susceptibles is s_0 and there was initially only one infective.*

(b) Draw a sketch of typical phase-plane trajectories. Deduce the direction of travel along the trajectories providing reasons.

(c) Using the phase-plane diagram describe how the number of infectives changes with time.

14. Fatal disease. The following model is for a fatal disease, where all infectives die from the disease, with death rate α:

$$\frac{dS}{dt} = -\frac{\lambda}{N}SC, \qquad \frac{dC}{dt} = \frac{\lambda}{N}SC - \alpha C,$$

where $N = S + C$ is the total population size and λ is the (constant) per-capita contact rate.

(a) Write the system in terms of the variables N and C (i.e. obtain a differential equation for N) and hence show that

$$\frac{dC}{dN} = \frac{\lambda}{\alpha}\frac{C}{N} + \left(1 - \frac{\lambda}{\alpha}\right).$$

(b) Solve the differential equation in (a) to obtain

$$C = N + KN^{\lambda/\alpha}$$

where K is the arbitrary constant of integration.

(c) Discuss any qualitative differences between this model and the one studied in Section 6.4. (Hint: Substitute $S = N - C$ for a meaningful comparison, and graph C against S.)

15. Battle model with desertion. The following battle model represents two armies where both are exposed to aimed fire, and for one of the armies there is significant loss due to desertion (at a constant rate). The numbers of soldiers, R and B, satisfy the differential equations

$$\frac{dR}{dt} = -a_1 B - c, \qquad \frac{dB}{dt} = -a_2 R,$$

where a_1, a_2 and c are positive constants.

(a) If the initial number of red soldiers is r_0 and the initial number of blue soldiers is b_0, use the chain rule to find a relationship between B and R.

(b) For $a_1 = a_2 = c = 0.01$ give a sketch of typical phase-plane trajectories and deduce the direction of travel along the trajectories.

16. Jungle warfare. *A simple mathematical model describing a jungle warfare, with one army exposed to random fire and the other to aimed fire, is given by the coupled differential equations*

$$\frac{dR}{dt} = -c_1 RB, \qquad \frac{dB}{dt} = -a_2 R,$$

where c_1 and a_2 are positive constants.

(a) *Use the chain rule to find a relation between R and B, given initial numbers of soldiers r_0 and b_0. Hence sketch some typical phase-plane trajectories. Give directions of travel along the trajectories, providing reasons for your choice.*

(b) *Given that, initially, both the red and blue armies have 1000 soldiers, and the constants c_1 and a_2 are 10^{-4} and 10^{-1}, respectively, determine how many soldiers are left if the battle is fought so that all the soldiers of one army are killed.*

(c) *In this model, one of the armies is hidden whereas the other is visible to their enemy. Which is the hidden army? Give reasons for your answer.*

17. Battle with long range weapons. *In a long range battle, neither army can see the other, but fires into a given area. A simple mathematical model describing this battle is given by the coupled differential equations*

$$\frac{dR}{dt} = -c_1 RB, \qquad \frac{dB}{dt} = -c_2 RB$$

where c_1 and c_2 are positive constants.

(a) *Use the chain rule to find a relationship between R and B, given the initial numbers of soldiers for the two armies are r_0 and b_0, respectively.*

(b) *Draw a sketch of typical phase-plane trajectories.*

(c) *Explain how to estimate the parameter c_1 given that the blue army fires into a region of area A.*

Chapter 7

Linearisation analysis

In the following chapter we develop some powerful theory, which often allows us to predict the dynamics of a system in general terms. It provides the means by which we can establish the phase-plane behaviour of a system and predict the outcome for any possible parameter combination. The theory is developed for both linear and nonlinear systems, and is applied to some of the nonlinear systems studied in Chapters 5 and 6.

7.1   Introduction

motivation   From the small sample of examples we have examined in the previous chapters, we have seen that systems of equations can result in many different types of behaviour. Depending on the initial conditions or the chosen parameter values, the outcome may be stable or unstable, cyclic or divergent.

overview   In this chapter we develop some powerful theory which allows us to predict the dynamical behaviour of a system. In the first case, we consider only linear systems of equations. However, since many of the interacting population models we have met are nonlinear (and in fact most natural systems are nonlinear), we then show how this theory for the linear case may be extended to nonlinear systems as well. In the final section we apply the process to the nonlinear models of population interactions, which we developed in Chapters 5 and 6, and shall extend in Chapter 8.

7.2   Linear theory

observed dynamics   So far, in our phase-plane analysis of systems of equations, we have encountered a variety of behaviours of trajectories near equilibrium points. For example we saw trajectories approaching some of these points, trajectories being repelled by others, as well as spiralling trajectories and closed loops.

overview   In the following theory we develop techniques which allow us to predict, for each equilibrium point, the behaviour of the trajectories close to that point. From this we can establish a complete picture of the system phase-plane. Initially, we consider the linear case (a pair of coupled linear equations in two unknowns) and then show how this can be extended to the nonlinear case. (Whilst we restrict our analysis to two equations in two unknowns, the theory is applicable to larger systems with many unknowns.)

The general linear system

general linear system   We start by considering the following general form of a pair of coupled linear equations:

$$X' = a_1 X + b_1 Y$$
$$Y' = a_2 X + b_2 Y$$

where differentiation is with respect to time t (i.e. $X' = dX/dt$, $Y' = dY/dt$) and a_1, a_2, b_1 and b_2 are constant.

We denote an equilibrium point (critical point or steady-state) for the general system by (x_e, y_e). Thus $a_1 x_e + b_1 y_e = 0$ and $a_2 x_e + b_2 y_e = 0$.

equilibrium point

Linear algebra notation

The system above can be written in terms of matrices and vectors in the following way. Let

general system in vector notation

$$\mathbf{x} = \begin{bmatrix} X \\ Y \end{bmatrix} \quad \text{and} \quad \mathbf{x}' = \begin{bmatrix} X' \\ Y' \end{bmatrix}$$

where $\mathbf{x}$ is a vector. Let

$$A = \begin{bmatrix} a_1 & b_1 \\ a_2 & b_2 \end{bmatrix}$$

where A is a matrix. Then the above system of two equations can be written as

$$\mathbf{x}' = A\mathbf{x}.$$

This means that

$$\begin{bmatrix} X' \\ Y' \end{bmatrix} = \begin{bmatrix} a_1 & b_1 \\ a_2 & b_2 \end{bmatrix} \begin{bmatrix} X \\ Y \end{bmatrix} = \begin{bmatrix} a_1 X + b_1 Y \\ a_2 X + b_2 Y \end{bmatrix} \tag{1}$$

using normal matrix multiplication.

What needs to be understood in general, is the effect of multiplying a vector by a matrix A. We take an example to establish this.

effect of multiplying by a matrix

Example 1: *Carry out the multiplication $A\mathbf{x}_1$ and $A\mathbf{x}_2$ where*

$$A = \begin{bmatrix} 3 & -2 \\ 1 & 0 \end{bmatrix}, \quad \mathbf{x}_1 = \begin{bmatrix} -1 \\ 1 \end{bmatrix}, \quad \mathbf{x}_2 = \begin{bmatrix} 2 \\ 1 \end{bmatrix}.$$

Solution:

$$A\mathbf{x}_1 = \begin{bmatrix} -3 - 2 \\ -1 \end{bmatrix} = \begin{bmatrix} -5 \\ -1 \end{bmatrix}$$

and

$$A\mathbf{x}_2 = \begin{bmatrix} 6 - 2 \\ 2 \end{bmatrix} = \begin{bmatrix} 4 \\ 2 \end{bmatrix} = 2 \begin{bmatrix} 2 \\ 1 \end{bmatrix}.$$

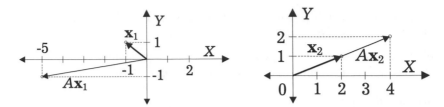

Figure 7.2.1: The effect of multiplying vectors $\mathbf{x}_1$ and $\mathbf{x}_2$ by a matrix A as defined in Example 1.

interpret-
ation

These results are illustrated in Figure 7.2.1. So multiplication by a matrix A maps a vector onto another vector. In the case of $\mathbf{x}_2$ we have that

$$Ax_2 = 2\mathbf{x}_2,$$

so that the effect of multiplying by A is the same as multiplying by a scalar or number (which is 2 in this case). We use this notion of the 'equivalence' of multiplication by a matrix and a scalar (number) in the process of finding eigenvalues and eigenvectors: the latter will be the vectors for which there is a nontrivial solution to the equation, and the former will be the associated scalars. These values turn out to be essential in predicting the behaviour of trajectories in the phase-plane associated with the system. For further details of matrix algebra see Appendix B.1.

Outline of method to solve equations

method
outline

In order to solve a general system of equations, as in (1), we can transform the equations onto a different system of axes. In choosing the new axes carefully, the equations transform to simple differential equations of exponential growth or decay, and thus are simple to solve. Finally, these solutions can be transformed back to the original system of axes to provide a solution in the required form.

The eigenvalues and eigenvectors

the
process

In order to transform a general pair of differential equations, which may be hard to solve, into a system which is easy to solve, we find and use the eigenvalues and eigenvectors. (These are described fully in Appendix B.1.) Essentially, *eigenvectors* $\mathbf{x}$ are the non-zero solutions of the matrix equation

$$A\mathbf{x} = \lambda\mathbf{x}$$

where the *eigenvalues* λ are the values for which these non-zero solutions exist.

Rewriting the equation as $A\mathbf{x} - \lambda\mathbf{x} = 0$ and then expanding using matrix multiplication we get

character-
istic
equation

$$A\mathbf{x} - \lambda\mathbf{x} = (A - \lambda I)\mathbf{x} \begin{bmatrix} a_1 - \lambda & b_1 \\ a_2 & b_2 - \lambda \end{bmatrix} \mathbf{x},$$

where I is the identity matrix

$$I = \begin{bmatrix} 1 & 0 \\ 0 & 1 \end{bmatrix}.$$

From the theory of linear algebra (see Appendix B.1) it follows that for non-zero solutions of this equation to exist, the determinant of the expression must be zero, and thus we get the *characteristic equation* of matrix A

$$|A - \lambda I| = \lambda^2 - \lambda(a_1 + b_2) + (a_1 b_2 - a_2 b_1).$$

This equation is central to the theory which is developed here. Note that the coefficient of λ is the sum of the diagonal elements of the matrix A, namely the *trace* of A, and also that the last term is the *determinant* of matrix A. We use these values extensively when applying the theory which we now develop.

We can establish the eigenvalues as the solutions to this characteristic equation. (We could also then solve for the associated pair of eigenvectors from the vector equation, but do not need these values in the applications.) From here the trajectory behaviour can be determined since, as will become apparent, it is dependent entirely on the eigenvalues.

getting
the eigen-
values and
eigenvec-
tors

Establishing the trajectory behaviour

We return to the general pair of linear first-order equations

the linear
system

$$X' = a_1 X + b_1 Y,$$
$$Y' = a_2 X + b_2 Y,$$

which has an equilibrium point at the origin, $(x_e, y_e) = (0,0)$. In vector notation then

$$\mathbf{x}' = A\mathbf{x}.$$

Suppose we have found the eigenvalues, λ_1 and λ_2, as well as the associated eigenvectors for A, namely

the matrix
of eigen-
vectors

$$\mathbf{u} = \begin{bmatrix} u_1 \\ u_2 \end{bmatrix} \quad \text{and} \quad \mathbf{v} = \begin{bmatrix} v_1 \\ v_2 \end{bmatrix}.$$

$$U = \begin{bmatrix} \mathbf{u} & \mathbf{v} \end{bmatrix} = \begin{bmatrix} u_1 & v_1 \\ u_2 & v_2 \end{bmatrix}.$$

the diagonal matrix of eigenvalues

From the definition of eigenvectors and eigenvalues we have

$$A\mathbf{u} = \lambda_1 \mathbf{u} \quad \text{and} \quad A\mathbf{v} = \lambda_2 \mathbf{v},$$

which implies that

$$A \begin{bmatrix} \mathbf{u} & \mathbf{v} \end{bmatrix} = \begin{bmatrix} \lambda_1 \mathbf{u} & \lambda_2 \mathbf{v} \end{bmatrix} = \begin{bmatrix} \mathbf{u} & \mathbf{v} \end{bmatrix} \begin{bmatrix} \lambda_1 & 0 \\ 0 & \lambda_2 \end{bmatrix} \quad \text{or} \quad AU = UD$$

with

$$D = \begin{bmatrix} \lambda_1 & 0 \\ 0 & \lambda_2 \end{bmatrix}.$$

Assuming that U is invertible, we can write

$$U^{-1}AU = D. \tag{2}$$

We use this equation below.

x as a linear combination of the eigenvectors

First we express $\mathbf{x}$ as a linear combination of the eigenvectors and, assuming this is possible, we have

$$\mathbf{x} = z_1 \mathbf{u} + z_2 \mathbf{v}.$$

Letting

$$\mathbf{z} = \begin{bmatrix} z_1 \\ z_2 \end{bmatrix} \quad \text{then} \quad \mathbf{x} = U\mathbf{z}.$$

Since X and Y are functions of time, and the eigenvectors are not (since A is not a function of time), then z_1 and z_2 must also be functions of time. We now establish two expressions for $\mathbf{x}'$:

$$\mathbf{x} = U\mathbf{z} \quad \text{so} \quad \mathbf{x}' = U\mathbf{z}',$$

and also

$$\mathbf{x}' = A\mathbf{x} \quad \text{so} \quad \mathbf{x}' = AU\mathbf{z}.$$

Equating these two expressions for $\mathbf{x}'$ and then using (2) gives

$$U\mathbf{z}' = AU\mathbf{z},$$

and then

$$\mathbf{z}' = U^{-1}AU\mathbf{z}$$
$$= D\mathbf{z}.$$

We are now in a position to solve the differential equations easily. Expanding $\mathbf{z}' = D\mathbf{z}$ *behaviour of trajectories*

$$z_1' = \lambda_1 z_1,$$
$$z_2' = \lambda_2 z_2,$$

we obtain two equations which are easy to solve. They are the equations for exponential growth and decay with which, by now, we are familiar. We have as solutions $z_1 = k_1 e^{\lambda_1 t}$ and $z_2 = k_2 e^{\lambda_2 t}$ where k_1 and k_2 are arbitrary constants.

Using these we can find solutions for X and Y by retracing our steps through this process and carrying the solutions with us. We have *solutions for X and Y*

$$\mathbf{x} = k_1 e^{\lambda_1 t}\mathbf{u} + k_2 e^{\lambda_2 t}\mathbf{v}$$
$$= e^{\lambda_1 t}\hat{\mathbf{u}} + e^{\lambda_2 t}\hat{\mathbf{v}},$$

where $\hat{\mathbf{u}} = k_1\mathbf{u}$ and $\hat{\mathbf{v}} = k_2\mathbf{v}$ are two eigenvectors (as any scalar multiple of an eigenvector is again an eigenvector) and so

$$X = e^{\lambda_1 t}\hat{u}_1 + e^{\lambda_2 t}\hat{v}_1,$$
$$Y = e^{\lambda_1 t}\hat{u}_2 + e^{\lambda_2 t}\hat{v}_2.$$

Geometric interpretation

What have we done? *process summary*

- We start with a set of axes, X and Y.
- We find eigenvectors $\mathbf{u}$ and $\mathbf{v}$ which give us the directions of a new system of axes z_1 and z_2 along which the effect of multiplying by A is the same as multiplying by a scalar λ. (It is clear from the diagram that any scalar multiple of the eigenvectors will suffice as an eigenvector.)
- We consider the plane described by the axes z_1 and z_2 and examine the behaviour of the solutions (which are now easy to find) $z_1 = k_1 e^{\lambda_1 t}$ and $z_2 = k_2 e^{\lambda_2 t}$ in this phase-plane. How the trajectories behave depends entirely on λ_1 and λ_2, the eigenvalues. Suppose they are as illustrated in the left-hand diagram of Figure 7.2.2.

- Next we return to the (X, Y) phase-plane with the transformation of this solution, which is distorted (stretched or contracted) in some way since we have changed axes again: note that it retains the main features, or dynamics, of the simpler system.

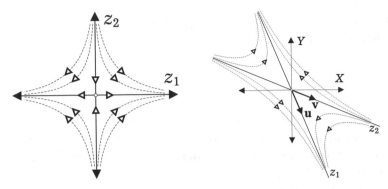

Figure 7.2.2: Sample saddle solution in the eigenvector phase-plane (left figure) and translated to the original phase-plane (right figure).

Equilibrium point classifications

eigenval-
ues carry
the infor-
mation
For the systems described above, we had the origin $(0,0)$ as the equilibrium (or critical) point. What we have found, using the techniques of eigenvalues and eigenvectors, is the behaviour of the trajectories in the phase-plane close to this point. The behaviour depends on the eigenvalues (λ_1 and λ_2) since the trajectories can be described by $z_1 = k_1 e^{\lambda_1 t}$ and $z_2 = k_2 e^{\lambda_2 t}$. Clearly different values of λ_1 and λ_2 may result in very different behaviours, and thus each case is dealt with separately below.

the effect
of the
eigenval-
ues on the
trajector-
ies
The following summarises the relationship between the eigenvalues and the forms of the trajectories:

- Case $\lambda_1 < 0$ and $\lambda_2 < 0$ (eigenvalues real and negative): We have

$$\lim_{t \to \infty} k_1 e^{\lambda_1 t} = 0 \quad \text{and} \quad \lim_{t \to \infty} k_2 e^{\lambda_2 t} = 0$$

  and thus all trajectories approach the equilibrium point at the origin. Such a point is called a *stable node* and is illustrated in Figure 7.2.3.
- Case $\lambda_1 > 0$ and $\lambda_2 > 0$ (eigenvalues real and positive):
  We have both z_1 and z_2 approaching ∞ (diverging) as t increases and thus all trajectories diverge from the equilibrium point. Such a point is called an *unstable node* (see Figure 7.2.3).

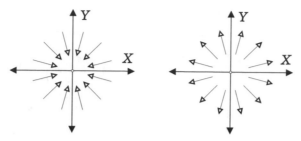

Figure 7.2.3: Trajectory behaviour close to a stable node (left) and an unstable node (right).

- Case $\lambda_1 > 0$ and $\lambda_2 < 0$ (eigenvalues real and of different sign): We have that $z_1 = k_1 e^{\lambda_1 t}$ and $z_2 = k_1 e^{\lambda_2 t}$, so $z_2 \to 0$ and $z_1 \to \infty$ as time increases. The trajectories approach zero along one axis and approach ∞ along the other. Such a point is called a *saddle* or an unstable saddle point and is illustrated in Figure 7.2.4.

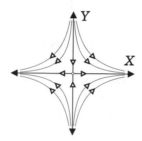

Figure 7.2.4: Trajectory behaviour close to an unstable saddle point.

- Case $\lambda_1 = \alpha + i\beta$ and $\lambda_2 = \alpha - i\beta$ (complex conjugate eigenvalues with $\alpha \neq 0$ and $\beta \neq 0$): In this case, the solutions can be written in the form $z_1 = e^{\alpha t} \cos \beta t$, $z_2 = e^{\alpha t} \sin \beta t$ and the trajectories spiral around the equilibrium point. If $\alpha < 0$ then they spiral inwards towards the equilibrium point. Such a point is called a *stable focus*. If $\alpha > 0$ then they spiral outwards and away from the equilibrium point. Such a point is called an *unstable focus*. A stable and unstable focus are illustrated in Figure 7.2.5.
- Case λ_1 and λ_2 purely imaginary. In this case, the solutions can be written in the form $z_1 = \cos \beta t$ and $z_2 = \sin \beta t$ and the trajectories form closed loops enclosing the equilibrium point. Such a point is called a *centre* and the solutions are called periodic. A centre is illustrated in Figure 7.2.6.

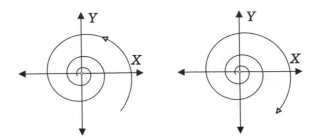

Figure 7.2.5: Trajectory behaviour close to a stable focus (left) and an unstable focus (right).

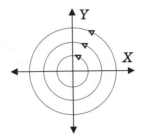

Figure 7.2.6: Trajectory behaviour close to a centre.

seen it all
before
We have already come across examples of most of these types of equilibrium points in Chapter 6, where we considered the behaviour of the trajectories in the phase-plane for some basic population models. The power of the above results is that, having located an equilibrium point, we now have the means to predict its type, once we have found the associated eigenvalues.

Summary

- We start with the characteristic equation

$$\lambda^2 - \lambda(a_1 + b_2) + (a_1b_2 - a_2b_1) = 0$$

and solve for λ. This provides us with the eigenvalues.
- Solving the quadratic characteristic equation gives

$$\lambda_1 = \frac{1}{2}p + \frac{1}{2}\sqrt{\Delta}, \qquad \lambda_2 = \frac{1}{2}p - \frac{1}{2}\sqrt{\Delta}$$

where $p = a_1 + b_2$ is the trace of matrix A, $q = a_1b_2 - a_2b_1$ is the determinant of A and $\Delta = p^2 - 4q$ the discriminant of the characteristic equation. The different possible classifications of the equilibrium points are given in Table 7.2.1.

- This can all be displayed in a diagram, Figure 7.2.7, which illustrates the general classifications.

Table 7.2.1: Table showing different classifications of equilibrium points.

Δ	p	q		Equilibrium point
$\Delta > 0$	$p < 0$		$\Rightarrow$	stable node
$\Delta > 0$	$p > 0$		$\Rightarrow$	unstable node
		$q < 0$	$\Rightarrow$	saddle point
$\Delta < 0$	$p < 0$		$\Rightarrow$	stable spiral
$\Delta < 0$	$p = 0$		$\Rightarrow$	centre
$\Delta < 0$	$p > 0$		$\Rightarrow$	unstable spiral

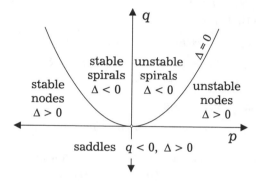

Figure 7.2.7: General classification diagram for equilibrium points using p, q and Δ from the characteristic equation.

Discussion

Note, we have not included the cases where $\Delta = 0$ or $q = 0$. When details $\Delta = 0$ the roots are equal and the equilibrium point is a stable/unstable inflected node. Such equilibrium points are called degenerate. When $q = 0$ the system 'equilibrium point' consists of a line and the trajectories are parallel. These special cases are illustrated in Figure 7.2.8.

Figure 7.2.8: Trajectory behaviour when $\Delta = 0$ (left) and $q = 0$ (right), illustrated for the stable case when $p < 0$.

attractors
and
repellers

Stable nodes and spirals are known as attractors, while unstable nodes and spirals are known as repellers. It is important to note that changes in the parameters a_1, a_2, b_1, b_2 can result in very different dynamics. Thus, it is possible to choose the dynamics through controlled variation of the parameters.

dealing
with
non-zero
equilib-
rium
points

We have considered only the case of an equilibrium point at the origin, i.e. $(x_e, y_e) = (0, 0)$, but in general this is not so. Suppose we have a system as before, but with $(x_e, y_e) \neq (0, 0)$,

$$X' = a_1 X + b_1 Y + c_1,$$
$$Y' = a_2 X + b_2 Y + c_2.$$

Let $\xi = X - x_e$ and $\eta = Y - y_e$ (so that $X = x_e$ implies that $\xi = 0$ and $Y = y_e$ implies that $\eta = 0$). Then

$$\xi' = X' = a_1(\xi + x_e) + b_1(\eta + y_e) + c_1$$
$$= (a_1 x_e + b_1 y_e + c_1) + a_1\xi + b_1\eta$$
$$= a_1\xi + b_1\eta$$

since $a_1 x_e + b_1 y_e + c_1 = 0$. Similarly

$$\eta' = Y' = a_2(\xi + x_e) + b_2(\eta + y_e) + c_2 = a_2\xi + b_2\eta.$$

change of
variable

This process is called a change of variable and allows us to transform the original system in X and Y with an equilibrium at (x_e, y_e), to a system with variables ξ and η which has its equilibrium point at the origin. We can now apply the above theory to this system, which is equivalent to the original system of equations.

x is a
linear
combin-
ation of z_1
and z_2

Another assumption we made was that we could express $\mathbf{x}$ as a linear combination of the eigenvectors and the new axes z_1 and z_2. This is a result of the linear algebra presented in Appendix B.1.

U is
invertible

Furthermore, we assumed that $U = \begin{bmatrix} \mathbf{u} & \mathbf{v} \end{bmatrix}$ was invertible, that is, there exists a matrix U^{-1} such that $U^{-1}U = I = UU^{-1}$. This follows as a consequence of Theorem 5 in Appendix B.1.

An alternative approach

a second-
order
represen-
tation

The above approach requires some understanding of linear algebra theory, and this can be avoided if we make the assumption that the solution is an exponential of the form $ke^{\lambda t}$, with λ possibly imaginary. In this case we need to write the coupled linear system

$$X' = a_1 X + b_1 Y,$$
$$Y' = a_2 X + b_2 Y,$$

as a single second-order equation

$$X'' - (a_1 + b_2)X' + (a_1 b_2 - a_2 b_1)X = 0.$$

(This can be done by differentiating the first equation and then eliminating Y using substitution from the second equation. For further details see Appendix A.5.)

Assuming that there is a solution of the form $X = ke^{\lambda t}$, we can calculate X'' and X' and substitute them into the second-order equation to give

the characteristic equation again

$$k\lambda^2 e^{\lambda t} - (a_1 + b_2)k\lambda e^{\lambda t} + (a_1 b_2 - a_2 b_1)ke^{\lambda t} = 0,$$

and then

$$\lambda^2 - (a_1 + b_2)\lambda + (a_1 b_2 - a_2 b_1) = 0.$$

Note that this is the characteristic equation once again. The solutions of this equation are the eigenvalues (λ_1 and λ_2) and from here the general solution is a linear combination of $e^{\lambda_1 t}$ and $e^{\lambda_2 t}$, that is

$$X(t) = C_1 e^{\lambda_1 t} + C_2 e^{\lambda_2 t}.$$

We are now in a position to classify the equilibrium points, as before.

Skills developed here:

- Write a system of equations in vector notation, or convert a vector equation into a system of equations.
- Find the trace and determinant of a matrix.
- Calculate the eigenvalues of a matrix.
- From the eigenvalues of a matrix, and for a given equilibrium point, classify the equilibrium point and sketch the trajectory behaviour in the phase-plane close to this point.

7.3 Applications of linear theory

overview

What we are most interested in is how this theory can be applied to the models we have been studying. However, to start with we apply the linear theory developed above to the example of a simple harmonic oscillator. This consists of a mass attached to a spring, which oscillates back and forth. (For further details see, for example, Fulford et al. (1997).) One example is a regularly vibrating (oscillating) object such as an idealised skateboard in a ramp, equipped with perfect bearings to ensure that no friction damps the motion. Such a skateboard would continue oscillating indefinitely between the edges of the ramp.

Simple harmonic motion

an
example
problem

The following example examines an equation for simple harmonic motion.

Example 1: *Find all equilibrium points associated with the system*

$$X' = Y \qquad Y' = -\omega^2 X \tag{1}$$

and determine their classification(s).

Solution: *In matrix form, the system is*

$$\mathbf{x}' = A\mathbf{x} \qquad \text{where} \qquad A = \begin{bmatrix} 0 & 1 \\ -\omega^2 & 0 \end{bmatrix}.$$

The only equilibrium point is $(x_e, y_e) = (0, 0)$.

To classify this point we need to examine the characteristic equation

$$\lambda^2 - \lambda \operatorname{trace}(A) + \det(A) = 0 \qquad \text{whence} \quad \lambda^2 - \lambda \times 0 + \omega^2 = 0.$$

Solving this equation yields imaginary eigenvalues

$$\lambda = \pm i\omega.$$

Now, with the notation as in Section 7.2,

$$p = \operatorname{trace}(A) = 0,$$
$$\Delta = p^2 - 4q = 0 - 4\omega^2 < 0.$$

This implies that the equilibrium point at $(0, 0)$ is a centre, and all solutions are periodic, as could be expected.

By eliminating the variable Y, the system (1) can be written as a single **second-** second-order differential equation $X'' = -\omega^2 X$. This is the well-known **order** simple harmonic oscillator, and periodic solutions are exactly what we **equation** would expect.

We will now include damping, that is, some friction. In this case our **including** idealised skateboard will eventually come to rest if the skateboarder does **damping** no work to keep it going. We consider what the changes will be. The differential equation is

$$X'' = -cX' - \omega^2 X$$

where c is a positive constant. Setting $Y = X'$, this can be written as the system

$$X' = Y, \qquad Y' = -cY - \omega^2 X. \tag{2}$$

The term $-cY$ describes the damping. In the following example we classify the equilibrium point.

Example 2: *Find and classify the equilibrium point(s) of the system (2).*

Solution: *The system can be written in matrix form*

$$\mathbf{x}' = A\mathbf{x} \quad \text{where} \quad A = \begin{bmatrix} 0 & 1 \\ -\omega^2 & -c \end{bmatrix}.$$

We still have $(0,0)$ as the only equilibrium point.

For its classification we consider the characteristic equation

$$\lambda^2 - \lambda\,\mathrm{trace}(A) + \det(A) = 0 \quad \text{or} \quad \lambda^2 - \lambda(-c) + \omega^2 = 0.$$

Hence

$$\lambda = \frac{-c}{2} \pm \frac{1}{2}\sqrt{c^2 - 4\omega^2}.$$

Again, with the notation of Section 7.2, we have $p = \mathrm{tr}\,A = -c < 0$, since $c > 0$, and $\Delta = c^2 - 4\omega^2$. There are three cases to consider:

- *If $\Delta > 0$, i.e. $c > 2\omega$ then $(0,0)$ is a stable node. This case is known as strong damping.*

- *If $\Delta < 0$, i.e. $c < 2\omega$ then this implies that $(0,0)$ is a stable focus. This case is known as weak damping.*

- *If $\Delta = 0$, i.e. $c = 2\omega$ then $(0,0)$ is astable inflected node. This case is known as critical damping.*

Battle model

review of
the model

In Section 5.7, we introduced a simple model of a battle between two armies using aimed fire against each other. The resulting coupled system of differential equations for the numbers of soldiers of the two armies were the linear Lanchester equations

$$\frac{dR}{dt} = -a_1 B, \qquad \frac{dB}{dt} = -a_2 R, \tag{3}$$

where the attrition coefficients a_1 and a_2 are positive constants.

Example 3: *Classify the equilibrium points for the system* (3).

Solution: *The system* (3) *can be written in matrix form*

$$\mathbf{x}' = A\mathbf{x}, \qquad A = \begin{bmatrix} 0 & -a_1 \\ -a_2 & 0 \end{bmatrix},$$

and the only equilibrium point is $(0,0)$.

The characteristic equation is

$$\lambda^2 - \text{trace}(A) + \det(A) = 0 \quad or \quad \lambda^2 + (-a_1 a_2) = 0.$$

With the notation of Section 7.2,

$$q = \det(A) < 0$$

which implies the equilibrium point is a saddle point. This is consistent with what we found in Section 6.5.

7.4   Nonlinear theory

overview

In Chapters 5 and 6, we examined coupled systems of nonlinear equations to describe interacting populations. We found equilibrium points and, at times, it was possible to establish the long-term behaviour of the system using graphical techniques such as the phase-plane and time-dependent diagrams. However, it would be extremely useful if we could extend the system of classification of the equilibrium points for linear systems, to nonlinear systems as well.

applying
linear
theory to
nonlinear
systems

We can do this by approximating the nonlinear system with a linear system close to the equilibrium point: this process is called *linearisation*. We then examine the dynamics of the linearised system close to an equilibrium point, and use the classifications from linear theory to predict the dynamics in the nonlinear system. It can be shown that in all classification cases (except that of a centre) the predictions of linear theory extend to the nonlinear system. (For proof see, for example, Hurewicz (1990).)

Linearisation

Our first task is to establish a linear approximation to the nonlinear equations. Consider a general system of two nonlinear differential equations

(margin note: general nonlinear system)

$$\frac{dX}{dt} = F(X, Y),$$

$$\frac{dY}{dt} = G(X, Y).$$

Let (x_e, y_e) be any equilibrium point for this system, not necessarily at $(0, 0)$, and then

$$F(x_e, y_e) = 0 \quad \text{and} \quad G(x_e, y_e) = 0.$$

Consider solutions close to the steady state (equilibrium) solutions

$$X(t) = x_e + \xi(t), \qquad Y(t) = y_e + \eta(t)$$

where $\xi(t)$ and $\eta(t)$ are small, and approach zero when X and Y approach the equilibrium point. (ξ and η are called *perturbations* of the steady state.)

We now change the variables in the system from X and Y to ξ and η. Then

(margin note: change of variable)

$$\frac{d(x_e + \xi)}{dt} = F(x_e + \xi, y_e + \eta),$$

$$\frac{d(y_e + \eta)}{dt} = G(x_e + \xi, y_e + \eta),$$

where ξ and η are functions of t. But we have, since x_e and y_e are constant,

$$\frac{dX}{dt} = \frac{d(x_e + \xi)}{dt} = \frac{d\xi}{dt},$$

$$\frac{dY}{dt} = \frac{d(y_e + \eta)}{dt} = \frac{d\eta}{dt}.$$

So

$$\frac{dX}{dt} = \frac{d\xi}{dt} = F(x_e + \xi, y_e + \eta),$$

$$\frac{dY}{dt} = \frac{d\eta}{dt} = G(x_e + \xi, y_e + \eta).$$

take a
linear
approxi-
mation

We now apply the Taylor series expansion in two variables to expand $F(x_e + \xi, y_e + \eta)$ and $G(x_e + \xi, y_e + \eta)$ and then take a linear approximation for each: a *tangent-plane approximation* for the system. (For details of partial derivatives and the Taylor series expansion, refer to Appendix B.2.) Applying the Taylor series expansion in two variables we find

$$\xi' = F(x_e, y_e) + F_\xi(x_e, y_e)\,\xi + F_\eta(x_e, y_e)\,\eta$$
$$+ \text{ (terms of higher order)},$$

$$\eta' = G(x_e, y_e) + G_\xi(x_e, y_e)\,\xi + G_\eta(x_e, y_e)\eta$$
$$+ \text{ (terms of higher order)},$$

where $F_\xi = \partial F/\partial \xi$, $F_\eta = \partial F/\partial \eta$ and likewise for G. Now $F(x_e, y_e) = 0$, since (x_e, y_e) is an equilibrium point, and similarly $G(x_e, y_e) = 0$. Then, taking the linear approximation of each Taylor series expansion (i.e. ignoring all terms of higher order) we are left with the tangent-plane approximation

$$\xi' = F_\xi(x_e, y_e)\,\xi + F_\eta(x_e, y_e)\,\eta,$$
$$\eta' = G_\xi(x_e, y_e)\,\xi + G_\eta(x_e, y_e)\,\eta.$$

the
linearised
system

In vector notation the linearised system is

$$\begin{bmatrix} \xi' \\ \eta' \end{bmatrix} = \begin{bmatrix} F_\xi & F_\eta \\ G_\xi & G_\eta \end{bmatrix} \begin{bmatrix} \xi \\ \eta \end{bmatrix}. \tag{1}$$

Equilibrium point classification

dynamics
from the
eigenval-
ues

Recall that we can predict the behaviour (the dynamics) of a linear system $\mathbf{x}' = A\mathbf{x}$ merely by considering the eigenvalues of A. Further, these eigenvalues are the solutions to the characteristic equation $\lambda^2 - \lambda(\text{trace}A) + |A| = 0$. Clearly from (1)

$$\text{trace}(A) = F_\xi + G_\eta,$$
$$|A| = \det(A) = F_\xi\,G_\eta - F_\eta\,G_\xi,$$

and λ_1, λ_2 can be found for any equilibrium point.

equating
terms
between
the
systems

Recall that ξ and η were not variables of the original equation. However, $X = x_e + \xi$ and $Y = y_e + \eta$ so that

$$\frac{\partial F}{\partial \xi} = \frac{\partial F}{\partial X}\frac{dX}{d\xi} = \frac{\partial F}{\partial X}\frac{d(x_e + \xi)}{d\xi} = \frac{\partial F}{\partial X},$$

and similarly

$$\frac{\partial F}{\partial \eta} = \frac{\partial F}{\partial Y}, \qquad \frac{\partial G}{\partial \xi} = \frac{\partial G}{\partial X} \qquad \text{and} \qquad \frac{\partial G}{\partial \eta} = \frac{\partial G}{\partial Y}.$$

This means that we have

classifica-
tion
details
from the
original
system

$$\begin{bmatrix} F_\xi & F_\eta \\ G_\xi & G_\eta \end{bmatrix} = \begin{bmatrix} F_X & F_Y \\ G_X & G_Y \end{bmatrix}$$

and in vector notation the linearised system is

$$\begin{bmatrix} \xi' \\ \eta' \end{bmatrix} = \begin{bmatrix} F_\xi & F_\eta \\ G_\xi & G_\eta \end{bmatrix} \begin{bmatrix} \xi \\ \eta \end{bmatrix} \qquad \text{or} \qquad \begin{bmatrix} X' \\ Y' \end{bmatrix} = \begin{bmatrix} F_X & F_Y \\ G_X & G_Y \end{bmatrix} \begin{bmatrix} X - x_e \\ Y - y_e \end{bmatrix}. \quad (2)$$

This is in the form $\mathbf{x}' = J\mathbf{x}$. Here J is the matrix of first order partial derivatives which is called the *Jacobian* matrix of the system above. Clearly, from this Jacobian matrix, which follows directly from the original system, the characteristic equation and eigenvalues can be calculated immediately.

Summary

In summary, if we start with a general nonlinear system

summaris-
ing the
process

$$\frac{dX}{dt} = F(X, Y),$$
$$\frac{dY}{dt} = G(X, Y),$$

with any equilibrium point (x_e, y_e) we can establish the dynamics of the system close to this point in the following way:

- First linearise the system to get

$$\begin{bmatrix} \xi' \\ \eta' \end{bmatrix} = \begin{bmatrix} F_X & F_Y \\ G_X & G_Y \end{bmatrix} \begin{bmatrix} \xi \\ \eta \end{bmatrix}.$$

- Evaluate the Jacobian matrix at the equilibrium point of interest. Each equilibrium point will produce different values of the Jacobian matrix.
- Then apply the linear theory of classification to this system.

It can be proved (see, for example, Hurewicz (1990)) that, with the exception of the prediction of a centre, the classification established for the linearised system can be extended to the nonlinear system. Furthermore, if an equilibrium point is a centre in the linearised system then it is either

the excep-
tional
case

a focus or a centre in the nonlinear system. In this case, it is necessary to consider the nonlinear terms to establish the exact dynamics. However, such analysis is beyond the scope of this book and we refer the interested reader to, for example, Hurewicz (1990).

Skills developed here:

- Find the Jacobian of a matrix.
- Linearise a system of nonlinear equations and derive the linearised system.
- From the eigenvalues of the Jacobian matrix classify any equilibrium point and sketch the trajectory behaviour in the phase-plane close to this point.

7.5   *Applications of nonlinear theory*

overview   We now apply this linearisation technique to some of the models of population dynamics we developed in Chapter 5.

Predator–prey model

the differential equations   Recall the model of the predator–prey interaction from Chapter 5,

$$X' = \beta_1 X - c_1 XY, \qquad\qquad Y' = -\alpha_2 Y + c_2 XY, \tag{1}$$

for which we found two equilibrium points $(0,0)$ and $(\alpha_2/c_2, \beta_1/c_1)$ (see Section 6.2).

Example 1: *Find the linearised system for model (1), and hence classify all equilibrium points of the basic predator–prey model.*

Solution: *From (1) we label $F(X,Y) = \beta_1 X - c_1 XY$ and $G(X,Y) = -\alpha_2 Y + c_2 XY$. The Jacobian matrix is*

$$J = \begin{bmatrix} F_X & F_Y \\ G_X & G_Y \end{bmatrix} = \begin{bmatrix} \beta_1 - c_1 Y & -c_1 X \\ c_2 Y & -\alpha_2 + c_2 X \end{bmatrix}.$$

There were two equilibrium points for this system, and we consider each in turn.

For the case where the equilibrium point is $(x_e, y_e) = (0,0)$ we have that

$$J = \begin{bmatrix} \beta_1 & 0 \\ 0 & -\alpha_2 \end{bmatrix}.$$

The trace is then $\beta_1 - \alpha_2$ and the determinant is given by $\det(J) = -\beta_1\alpha_2$. Using the notation of Section 7.2, the characteristic equation follows as $\lambda^2 - (\beta_1 - \alpha_2)\lambda + (-\beta_1\alpha_2) = 0$ and we have $q = -\beta_1\alpha_2 < 0$ which implies that the equilibrium point is a saddle point.

In the second case where the equilibrium point is

$$(x_e, y_e) = \left(\frac{\alpha_2}{c_2}, \frac{\beta_1}{c_1} \right)$$

we have that

$$J = \begin{bmatrix} 0 & -\dfrac{c_1\alpha_2}{c_2} \\ \dfrac{\beta_1 c_2}{c_1} & 0 \end{bmatrix}.$$

The trace of this matrix is 0 and its determinant is given by $\det(J) = \beta_1\alpha_2$. As above, with the notation of Section 7.2, the characteristic equation in this case $\lambda^2 - 0\lambda + \beta_1\alpha_2 = 0$ and we have $p = 0$ and $\Delta < 0$ which implies that this equilibrium point is a centre.

The results from this example are those we might have expected from the numerical solutions obtained earlier using **Maple**. However, the latter critical point is predicted as a centre and, as we have stated, linear theory is not adequate in this case where it is necessary to consider the nonlinear terms to establish the true dynamics. On the other hand, we were able, for this example, to solve for the phase-trajectories exactly in Section 6.6. The solution obtained there shows that the critical point is indeed a centre.

results for linearised system

Epidemic model

Recall the basic model for the spread of a disease in a population from Section 5.6. The differential equations were

epidemic model example

$$\frac{dS}{dt} = -bSC, \qquad \frac{dC}{dt} = bSC - rC. \tag{2}$$

Here $S(t)$ describes the susceptible population and $C(t)$ describes contagious infectives, as functions of time. Parameter b is the disease transmission rate (transmission coefficient) and r is the removal rate.

Example 2: *Classify the equilibrium points of the epidemic model (2).*

Solution: *We established in Section 6.4 that (2) has an infinite set of equilibrium points, $(S, C) = (s_e, 0)$, where s_e is an arbitrary positive real number.*

From (2) we label $F(S, C) = -bSC$ and $G(S, C) = bSC - rC$. The Jacobian matrix is

$$J = \begin{bmatrix} F_S & F_C \\ G_S & G_C \end{bmatrix} = \begin{bmatrix} -bC & -bS \\ bC & bS - r \end{bmatrix}.$$

At any given equilibrium point, $(S, C) = (s_e, 0)$, we have the linearised system

$$\begin{bmatrix} S' \\ C' \end{bmatrix} = \begin{bmatrix} 0 & -bs_e \\ 0 & bs_e - r \end{bmatrix} \begin{bmatrix} S - s_e \\ C - 0 \end{bmatrix}.$$

At these equilibrium points the trace $(J) = bs_e - r$ is negative if $s_e < r/b$ and positive if $s_e > r/b$. This implies stable points for $S < r/b$ and unstable points for $S > r/b$. With the notation of Section 7.2, we have that $q = \det(J) = -bC(bS - r) + b^2 SC = 0$ at $(S, C) = (s_e, 0)$. This is a case of parallel trajectories as discussed in Section 7.2.

We have a line of stable nodes for $S < r/b$ and a line of unstable nodes for $S > r/b$. The same is true for the original nonlinear system (2).

These results are consistent with the phase-plane analysis of Section 6.4.

Using Maple

Maple code Most mathematics software packages will have inbuilt routines to find equilibrium solutions, Jacobian matrices, characteristic polynomials and eigenvalues. What follows is some Maple code, which has been used to solve the basic predator–prey example at the beginning of Section 7.5.

```
> restart:with(linalg):
> eqn1:=beta[1]*X-c[1]*X*Y;
> eqn2:=-alpha[2]*Y+c[2]*X*Y;
Critical/equilibrium points
> critpts:=solve({eqn1=0,eqn2=0},{X,Y});
The Jacobian matrix
> sys:=vector([eqn1,eqn2]);
> J:=jacobian(sys,[X,Y]);
The characteristic equation
> ch:=charpoly(J,lambda);
Eigenvalues
> ch1:=subs(critpts[1],ch);
> ev1:=solve({ch1},{lambda});
> ch2:=subs(critpts[2],ch);
> ev2:=solve({ch2},{lambda});
Using the p, q , Delta approach for classification
> p:=trace(J);
> q:=det(J);
```

```
> delta:=p^2-4*q;
> critpts[1];
> p1:=subs(critpts[1],p);
> q1:=subs(critpts[1],q);
> delta1:=subs(critpts[1],delta);
> critpts[2];
> p2:=subs(critpts[2],p);
> q2:=subs(critpts[2],q);
> delta2:=subs(critpts[2],delta);
Calculating the eigenvalues from p, q  and Delta
> lambda1a:=p1*1/2 +1/2*sqrt(delta1);
> lambda1b:=p1*1/2 -1/2*sqrt(delta1);
> lambda2a:=p2*1/2 +1/2*sqrt(delta2);
> lambda2b:=p2*1/2 -1/2*sqrt(delta2);
Calculating the eigenvalues directly from Maple
> eigs:=eigenvals(J);eigs[1];
> eig1a:=subs(critpts[1],eigs[1]);
> eig1b:=subs(critpts[1],eigs[2]);
> eig2a:=subs(critpts[2],eigs[1]);
> eig2b:=subs(critpts[2],eigs[2]);
```

Further /Maple commands, such as `simplify(X)` and `factor(X)`, can reduce complex expressions of X to a more manageable form.

Discussion

The above applications give some indication of the power of linearisation theory, and the equilibrium point classification process developed. It should be noted that changes in the parameter values in each case may alter the equilibrium point classification and predict different dynamics. In this way it is not only possible to predict the system dynamics, but also to control the outcome in certain dynamical systems where such control parameters determine changes in the classification. This is an extremely important aspect in the study of dynamical systems.

strengths of linearisation

Skills developed here:

- For a nonlinear system of equations, find all the equilibrium points and establish the trajectory behaviour close to each of them.
- Sketch a picture of the entire phase-plane indicating the direction of the trajectories in any part of the plane.
- Identify how a parameter might be changed to alter the dynamics in the phase-plane.

7.6 Exercises for Chapter 7

1. Constructing a phase-plane. *Consider the linear system*

$$x' = x - y, \qquad y' = x + y,$$

where differentiation is with respect to time.

(a) *Find the equilibrium point(s) and establish what type of equilibrium it is (they are).*

(b) *Using the above results draw a sketch of how you would expect the phase-plane trajectories to behave. (Note that, along the line $x = y$ (or equivalently $x' = 0$) there is no change in x, and along the line $x = -y$ there is no change in y.)*

(c) *Use* `Maple` *to draw the phase-plane to check your results.*

2. Defining the trajectories. *Consider the system*

$$x' = x - 5y, \qquad y' = x - y,$$

where differentiation is with respect to time.

(a) *Confirm there is only one equilibrium point which is a centre.*

(b) *Confirm that the family of ellipses given by*

$$x^2 - 2xy + 5y^2 = K,$$

with K some constant, describe the solution trajectories in the (x, y)-plane. (Hint: One way to do this is to use direct substitution.)

3. Linearisation. *Consider the nonlinear system of equations*

$$x' = x - y, \qquad y' = 1 - xy,$$

where differentiation is with respect to time.

(a) *Find all equilibrium points for the system.*

(b) *Linearise the system and establish the classification of each equilibrium point.*

(c) *From the above results sketch the trajectory behaviour in the phase-plane. (Note that along $x = y$ we have $x' = 0$ and along $y = 1/x$ we have $y' = 0$.)*

(d) *Using* `Maple` *check your results.*

4. Linearisation. *Given the system of equations*

$$x' = 3 + q(x - y), \qquad y' = 4 - xy,$$

where differentiation is with respect to time.

(a) *Assume $q = 1$. Give the linearised system of equations.*

(b) *Assume $q = 1$. Find all equilibrium points for the system and classify each point. Hence sketch the phase-plane.*

(c) *Consider q to be a variable parameter. Prove that one equilibrium point will always be a saddle point. Give full reasons for your answer. (Hint: Show graphically, or otherwise, that for each such equilibrium point (x_e, y_e), the x and y coordinates will have the same sign.)*

5. Linearisation of competition model. *Consider a competing species model, species X competing with species Y, including logistic growth for both species*

$$X' = \beta_1 X - d_1 X^2 - c_1 XY, \qquad Y' = \beta_2 Y - d_2 Y^2 - c_2 XY.$$

Suppose that the parameter values are $\beta_1 = 3$, $\beta_2 = 2.5$, $d_1 = 2.8$, $d_2 = 2$, $c_1 = 2$ and $c_2 = 1$. The equilibrium values are

$$(0,0), \quad \left(0, \frac{\beta_2}{d_2}\right), \quad \left(\frac{\beta_1}{d_1}, 0\right), \quad \left(\frac{c_1\beta_2 - d_2\beta_1}{c_1 c_2 - d_1 d_2}, \frac{c_2\beta_1 - d_1\beta_2}{c_1 c_2 - d_1 d_2}\right)$$

and the phase-plane indicates that, with time, the populations will settle to coexistence of both species.

(a) *Linearise the system, giving the general form of the linearised equations and substituting in the parameter values as given.*

(b) *Use the linearised system to classify each equilibrium point. Sketch the phase-plane and illustrate the trajectory behaviour close to each equilibrium point from your classifications. Verify that these classifications agree with coexistence of the species.*

(c) *How, by varying the interaction parameters, could you change the phase-plane predictions to ensure that X will always die out?*

6. Comparison of nonlinear and linearised systems. *In order to compare the nonlinear system with the linearised system we can draw them both on the same system of axes and compare them directly.*

(a) For the predator–prey model

$$X' = X(\beta_1 - c_1 Y), \qquad Y' = Y(-\alpha_2 + c_2 X),$$

taking $\beta_1 = 1.3$, $c_1 = 0.01$, $\alpha_2 = 1$ and $c_2 = 0.01$, and considering the trajectories close to the non-zero equilibrium, draw on the same system of axes the time-dependent plots for the nonlinear and the linearised systems. (Use `Maple` to solve the systems numerically).

(b) Also compare the trajectories in the phase-plane of the nonlinear and the linearised systems.

(Hint: Set up the equations for the linearised system, and then write them in terms of the X and Y coordinate system of the given nonlinear equations. For example $X = x_e + \xi$, or $\xi = X - x_e$, where x_e is the equilibrium value. Then solve both systems of equations using `Maple` and display the results in the same diagram.)

7. Linearisation of an epidemic model. *Consider the following model for the spread of a disease, with susceptibles constantly renewed by births at a constant overall birth rate of μ.*

$$S' = -bSC + \mu, \qquad C' = bSC - rC.$$

(a) Show that the only equilibrium point is $(r/b, \mu/r)$.

(b) Linearise the system and determine the behaviour of the trajectories near the equilibrium point.

(c) Discuss, briefly, the implications for the time-dependent behaviour of both $S(t)$ and $C(t)$.

8. Linearisation of jungle warfare model. *A model for a battle between red soldiers and blue soldiers, where one army uses random fire and the other uses aimed fire, is*

$$R' = -a_1 RB, \qquad B' = -a_2 R.$$

Linearise the system about the only equilibrium point, and hence determine the behaviour of trajectories near that point.

Chapter 8

Some improved population models

In this chapter, we present a variety of case studies based on research papers or articles. Each study significantly extends one or more of the simple models we have developed in Chapter 5 and applies the techniques of analysis developed in Chapter 6 and particularly Chapter 7. It serves to illustrate the relevance of the theory presented in this text and its current use in practice.

8.1   Introduction

modelling
is
applicable
There are many interesting problems requiring the use of population models, and we present a sample of them here with the aim of illustrating the versatility, diversity and applicability of these models to current issues.

overview
In previous chapters, we developed the simplest type of population models. Each of those may be extended or combined with others to reflect the particular features of the population under study. As we see in the case studies which follow, the models examined so far form the basis on which the further models proposed are built. The concept of the balance law and the compartmental structure are integral in each case.

chapter
diagrams
The references from which the following case studies are adapted are provided, and all figures presented have been generated using `Maple`. (The code has not been included here, although all figures can be generated from code integrated into the text of earlier chapters.) Code building and model exploration are encouraged in the exercises.

8.2   Case Study: Competition, predation and diversity

In Section 5.4, we developed simple models for predator–prey interactions and for competing species. For the latter we found that the model predicted, in most cases, the survival of one species and the extinction of all others. We noted that this was a limitation of the model, as in many instances observations in nature confirm the stability and coexistence of a number of competing species.

One notion which would explain coexistence as well as diversity is that of cooperation, as opposed to competition. In the infertile country of the arid regions of Australia, this is one hypothesis for the substantial diversity and coexistence of species relying on the same resources for survival.

In the following case study we examine a different hypothesis. A combination of the basic competition model together with the predator–prey model provides a mathematical argument for the stability and coexistence of competing species within an ecosystem. The case study supports the earlier comment in Section 5.1, that systems operating in isolation may be unstable, but when incorporated into more complex systems they typically display stability.

Note here the use of the word stability. A stable system in this sense is a system in which species coexist, whereas if one species suffers extinction, the system is referred to as unstable. This use appears in practice, but does not satisfy the strict definition we have met previously in the text.

The following is adapted from Cramer and May (1972).

Observations have confirmed that whilst simple systems (for example, a single trophic level) are likely to be unstable in isolation, more complex systems comprising a collection of interacting trophic levels are often extremely stable.

Parrish and Saila (1970) attempted to validate this idea with some computer experiments. They proposed that while a two species competition model might be unstable, the introduction of a predator could allow the system to stabilise, thus permitting all three species to coexist.

For the two species of prey, N_1 and N_2, and a predator N_3, the model they constructed was a simple competition model between N_1 and N_2 (both populations following a logistic growth pattern) with a predator N_3 included, whose birth rate was dependent on both N_1 and N_2, namely

$$\begin{aligned}
N_1' &= N_1(\epsilon_1 - \alpha_{11}N_1 - \alpha_{12}N_2 - \alpha_{13}N_3) \\
N_2' &= N_2(\epsilon_2 - \alpha_{21}N_1 - \alpha_{22}N_2 - \alpha_{23}N_3) \\
N_3' &= N_3(-\epsilon_3 + \alpha_{31}N_1 + \alpha_{32}N_2).
\end{aligned} \tag{1}$$

In the absence of the predator $(N_3 = 0)$ we have a simple competition model which predicts the extinction of all but one species in most cases: that is, $\alpha_{11}\alpha_{22} - \alpha_{12}\alpha_{21} \leq 0$. Now starting with parameter values which ensure instability for the competition model, the predator is included. The new system can be stable, even when we do not have equal predation (that is, $\alpha_{13} = \alpha_{23}$ and $\alpha_{31} = \alpha_{32}$).

Figure 8.2.1 illustrates the case of the unstable competition model without predation. In this case N_1 dies out and $N_2 \to 3.5 \times 10^5$. Figure 8.2.2 corresponds to the case where a predator is included and predation of each of the species is the same, that is, $\alpha_{13} = \alpha_{23}$). In this case all populations approach fixed positive population sizes with $N_1 \to 11.66 \times 10^3$, $N_2 \to 8.34 \times 10^3$ and $N_3 \to 11.3$. Finally, Figure 8.2.3 illustrates the case of a stable system where all three populations coexist, but without the condition of equal predation. The results are that $N_1 \to 69.4 \times 10^3$, $N_2 \to 16.7 \times 10^3$ and $N_3 \to 11.1$.

The example serves to illustrate that while a single two species population model may be unstable, the introduction of a further species (in this case a

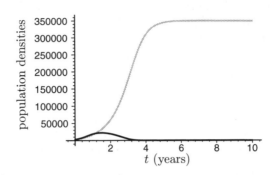

Figure 8.2.1: Unstable competition model without predation, with N_1 (black) and N_2(grey). Parameter values: $\alpha_{11} = 9 \times 10^{-5}$, $\alpha_{12} = 3 \times 10^{-5}$, $\alpha_{13} = 0$, $\alpha_{21} = 3 \times 10^{-5}$, $\alpha_{22} = 0.6 \times 10^{-5}$, $\alpha_{23} = 0$, $\epsilon_1 = 3$ and $\epsilon_2 = 2.1$. The initial populations are $N_1(0) = N_2(0) = 3 \times 10^3$.

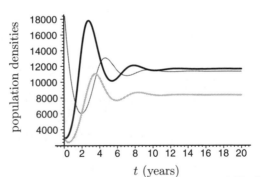

t (years)

Figure 8.2.2: Unstable competition model with predation and N_1 (black), N_2 (grey) and N_3 (fine black). Parameter values are as in Figure 8.2.1 but with $\alpha_{13} = 0.15$, $\alpha_{23} = 0.15$, $\alpha_{31} = \alpha_{32} = 0.6 \times 10^{-4}$ and $\epsilon_3 = 1.2$. The initial populations are $N_1(0) = N_2(0) = 3 \times 10^3$ and $N_3(0) = 20$. In the displayed results N_3 has been multiplied by 10^3.

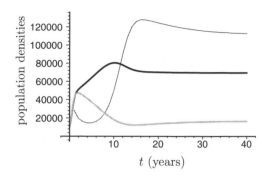

Figure 8.2.3: Unstable competition model with predation and N_1 (black), N_2 (grey) and N_3 (fine black). Parameter values: $\alpha_{11} = 0.9 \times 3.22 \times 10^{-5}$, $\alpha_{12} = 3.22 \times 10^{-5}$, $\alpha_{13} = 0.06$, $\alpha_{21} = 3.22 \times 10^{-5}$, $\alpha_{22} = 3.22 \times 10^{-5}$, $\alpha_{23} = 0.04$, $\alpha_{31} = 3 \times 10^{-5}$, $\alpha_{32} = 10^{-5}$, $\epsilon_1 = 3.22$, $\epsilon_2 = 3.22$ and $\epsilon_3 = 2.25$. The initial populations are $N_1(0) = N_2(0) = 10^4$ and $N_3(0) = 10$. In the displayed results N_3 has been multiplied by 10^4.

predator of both other species) may result in a stable system with the persistence and coexistence of all species. This provides an argument for the existence of a rich diversity of species, some of which may rely on the same resources, as is apparent in any natural ecosystem.

8.3   Extended predator–prey model

In this section we improve on the predator–prey model by incorporating further biological observations. The new model also exhibits oscillatory behaviour, which means that in the phase-plane we have a limit cycle or periodic solution.

overview

Review of previous models

In Section 5.3, we developed a predator–prey model which predicted oscillations in the population densities, and we saw, in Section 6.6, how this led to closed curves, or a set of periodic solutions, in the phase-plane.

predator–prey model revisited

We then improved the model to include logistic growth and noted that the oscillations died out and the populations stabilised to some fixed equilibrium point. In the phase-plane this translated to an inwardly moving spiral approaching this equilibrium point. The differential equations for

logistic growth

this model were

$$\frac{dX}{dt} = r_1 X \left(1 - \frac{X}{K}\right) - c_1 XY, \qquad \frac{dY}{dt} = c_2 XY - \alpha_2 Y. \qquad (1)$$

Here $X(t)$ is the prey density and $Y(t)$ is the predator density. The constant K is the prey carrying capacity, r_1 is the prey per-capita intrinsic growth rate (the per-capita growth rate for small populations), α_2 is the predator per-capita death rate and c_1 and c_2 are positive constants which may be interpreted in terms of the predator searching rate (see Section 5.3).

model limitations   However, oscillating predator–prey pairs have been observed in nature, and this 'improved' model does not predict periodic solutions. This is a fundamental flaw, and we now suggest possible extensions which make the model more realistic.

Extending the model

functional and numerical response   The basic problem with the Lotka–Volterra predator–prey model is that the assumptions made regarding prey deaths and predator births and deaths are not sufficiently realistic. In practice, field ecologists studying predator populations measure two quantities: the *functional response*, $F(X)$, and the *numerical response* $N(X,Y)$. The functional response is the rate at which a single predator kills prey, as a function of the prey population. The numerical response is the per-capita growth rate of the predators, which may be dependent on both prey and predator populations.

general predator–prey model   A general model for a predator–prey system, based upon functional and numerical response functions, and with X and Y the population densities, is

$$\frac{dX}{dt} = r_1 X \left(1 - \frac{X}{K}\right) - F(X)Y, \qquad \frac{dY}{dt} = N(X,Y)Y. \qquad (2)$$

interpretation of original model   In equations (1) the functional response is $F(X) = c_1 X$. This states that a single predator continues to increase its rate of killing prey as the prey density increases. In practice, it is observed that this function levels off to a constant amount as X increases, representing the maximum that a single predator will want to eat (in a given time). Another way of thinking about this is that if the prey are plentiful, the predator will not have to spend much time hunting for them, but will spend more time waiting until they need to hunt again. The following example uses the idea of a handling time to formulate a suitable functional response function.

Example 1: *Assuming that a single predator takes a time t_h to handle each prey find a suitable functional response.*

Solution: *We modify the argument given in Section 5.3 for the interpretation of one of the parameters in terms of a searching rate a'. Let T denote the time left available to the predator for searching (hunting prey). Consider a time interval Δt.*

The number of prey eaten is determined from the searching rate a' multiplied by the prey density $X(t)$

$$\left\{ \begin{array}{c} \text{no. prey eaten} \\ \text{in time } \Delta t \\ \text{by one predator} \end{array} \right\} = a'TX(t). \tag{3}$$

The rate of prey eaten by a single predator, is then $a'TX(t)/\Delta t$. Now

$$\Delta t = T + t_h \times \left\{ \begin{array}{c} \text{no. prey eaten} \\ \text{in time } \Delta t \\ \text{by one predator} \end{array} \right\} = T(1 + t_h a' X(t)). \tag{4}$$

The functional response $F(X)$ is the rate of prey eaten by a single predator, which is given by dividing equation (3) by Δt. Using (4) and writing $c_1 = a'$ we obtain

$$F(X) = \frac{c_1 X}{1 + t_h c_1 X}. \tag{5}$$

As $X \to 0$, we see that $F(X) \to 0$. For large prey densities, or as $X \to \infty$, then $F(X) \to t_h^{-1}$. This value is the maximum possible removal rate of prey: the functional response is *saturated* at this value. For large X, the prey are very easy for a predator to find, so this maximum rate is the reciprocal of the handling time t_h, for a single predator. This function also arises in the theory of reaction kinetics in chemistry, see Murray (1990), where it is known as the Michaelis–Menten function.

properties of F

This particular form of the functional response $F(X)$ is called a Holling type II response function and is commonly used for invertebrate predators May (1981). Some examples of fitting the Holling type II model to field data are given in Hassell (1978, 1976) and Gotelli (1995), together with data for handling times and searching rates. For vertebrate predators a Holling type III response is often used: this is obtained by setting the searching rate $c_1 = a'$ in (5) proportional to prey density. This could be thought of as reflecting the higher intelligence of vertebrates whose

model validity

searching rates improve with higher prey density, since they learn how to track the prey.

numerical response

For the numerical response, $N(X, Y)$, we assume a logistic growth function for the predator population, $r_2(1 - Y/K_2)$ but with the predator carrying capacity assumed to be proportional to the prey density, $K_2 = c_3 X$. Hence

$$N(X, Y) = r_2 \left(1 - \frac{Y}{c_3 X}\right), \tag{6}$$

where r_2 and c_3 are positive constants. Note that as X increases, the deaths due to overcrowding decrease, but an increase in Y implies an overall increase in the deaths proportional to Y^2.

improved system

Our model becomes

$$\frac{dX}{dt} = r_1 X \left(1 - \frac{X}{K}\right) - \frac{c_1 XY}{1 + t_h c_1 X}, \qquad \frac{dY}{dt} = r_2 Y \left(1 - \frac{Y}{c_3 X}\right). \tag{7}$$

This model is sometimes called the Holling–Tanner predator–prey model.

Scaling the equations

a simpler system

Before using `Maple` to examine the time-dependent graphs and phase-plane we apply a technique often adopted in practice to simplify the analysis. We scale the equations. This process may offer several advantages.

purpose of scaling

After scaling, the parameters are in dimensionless form and provide a measure of the relative strengths of the interacting or competing effects within the model. This may provide important information towards improving the design of experimental processes, in turn creating a more applicable model. Also as a result of scaling, the mathematics may be simplified, or the number of parameters reduced. However, this is not necessarily true in all cases.

scaling the system

We scale the above system of equations by defining new variables x, y and τ, by

$$x = \frac{X}{K}, \qquad y = \frac{Y}{c_3 K}, \qquad \tau = r_2 t.$$

Then, using the chain rule, it can be shown that (see Exercises, Question 3), the system becomes

$$\frac{dx}{d\tau} = \lambda_1 x(1 - x) - \frac{\lambda_2 xy}{(\lambda_3 + x)}, \qquad \frac{dy}{d\tau} = y \left(1 - \frac{y}{x}\right), \tag{8}$$

with dimensionless parameters

$$\lambda_1 = \frac{r_1}{r_2}, \qquad \lambda_2 = \frac{c_3}{t_h r_2}, \qquad \lambda_3 = \frac{1}{t_h c_1 K}.$$

Interpretation of the parameters

The newly defined parameters can be interpreted in the following way. parameter
Since r_1 is the maximum specific growth rate of the prey with no preda- λ_1
tors and r_2 is the specific growth rate of predators with infinite prey
density, then λ_1 is the ratio of the specific growth rates under ideal conditions.

The quantity λ_2 measures the predators' maximum prey removal rate parameter
t_h^{-1} relative to its ideal specific growth rate r_2. Parameter c_3 is already λ_2
dimensionless as $c_3 X = K$.

Since t_h is the handling time per prey then the combination $t_h K$ rep- parameter
resents a time scale (per unit area) for the total time to handle the λ_3
maximum possible prey density (given by the carrying capacity K in the
absence of predation). Since c_1 is the searching rate (area per unit time)
then c_1^{-1} is a time scale for searching a unit area. Hence the parameter λ_3
represents a ratio of time scales (searching time to handling time). For
large λ_3 the handling time is insignificant, but for small values of λ_3 the
predators ability to remove prey is more likely to rise to its saturation
value of λ_2.

Numerical solution for scaled model

We use `Maple` to solve the system numerically with parameter values of example
$\lambda_1 = 5$, $\lambda_2 = 10$ and $\lambda_3 = 0.015$. Consider the phase-plane diagram as numerical
illustrated in Figure 8.3.1. Note that for all the initial conditions chosen solution
from the illustrated domain, the solution is a single periodic solution.

For different parameter values, $\lambda_1 = 5$, $\lambda_2 = 1$ and $\lambda_3 = 0.6$, as in different
Figure 8.3.2, the equilibrium solution is a stable node. Trajectories with behaviour
initial values close to the equilibrium are attracted to the equilibrium.

Discussion

Notice that this new system can give rise to a single periodic solution single
for certain parameters (Figure 8.3.1). In this case, the periodic cycle periodic
is always the same, regardless of the initial conditions chosen. This is solution
very different from the infinite series of closed curves, each associated
with a different set of initial values, which we obtained from our basic
predator–prey model in Section 6.2.

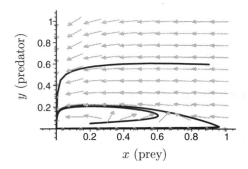

Figure 8.3.1: Phase-plane diagram for the scaled system (8) with $\lambda_1 = 5$, $\lambda_2 = 10$ and $\lambda_3 = 0.015$.

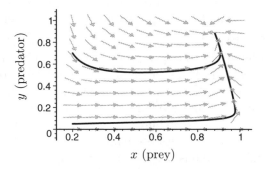

Figure 8.3.2: Phase-plane diagram for the scaled system (8) with $\lambda_1 = 5$, $\lambda_2 = 1$ and $\lambda_3 = 0.6$.

stable
node
solution

The system also predicts populations stabilising to fixed densities for other parameter combinations (Figure 8.3.2). It would be useful to establish a diagram of the parameter values which lead to periodic solutions and those which do not. This is left as an exercise for the interested reader, however, such a diagram, establishing the parameter regimes which result in different dynamics, is an important part of the system analysis process.

8.4   *Case Study: Lemming mass suicides?*

Dramatic fluctuations have been observed in the populations of several rodents. In the case of lemmings, several myths have evolved to account for these fluctuations, such as lemmings hurling themselves off cliffs in mass suicides. In the following case study, a simple model of rate equations is presented to describe these dramatic cycles.

The model proposed is a further extension of the basic predator–prey model and makes use of the concept of modelling a single population in two parts, each part having different behavioral characteristics. This is similar to the notion we made use of when modelling the spread of a disease in a population by dividing the population into two subpopulations of infectives and susceptibles.

Adapted from Dekker (1975).

Legend has it that lemmings periodically 'migrate' into the sea in an act of mass suicide. Early Scandinavians believed that lemmings fell from heaven in stormy weather! More recently it has been proposed by natural historians that epidemic disease, or possibly predation, intermittently wipe out large numbers of the lemming population. Each of the above attempts to explain the drastic fluctuations in the populations of lemmings, true also for certain other small rodents or voles, which have been observed.

It appears from collected data that the population cycles have a period of three to four years. Further, there appear to be no external forces causing these fluctuations, suggesting the existence of some internal control mechanism. To support this idea, the observations indicate changes in the population characteristics during a cycle, in terms of (amongst others) sex and age.

Following the work of Myers and Krebs (1974), populations can be divided into two main genotypes: one with a high reproduction rate and intolerant of high population density living conditions, and the other with a low reproduction rate and well adapted to crowded living conditions. The former population, which deals with overcrowding by migrating (not necessarily off a cliff into the ocean) are called the emigrants and their density will be denoted by n_1, while the latter population are known as the tolerants and their density is denoted as n_2. The mathematical model proposed to describe these interacting populations is given as

$$\frac{dn_1}{dt} = n_1 \left[a_1 - (b_1 - c_1)n_2 - c_1(n_1 + n_2) \right],$$
$$\frac{dn_2}{dt} = n_2 \left[-a_2 + b_2 n_1 \right],$$

$$(1)$$

where a_1, a_2, b_1, b_2 and c_1 are parameters which will be described below.

First, consider the growth of the emigrants in the absence of the tolerants. In that case the emigrants follow a logistic growth pattern and, since emigration depends on the total population density, the term which curbs the infinite exponential growth is $-n_1^2 c_1$. The individuals which migrate are those with a high reproductive potential, largely the young, lactating females.

On the other hand the tolerants, in the absence of emigrants, will die off exponentially. They have an average age far greater than that of the emigrant population, but a far lower reproductive potential. The tolerants interact with the emigrants in both a social and sexual manner, which leads to an improved birth rate for the tolerants.

Thus the parameter b_2 can be split into parts: b_2^0 governing the sexual creation of individuals, and b_2^1 governing the possible creation of a tolerant from an emigrant through a social encounter. The latter process allows a flow of individuals from the emigrant to the tolerant population. Similarly, the parameter $(b_1 - c_1)$ describing the reduction in emigrant population consists of three parts: a social contribution where individuals flow to the tolerants, a sexual contribution where individuals are created, and a dispersive contribution where individuals leave the population (largely young females and usually when the total density of the populations is large).

A further consideration is that of seasonal fluctuations in the reproductive rates, a_1 and a_2, which reflect the observed maximum rate in late summer and minimum in late winter. Assuming a simple sinusoidal rate, the time dependent parameters are taken to be

$$a_1(t) = a_1^0 + a_1^1 \sin\left(\frac{\pi}{6}t\right), \qquad a_2(t) = a_2^0 - a_2^1 \sin\left(\frac{\pi}{6}t\right). \tag{2}$$

The following figures (Figures 8.4.1 and 8.4.2) are constructed applying the above model and estimating the parameters from data collected by Krebs et al. (1973) and Myers and Krebs (1974) for the rodent voles Microtus pennsylvanicus (or meadow mouse).

For the figures, a monthly time step was taken and the parameter values are listed in the associated captions. From Figure 8.4.1 (following Dekker (1975)) it should be noted that the total population constitution of the two genotypes is very different in different phases of the first 48-month (four year) cycle. Furthermore, for the total rodent population the dramatic fluctuations are observed in the initial four year period. While the model is simple, easy to analyse, and provides a close resemblance to the observations in terms of the initial main cycles, it is not robust over long time periods.

If the time-dependent graph is considered over long periods (more than 20 years) then the population settles to a simple 1-year oscillation. This is illustrated in the time-dependent plot and phase-plane of Figure 8.4.2, and is not included or discussed in Dekker (1975).

However, if disease or predation (or some other mechanism) forced a reduction in the population of the emigrants in particular, then this model would

Figure 8.4.1: Population cycles for the meadow mouse. Parameters $a_1^0 = 1$, $a_1^1 = 0.35$, $b_1 = 1.75 \times 10^{-2}$, $c_1 = 1 \times 10^{-2}$, $a_2^0 = 0.140$, $a_2^1 = 0.075$ and $b_2 = 1.5 \times 10^{-2}$ with initial populations of $n_1 = 0.5$ and $n_2 = 0.5$. Emigrants n_1 (black), tolerants n_2 (grey) and total population (fine black).

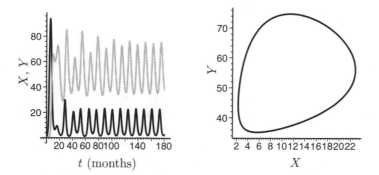

Figure 8.4.2: The time-dependent (first graph) and phase-plane (second graph) plots for the population cycles of the meadow mouse. Parameters as for Figure 8.4.1, but the model is run over a longer time period. (n_1 represented with black, and n_2 with grey in the first graph.)

predict dramatic fluctuations over the following decade. Thus the model re-quires further refinement, which might include different sex groups and an age distribution, as well as a better representation of the rodent dispersion with respect to distance and location, namely their spatial distribution. Fur-thermore, some statistics on the likelihood of interspecies and between-species interactions, such as predation, would provide an enhancement.

8.5 Case Study: Prickly-pear meets its moth

In Chapter 5, we studied a predator–prey model describing a case where the introduction of an insect species to a region, to eradicate a previ-ously introduced pest was unsuccessful. Many such introductions have been and are currently being carried out, with many disasters and un-foreseen ramifications for the environment, as we saw in Section 5.2 with the introduction of the Nile Perch to Lake Victoria. However, there have also been success stories and the following case study examines one spec-tacular success.

The model proposed in this case study is a plant-herbivore system, which is similar to the predator–prey model. It includes extensions of the basic form, and the inclusion of a Michaelis–Menten type function.

Adapted from a discussion in May (1981) and a description in Rolls (1969).

In 1864 Captain Phillip, on his way to Australia with the First Fleet, stopped off in Brazil to collect the cochineal insect and their host plant the prickly-pear (Opuntia inermis and Opuntia stricta). He was to introduce them to Australia to ensure a plentiful supply of red dye (from the cochineal insects) for his soldiers' coats. The insects did not fare well and several more introductions were tried but failed. However the cacti, which were also planted as hedges for additional stock feed as well as garden plants, adapted well: far too well in fact. They ran wild. Extensive, dense stands of the cacti spread into the farmland of northern New South Wales and southern Queensland, averaging about 500 plants per acre (or 1250 plants per hectare). They walled in homesteads with growth of impenetrable density, and destroyed the viability of thousands of square kilometres of farmland.

Attempts were made to eradicate the cacti by spraying them with arsenic pentoxide and sulphuric acid from horseback. Whilst this destroyed the cacti it came into contact with, it was a drop in the ocean and, furthermore, it destroyed the mens' clothing, their boots and saddles and finally their horses, which lost their hair and developed sores which would not heal.

In 1925 a moth *Cactoblastis cactorum* was introduced from Argentina to combat the cactus growth. Its larvae bore through the plant, their breeding being restricted solely to the cacti. They were bred up in Brisbane, from where some 3000 million eggs were distributed to farmers. Within only two years they had virtually wiped out the cacti.

The eggs of the moth are not laid at random, but in clumps of egg-sticks on the plants with each egg-stick comprising about 80 eggs. Thus some plants were hit with large numbers of the moths while others escaped completely. Since about 1.5 sticks of eggs produce sufficient larvae to destroy a plant, many plants received many more larvae than were required to destroy them.

A mathematical predator–prey type model is proposed (May, 1981) to describe this plant-herbivore system. With V the plant mass (in plants per acre) and H the herbivore (moth) population size (in egg-sticks), the model is

$$\frac{dV}{dt} = r_1 V \left(1 - \frac{V}{K} \right) - c_1 H \left(\frac{V}{V + D} \right)$$
$$\frac{dH}{dt} = r_2 H \left(1 - \frac{JH}{V} \right). \tag{1}$$

Here K is the maximum biomass, r_1 is the intrinsic rate of increase in the cacti biomass, c_1 is the per-capita maximum rate of food intake by the moths, D is a grazing efficiency term at low plant density, J is the proportionality constant associated with the biomass required to sustain a moth at equilibrium, and r_2 is the per-capita rate of increase of the moth population.

Values for these parameters have been estimated from data collected by Dodd (1940) and Monro (1967), and the older unit of measurement, the acre, is retained to make use of the whole numbers from this collected data. (The conversion is 1 ha = 2.471 acres.) Since the root stock of the cacti can increase to 250 tons per acre in 2 years, r_1 is close to 2.7, however, this is only based on vegetative growth, and if we consider sexual reproduction r_1 is estimated to be 2, at maximum. The carrying capacity K is taken to be 5000 plants per acre. Laboratory experiments were used to get an estimate for r_2 in which 2750 eggs increased to 100 605 eggs in the following cycle, providing an estimate of 3.6. This compares with field data where 5 000 larvae could multiply to 10 000 000 in two years, suggesting a larger r_2 of 3.8 a year. Both figures may underestimate the rate and so r_2 was taken as 4.

From the work of Monro (1967) J was estimated as 2.23 in units of cactus plants per egg-stick. The parameters c_1 and D were more difficult to estimate. The former should be large, reflecting the damage to the plants by the feeding larvae, while the latter should be small to reflect the efficiency with which the female moths choose plants on which to lay eggs. Following the near total

*destruction of the cacti in a period of only two years, the model should predict
this crash and then settle to a stable 11 plants per acre, as was observed . For
this result $c_1 = 6.2$ and $D = 4$.*

*Figure 8.5.1 plots the results of applying the proposed model with the param-
eter values as above.*

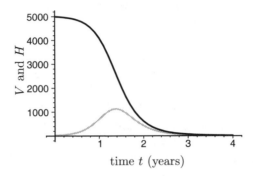

Figure 8.5.1: The crash of the prickly-pear cactus in just two years. The plants per
acre V (black) and the moth population H in egg-sticks per acre (grey).

*Clearly the destruction after two years is apparent, albeit with caution we
should accept the model as it does not incorporate many of the features of the
system under study.*

*Today, the cacti can still be found in certain small regions of southern Queens-
land, but they are well under control and no longer pose a threat to the
environment. As a token of their gratitude to the moths, affected communities constructed memorial halls, still in current use, in honour of Cactoblastis.*

8.6   Case Study: Geese defy mathematical convention

So far we have followed the convention that more available food or prey
means more animals or predators: the result of the top-down approach
of control theory. However, observations do not always support this
theory, and the following case study provides an argument and model for
a slightly different approach.

The model presented below is based on the predator–prey model, in this
case a plant-herbivore system, with a Michaelis–Menten type function
to allow for an increase in predator (herbivore) numbers to a maximum,
and then a reduction as the amount of prey (grass) is increased.

The case study is adapted from a model considered by van de Koppel et al. (1996).

Classical exploitation theory has led us to believe that where more grazing is available, more herbivores will be found inflicting more pressure on the standing crop and thus keeping the crop height low. A top-down control in productive environments. This approach is a hotly debated topic, although still widely accepted. van de Koppel et al. (1996) contest the standard plant-herbivore models, arguing that herbivores do not 'control' the plant growth entirely.

Their models, supported by empirical evidence, include the possible coexistence of dense vegetation and low herbivore grazing pressure, all in the absence of any predator.

Field data was collected from a salt marsh on the Island of Schiermonnikoog in The Netherlands. It provided an ideal location as the island supports all types of vegetation (sparse, intermediate, and dense) as well as a collection of herbivores (rabbits, hares and two species of geese). Also, there are no predators for these herbivores, other than the occasional raptor or human.

The observation of the maximal grazing pressure by rabbits, hares and geese at the intermediate level of the crop can be explained by several hypotheses. For example, in the denser regions, lack of light may impact on the stem development of the plants, robbing the leaves of the plant protein content and making them less attractive to the herbivores. Alternatively, because of the greater density, the herbivores in these regions may be under greater potential danger from the few predators on the island, and so may avoid them. Another theory is that the herbivores may have greater difficulty penetrating the thicker growth to forage. Whatever the reason, reduced foraging efficiency is observed in the dense regions of vegetation, and typically plant-herbivore models have not incorporated this aspect.

Let P and H be the plant and herbivore densities, respectively, with P measuring the plant mass per unit of area (g/m^2), and H the number of herbivores in the unit of area ($number/m^2$) calculated from the number of droppings found. The standard form of a plant-herbivore model is

$$\frac{dP}{dt} = G(P) - F(P)H, \qquad \frac{dH}{dt} = N(P)H,$$

where $G(P)$ is the plant growth function dependent on density, $F(P)$ is the per-capita consumption rate of a herbivore (the functional response), and $N(P)$ is the per-capita growth rate of herbivores (numerical response). Time t is measured in years. Typically $F(P)$ and $N(P)$ are monotonically increasing functions with increasing P. The proposed models now incorporate the reduced foraging efficiency.

Consider first the case of reduced digestion efficiency. This model assumes that the herbivore digestion is hampered by the change in stem structure of plants in regions of vegetation with greater density. To include this reduced foraging due to reduced digestion efficiency, the per-capita growth rate was taken to decline at high plant density. Let P_1^* and P_2^* be the two densities of vegetation for which herbivore growth is zero, so that below P_1^* there is too little vegetation to support a herbivore population, while above P_2^* there is too much vegetation for the herbivores to thrive and their density decreases.

The model proposed is

$$\frac{dP}{dt} = rP\left(1 - \frac{P}{K}\right) - c_{\max}\left(\frac{P}{a+P}\right)H,$$

$$\frac{dH}{dt} = \left(e_{\max}\left(\frac{P}{a+P}\right)e^{-bP} - d\right)H,$$

(1)

where clearly the growth rate for P is modelled as logistic. Parameter d is the herbivore mortality rate. The herbivore population growth rate (numerical response)

$$N(P) = \left(e_{\max}\left(\frac{P}{a+P}\right)e^{-bP} - d\right),$$

and the saturating functional response

$$F(P) = c_{\max}\left(\frac{P}{a+P}\right),$$

are illustrated in Figure 8.6.1 with parameter values $a = 10$, $b = 0.65$, $e_{\max} = 0.4$ and $d = 0.1$.

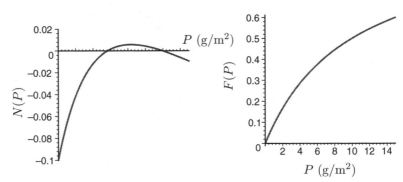

Figure 8.6.1: For model (1) the first figure illustrates the numerical response function (herbivore growth rate) and the second figure the functional response (herbivore consumption rate).

Note that the functional response is approximately $c_{\max}P/a$ when $a \gg P$, and is approximately $c_{\max}$ when $a \ll P$. Here it has been assumed that the *consumption rate of the herbivore is an increasing function of plant density but the negative effect of the plant density, when it increases beyond some upper bound, is translated to growth in the herbivores via the factor e^{-bP}.*

What is of interest is how this plant-herbivore system may behave with time. Does it settle to a situation of coexistence, or, are there conditions under which the herbivore population dies out? To establish this the phase-plane is examined.

Applying the above model, Figure 8.6.2 represents a typical phase-plane diagram with $K = 18$, $r = 1$, $c_{\max} = 1$, $a = 10$, $b = 0.065$, $e_{\max} = 0.4$ and $d = 0.1$.

Figure 8.6.2: Phase-plane diagram with $K = 18$, including the H and P nullclines and the equilibrium points at their intersection.

The nullclines are included and it is clear from the figure (see enlargements in Figure 8.6.3) that there are three equilibrium points away from the origin in this case, two stable and one unstable. In only one case is there the possibility of long-term coexistence.

Figure 8.6.3: The phase-plane of Figure 8.6.2, magnified about two of the equilibrium points.

Now, considering K (the measure of the plant carrying capacity) as a variable parameter $0 < K < 25$ with all other parameters as before ($r = 1$, $c_{max} = 1$, $a = 10$, $b = 0.065$, $e_{max} = 0.4$ and $d = 0.1$) several phase-planes are examined to see how this situation may change with changing maximum density of the plant crop. These are illustrated in Figure 8.6.4.

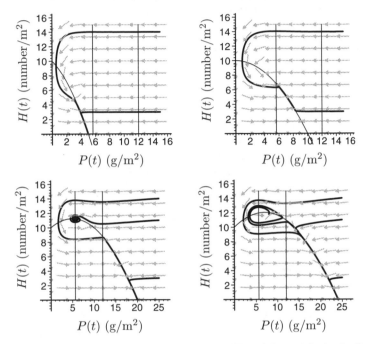

Figure 8.6.4: Phase-plane diagrams for varying K. From left to right in the first row, $K = 5$ and $K = 10$. In the second row, $K = 20$ and $K = 24$.

From Figure 8.6.4 it is clear that there is no herbivore population if $K < 5$ or $K > 25$. For $K = 10$ coexistence ensues from any initial densities. The more interesting situation occurs for $10 < K < 24$, where coexistence ensues from only a select number of initial conditions and the herbivores approach extinction otherwise, even with large initial densities.

A second scenario to consider is that of reduced consumption rate. In this case, the model can be further modified to include the increasing crop accompanied by a reduction in consumption rate. This is due to the possible situation where herbivores keep out of the regions of greater density for fear of predators. A further exponential is introduced into the term describing plant consumption as follows:

$$\frac{dP}{dt} = rP\left(1 - \frac{P}{K}\right) - c_{\max}\left(\frac{P}{a+P}\right)e^{-bP}H,$$

$$\frac{dH}{dt} = \left(e_{\max}\left(\frac{P}{a+P}\right)e^{-bP} - d\right)H. \tag{2}$$

This reduction in the consumption rate results in a feedback between plant growth of greater density and reduced grazing pressure, greater than in the earlier case of reduced digestion efficiency in system (1).

We can summarise the results from the models as follows. For K below P_1^* there is insufficient vegetation to sustain a herbivore population. For K between P_1^* and P_2^* the herbivores control the plant growth at a constant level (as is predicted in traditional plant-herbivore models where increases in plant mass result in increases of the herbivore population). For values of K above P_2^* there are two possible states in which the system can exist. In one there is intermediate plant density as well as herbivores, while in the other only dense vegetation exists. For certain initial conditions the system will tend to the coexistence stable state, while for others it will tend to the extinction of the herbivores. As K increases still further, the only possible stable state becomes that of herbivore extinction and dense vegetation: the vegetation 'escapes' herbivore 'control'.

One consequence of this perspective is that nutrient enrichment, such as fertiliser, is capable of triggering extra growth which may spark a relief from some herbivore control. Furthermore, there is the possibility of permanently destroying a plant-herbivore system; this may have its uses when certain herbivores are considered as pests and not desired as part of the equilibrium.

8.7 Case Study: Possums threaten New Zealand cows

In Chapters 5 and 6, we examined the spread of an influenza in a school, however, we did not consider any strategies to combat the spread. In the following case study we examine a simple disease model which describes the spread of bovine tuberculosis in possums, and the impact of three different strategies to reduce the spread. Note that, unlike epidemics in human communities, we are not concerned about the survival of the New Zealand possum population, in fact if it were wiped out that would be a bonus!

This model is adapted from the two articles: Wake and Roberts (1995) and Roberts (1992). The basic model is also discussed in the differential equations text of Borelli and Coleman (1996).

Sometime between 1830 and 1860 the brushtailed possum (Trichosurus vulpecula) was introduced into New Zealand from its native Australia. It is a cat-sized marsupial with large appealing eyes, a pink nose and a very warm fluffy brown coat. Back in the 1800s it was prized for this coat and thus it was introduced to New Zealand when the fur trade was booming. It thrived in its new environment, without predators of any sort, and while the fur trade died out the possum numbers grew to plague proportions. Its estimated population by 1995 was 70 million with only a few possum-free regions remaining.

Not only does this introduced species have an impact on the native vegetation, where it creates a conservation problem of huge dimensions (by eating approximately 4000 football fields of native forest each day!), it also threatens bird species by robbing their nests of eggs, and as a final straw it harbours and spreads bovine tuberculosis (Mycobacterium bovis). This threatens the primary industry of New Zealand, livestock, posing an enormous and extremely costly problem for New Zealand agriculture.

In New Zealand possums are thought to be the main source for bovine tuberculosis in cattle. While the actual mechanisms of transmission from possums to cattle are still being debated, one possibility is that they leave secretions containing the bovine tuberculosis bacteria on the pastures where cattle feed. This is a serious problem for the dairy industry where the disease may be passed through the milk and, without adequate pasteurisation, on to the human population. All animals are tested on a regular basis and any meat or milk from diseased animals cannot be sold. Animals testing positive are slaughtered and whole dairies suspended until they are clear of infection. The economic impact is enormous.

In order to curb the spread of the disease in cattle, it is necessary to examine the dynamics of its spread in the possum population, as well as the resulting changes in these dynamics when various strategies are implemented to reduce the threat. With such large numbers of animals, any feasible plan of action is costly and thus requires extremely careful consideration. We will consider a very simple model for the disease dynamics, and then include some strategies to reduce its spread: culling, sterilisation and vaccination. We are interested in examining the impact of these strategies on the population dynamics.

Consider the possum population to have density $N(t)$ per hectare. We will divide this into two mutually exclusive groups: those with the disease, the contagious infectives $C(t)$, and those who are without disease, the susceptibles $S(t)$. Then $N(t) = S(t) + C(t)$ at any time and subsequently $N' = S' + C'$.

We will assume a simple natural birth rate of β ($\beta > 0$) and a natural death rate of α ($\alpha > 0$). We will adopt α_d to describe the death rate attributable to the disease and b to describe the cross-infection rate, where again the parameters α_d and b are positive. These parameters are annual estimates and thus we will adopt a yearly time scale for the model. Furthermore, since the population of possums in New Zealand is so large and successful, we will assume that $\beta - \alpha > 0$.

Then, a system of differential equations describing the dynamics is

$$\frac{dS}{dt} = (\beta - \alpha)S + \beta C - bCS,$$

$$\frac{dC}{dt} = bCS - (\alpha_d + \alpha)C. \tag{1}$$

(Note that the symbols used here are different from those used by Roberts (1992), but are consistent with those used for disease models in this text.) It is easy to establish that there are at most two equilibrium points, the first (very attractive one for New Zealanders) at $(S, C) = (0, 0)$ and the other at

$$(S, C) = \left(\frac{\alpha_d + \alpha}{b}, \frac{(\beta - \alpha)(\alpha_d + \alpha)}{b\left(\alpha_d - (\beta - \alpha)\right)} \right).$$

We are interested in how the populations behave over time, and so we will employ linearisation analysis and examine the system dynamics through the phase-plane. From above, the non-zero equilibrium point becomes very large when α_d is close to, or approaches, $(\beta - \alpha)$. If this is a stable point, it is very bad news if we are hoping to eradicate the pest. We would like to alter this with the implementation of some strategy, but first we examine the dynamics of the system (1).

The Jacobian matrix is

$$J = \begin{bmatrix} \beta - \alpha - bC & \beta - bS \\ bC & bS - (\alpha_d + \alpha) \end{bmatrix}.$$

Thus for the $(0, 0)$ equilibrium point we have eigenvalues $\lambda_1 = (\beta - \alpha)$ and $\lambda_2 = -(\alpha_d + \alpha)$, and we can establish that this point is always an unstable saddle point. For the other equilibrium point the algebra is not quite as simple, however, the real parts of the eigenvalues determine the stability and they can be established from the sign of the trace of J:

$$\text{trace}(J) = \frac{-\beta(\beta - \alpha)}{\alpha_d - (\beta - \alpha)}.$$

Clearly this value is negative when $\alpha_d > \beta - \alpha$ and then the equilibrium is stable. Alternatively, when $\alpha_d < \beta - \alpha$ the point is unstable, but in that case the positive populations C and S increase without bound as can be deduced from the gradient of the trajectories in the phase-plane. (It is worth noting here that the measured values of these parameters pertaining to the New Zealand possums, as given by Roberts (1992), have $\alpha_d > \beta - \alpha$.)

Whether this point is a node or a focus can also be established from the Jacobian matrix, and by examining the discriminant of the characteristic polynomial it can be shown that the equilibrium point is a focus if

$$\frac{4(\alpha_d + \alpha)(\alpha_d + \alpha - \beta)}{\beta^2(\beta - \alpha)} > 1.$$

Alternatively it is a node.

So linear theory predicts that the equilibrium point at the origin is always unstable and a saddle point under our assumption that $\beta > \alpha$. That away from the origin is an unstable node (if $\alpha_d < \beta - \alpha$) with the trajectories approaching infinity, or a stable node/focus (if $\alpha_d > \beta - \alpha$) with the trajectories converging to this point. Adopting estimates for the parameter values in Roberts (1992) which relate to the possum population in New Zealand, we have taken $\beta = 0.305$, $\alpha = 0.105$, $\alpha_d = 3$ and $b = 0.7$. These have been incorporated into the time-dependent graphs and phase-plane diagram (generated with Maple) of Figure 8.7.1.

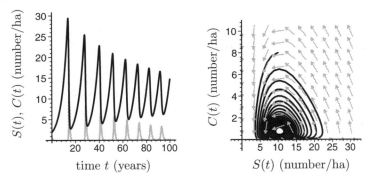

Figure 8.7.1: The time-dependent plot, (S (black line) and C (grey line) and the phase-plane diagram for the NZ possum populations.

Over time, the populations oscillate with decreasing amplitudes and approach a stable population with densities $S = 10.35$ and $C = 0.74$ per hectare, thus predicting a total population of ≈ 11 animals per hectare. This is in agreement with NZ possum population estimates from Roberts (1992).

We now include in the model a constant per-capita rate of culling per hectare, κ. The system of equations becomes

$$\frac{dS}{dt} = (\beta - \alpha - \kappa)S + \beta C - bCS,$$

$$\frac{dC}{dt} = bCS - (\alpha_d + \alpha + \kappa)C. \tag{2}$$

This is equivalent to increasing the value of parameter α so that we can use the results from above to predict the dynamics. However, note that now we may have the situation where $\beta < \alpha + \kappa$ so we need to include this in our interpretations. At this point it is informative to examine a graph of the nullclines to understand how the equilibrium points vary with parameter α. This is illustrated in Figure 8.7.2.

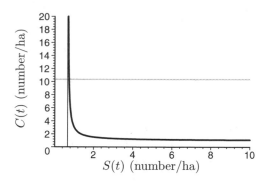

Figure 8.7.2: Graph of the nullclines (S-nullcline in black, C-nullcline in grey) to establish the impact of varying α. The vertical asymptote is at $(\beta - \alpha)/b$ (fine black).

As α increases, so the hyperbole (with its asymptote at $(\beta-\alpha)/b$) shifts to the left and the C-nullcline at $(\alpha-\alpha_d)/b$ increases. This means that at equilibrium the number of susceptibles decreases while the number of infectives increases. However, what is more relevant for the dynamics is that the stability of both equilibrium points changes when $\beta = \alpha + \kappa$ and we have extinction as the stable equilibrium. For the particular parameter combination given above we have extinction for a cull rate $\kappa > 20\%$.

Using Maple to plot the phase-plane with the parameters as before, but with $\kappa = 0.33$ included, we see that the dynamics match our theoretical predictions. The origin is now a stable equilibrium and the possum population rapidly approaches extinction as illustrated in Figure 8.7.3.

Instead of culling, we can consider the effect of sterilising a proportion of the population r_s. As a crude approximation, this can be considered to reduce the possum birth rate β to $\beta - r_s$. (See Roberts (1992) for an alternative model which involves introducing further dependent variables.) This case is similar to that of the culling strategy and the details are omitted.

The more interesting case (from a mathematical perspective) is the inclusion of a vaccination program. We will assume that a proportion of susceptible animals are vaccinated (γS), as obviously there is no gain in vaccinating infected possums. We will also assume that the vaccine will be permanent for that possum, so that once vaccinated, it cannot be infected during its life. (This is not necessarily true, with some vaccines requiring booster shots. In this

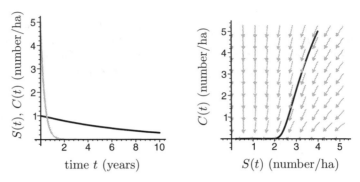

Figure 8.7.3: The time-dependent plot, S (black line) and C (grey line) and phase-plane diagram for the NZ possum populations with $\kappa = 0.33$.

case it is somewhat hypothetical to include vaccines as they are not currently available, however, we can still gain an understanding of the impact of such a strategy if implementation were possible.)

We introduce a new dependent variable Z to represent the population of vaccinated animals thus rendering three mutually exclusive groups: S the susceptibles, C the infectives and Z the vaccinated. The new model is

$$\frac{dS}{dt} = (\beta - \alpha)S + \beta C + \beta Z - bCS - \gamma S,$$
$$\frac{dC}{dt} = bCS - (\alpha_d + \alpha)C, \tag{3}$$
$$\frac{dZ}{dt} = \gamma S - \alpha Z.$$

We do not intend to carry out any further algebraic analysis on this system here. However, we aim to establish, through the use of Maple, some idea of how this vaccination program has affected the dynamics. Figure 8.7.4 illustrates the time-dependent graphs of the three populations, S, C and Z.

After a few initial oscillations the population settles to the equilibrium value of approximately 48 possums per hectare. Thus there is a substantial increase in the total population, as one might expect. Furthermore, the non-zero equilibrium point given as $(S, C, Z) = (10.35, 3.06, 32.53)$ (from Maple) is stable for the parameter combination of interest, and thus the infected population is in fact larger than that where there is no implemented strategy. Not a very attractive scenario. This shift is due to the initial response to fewer deaths from disease. The scenario is not very attractive from a conservation perspective either, with the destruction of native vegetation increasing substantially.

In Figure 8.7.5 we have presented a comparison of the three strategies on the possum infective population (C) over time, with the initial number of animals in each case given by the non-zero equilibrium of model (1).

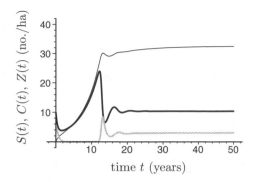

Figure 8.7.4: The possum population, S (black), C (grey) and Z (fine black), over time with the inclusion of a vaccination program with $\gamma = 0.33$. The initial condition is $(S(0), C(0), Z(0)) = (10, 5, 0)$.

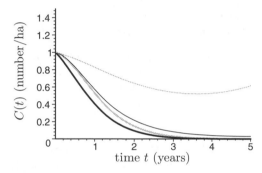

Figure 8.7.5: The infective possum population over time under a culling strategy (black), under a sterilisation program (grey), vaccinating a percentage of the population (fine black), and the infective population without any applied strategy (fine grey)

Clearly culling appears the most successful strategy in the short term. This conclusion is consistent with the results of more complex models examined by Roberts (1996), which include a density dependent contact rate.

We note that there are still some aspects of the possum and disease dynamics which are not included in these models, and which may have substantial impact. For example, the spatial distribution and movement of possums (see Louie et al. (1993) for an associated model), the rate at which cattle are infected, or the characteristics of a particular vaccine or sterilisation agent. Furthermore, the direct and indirect costs associated with the strategies were not considered, such as the costs in hard dollar terms, or the environmental impact of vaccination which does not reduce the population size. However, we do gain an understanding of the impact on a population of a variety of disease eradication strategies. For this particular case of New Zealand possums, we have some preliminary estimates for the size that these programs would need to be in order to hope for their success. From here, we would need to consider in greater detail the most cost effective strategy for New Zealand.

For more complex models see, for example, Roberts (1992) and Roberts (1996). In particular, Roberts (1996) includes a contact rate function to model the change in disease transmission as the population size changes.

8.8   *Exercises for Chapter 8*

1. Species diversity. *With reference to the case study, Competition, predation and diversity, of Section 8.2:*

(a) *Using* `Maple`, *generate the time-dependent graphs presented in the case study with the parameter values provided in the text.*

(b) *Show that for the simple competition model (when $N_3 = 0$) the condition $\alpha_{11}\alpha_{22} - \alpha_{12}\alpha_{21} \leq 0$ predicts extinction for all but one species in most cases.*

(c) *Find the equilibrium points for the system.*

2. Scaling. *Scaling can be used to simplify the mathematical analysis of a model by reducing the number of parameters, and writing the equation in a dimensionless form. Consider the logistic equation*

$$\frac{dX}{dt} = rX\left(1 - \frac{X}{K}\right), \qquad X(0) = x_0.$$

Since X and K are in the same units, we choose a new dimensionless variable Y as a measure of X in terms of the carrying capacity K, such that $X/K = Y$.

(a) *Show that, with this change of variable, the above logistic equation becomes $dY/dt = r(1 - Y)Y$, with $Y(0) = y_0 = x_0/K$.*

(b) *Furthermore, the independent variable can be scaled in units of $1/r$, using $t = s/r$. Recall, the chain rule gives*

$$\frac{dY}{ds} = \frac{dY}{dt}\frac{dt}{ds}.$$

Show, by using the chain rule, that with this scaling the logistic equation becomes

$$\frac{dY}{ds} = (1 - Y)Y, \qquad Y(0) = \lambda, \qquad \lambda = \frac{x_0}{K}.$$

Thus the model is reduced to a dimensionless form with only one parameter, λ.

3. **Scaling the extended predator–prey model.** *Consider the extended predator–prey model, the Holling–Tanner model developed in Section 8.3,*

$$\frac{dX}{dt} = r_1 X \left(1 - \frac{X}{K}\right) - \frac{c_1 XY}{1 + t_h c_1 X}, \qquad \frac{dY}{dt} = r_2 Y \left(1 - \frac{Y}{c_3 X}\right).$$

To simplify the analysis the equations can be scaled as follows: Define dependent and independent variables x, y and τ by

$$x = \frac{X}{K}, \qquad y = \frac{Y}{c_3 K}, \qquad \tau = r_2 t.$$

The chain rule can be used to change both the dependent and independent variables in the equations since we have that

$$\frac{dx}{dt} = \frac{dX}{dt}\frac{1}{K}, \qquad \frac{dx}{d\tau} = \frac{dx}{dt}\frac{dt}{d\tau}.$$

In order to find the derivative $dx/d\tau$, use the chain rule as above to get expressions for X' in terms of x and $dx/d\tau$. Then substitute these expressions into the original equation for X'. In this way, all terms involving X, X' and Y can be replaced with terms in x, x' and y. Similarly, the equation for $dy/d\tau$ can be obtained. Show that

$$\frac{dx}{d\tau} = \lambda_1 x (1 - x) - \frac{\lambda_2 xy}{(\lambda_3 + x)}, \qquad \frac{dy}{d\tau} = y \left(1 - \frac{y}{x}\right),$$

with

$$\lambda_1 = \frac{r_1}{r_2}, \qquad \lambda_2 = \frac{c_3}{t_h r_2}, \qquad \lambda_3 = \frac{1}{t_h c_1 K}.$$

4. **Extended predator–prey model.** *Use* `Maple` *to draw the phase-plane for the scaled system of Question 3 above, with parameter values $\lambda_1 = 5$, $\lambda_2 = 10$ and $\lambda_3 = 0.015$. Show that, for a variety of initial conditions, the system settles to a single periodic solution over time as illustrated in Figure 8.3.1.*

Now, by decreasing the ratio of searching time to handling time, estimate roughly for what value of λ_3 this periodic solution collapses to a stable/unstable node/focus. What changes, if any, do you observe in the nature of the periodic solution as the parameter is decreased? What does this mean for the time-dependent graphs of the populations?

Repeat this process for a variety of λ_2 values between 1 and 10. Then, using these results, plot the points (λ_3, λ_2), where the change in dynamics takes place, in the (λ_3, λ_2)-plane. Joining the points together provides the definition of a rough boundary in the parameter plane, which divides periodic behaviour from where all trajectories approach a stable point.

5. Lemmings. *Consider the case study on Lemming mass suicides of Section 8.4.*

(a) *The model presented claims to support a four yearly cycle. Using* `Maple` *generate the time-dependent plots over both short and long time periods, as well as a phase-plane. Do your results support the claim of the authors? Provide clear reasons for your answers.*

(b) *Find the equilibrium values for the system in the case with constant parameter values (not time dependent parameters).*

(c) *Linearise the system and classify these equilibrium points. Comment on whether your results support the claim made in (a) above.*

6. Prickly-pears. *Consider the plant-herbivore model presented to describe the spread of the prickly-pear in Australia, and its annihilation by an introduced moth, Cactoblastis: namely, Prickly pear meets its moth (Section 8.5).*

(a) *What growth processes (exponential or logistic) are adopted for the plant biomass and the moths in this model? What are the numerical response and the functional response terms for this system?*

(b) *The interaction term appearing in the rate equation for the plant biomass is given as*

$$-c_1 H \left(\frac{V}{V + D} \right).$$

Explain clearly why you think a function of this nature has been used, in terms of its changing contribution when the number of plants increases, or decreases, significantly.

(c) *In the case study, the claim is made that the system settles to a fixed and stable number of 11 plants per acre over a period of time. Using* `Maple`, *or otherwise, show that this is indeed the case for the suggested model with the parameter values given.*

7. Rabbits, hares and geese. *The case study of Section 8.6, Geese defy mathematical convention, presents improved plant-herbivore models appropriate to systems which do not exhibit the top-down control approach.*

(a) *Explain the suggested meaning of all parameters used in the models.*

(b) *Sketch graphs of the numerical response and the functional response terms for the general case, in each of the two models. When do the two functional response terms differ significantly and when are they similar? Explain how these differences and similarities reflect the different intentions of the two proposed models in the case study.*

(c) *Consider the reduced digestion efficiency model with parameter values as in the text, $K = 10$, $r = 1$, $c_{max} = 1$, $a = 10$, $b = 0.065$, $e_{max} = 0.04$ and $d = 0.1$. The case study claims that the model can be used to predict when and how, by changing a parameter, the herbivore could move towards extinction, or the plants die out. In the former case there is the possibility of eradicating a pest by encouraging plant growth beyond a threshold, or in the latter case a weed could be controlled by increasing herbivore numbers beyond a threshold.*

Use Maple *to generate the phase-plane, together with direction trajectories and system nullclines. Show, considering the phase-plane, how increasing or decreasing the herbivore carrying capacity can force a change in the dynamics so that the two thresholds discussed above can be crossed. Make rough estimates of the values of K for which this change occurs.*

(d) *Using the second model proposed, the reduced consumption rate model, with the same parameter values as above, do the threshold values estimated in (b) change? Give reasons for your answer and find the new thresholds if they are different.*

(e) *Identify a value of K for which there is a periodic solution in each of the models proposed, and choose appropriate initial conditions to illustrate them in a phase-plane, using* Maple.

8. Bovine tuberculosis. *A model for the bovine tuberculosis epidemic in NZ possums is presented in the case study, Possums threaten New Zealand cows, of Section 8.7:*

$$\frac{dS}{dt} = (\beta - \alpha)S + \beta C - bCS,$$

$$\frac{dC}{dt} = bCS - (\alpha_d + \alpha)C.$$

(a) *Show that the only two equilibrium points are at the origin and*

$$(S, C) = \left(\frac{\alpha_d + \alpha}{b}, \frac{(\beta - \alpha)(\alpha_d + \alpha)}{b(\alpha_d - (\beta - \alpha))} \right).$$

(b) *Construct the Jacobian matrix (J). Show that*

$$\text{trace}(J) = \frac{-\beta(\beta - \alpha)}{\alpha_d - (\beta - \alpha)},$$

and that the equilibrium point is a focus if

$$\frac{4(\alpha_d + \alpha)(\alpha_d + \alpha - \beta)}{\beta^2(\beta - \alpha)} > 1,$$

and alternatively that it is a node.

(c) *Establish the nullclines for the system and sketch a graph to indicate how these nullclines, and thus the equilibrium points, shift when culling is included in the model.*

(d) *Alternatively, if a sterilisation program is introduced how are the null-clines, and thus the equilibrium points, changed? Estimate the minimum number of sterilisations which ensures that the model predicts the possible extinction of possums.*

9. Bovine tuberculosis and a vaccination program. *Consider the equations for the possum bovine tuberculosis epidemic from Section 8.7, Possums threaten New Zealand cows, with vaccination included,*

$$\frac{dS}{dt} = (\beta - \alpha)S + \beta C + \beta Z - bCS - \gamma S,$$
$$\frac{dC}{dt} = bCS - (\alpha_d + \alpha)C,$$
$$\frac{dZ}{dt} = \gamma S - \alpha Z.$$

(a) *Show that there are only two equilibrium points: one at the origin and the other at*

$$(S, C, Z) = \left(\frac{\alpha_d + \alpha}{b}, \frac{(\alpha_d + \alpha)(\alpha + \gamma)(\beta - \alpha)}{\alpha b (\alpha_d + \alpha - \beta)}, \frac{\gamma(\alpha_d + \alpha)}{\alpha b} \right).$$

(b) *Construct the associated Jacobian matrix (at the non-zero equilibrium point) and show that it is given by*

$$J = \begin{bmatrix} \beta - \alpha - bC - \gamma & \beta - bS & \beta \\ bC & bS - \alpha_d - \alpha & 0 \\ \gamma & 0 & -\alpha \end{bmatrix}.$$

For known parameter values the trace and determinant can be calculated easily, but for general parameter values the algebra can get complicated and may yield little information.

(c) *With parameter values as in Section 8.7, draw time-dependent plots for the three sub-populations, using* `Maple`*.*

(d) *On the same system of axes, and parameter values, draw time-dependent plots for the infective population, comparing the three models (culling, sterilisation, vaccination) with the number of infectives in the population without any applied strategy.*

10. Density dependent contact rate. *As discussed at the end of Section 5.6, a better model for the spread of a disease in a population where the population density changes with time, is one which incorporates a density dependent contact rate function. One basic model, proposed for the spread of TB in possums, (Roberts (1996)) is*

$$\frac{dS}{dt} = (\beta - \alpha)S + \beta C - \frac{\lambda(N)}{N}CS,$$

$$\frac{dC}{dt} = \frac{\lambda(N)}{N}CS - (\alpha_d + \alpha)C,$$

(1)

where $N = S + C$ *and*

$$\lambda(N) = \frac{\lambda_m N}{K(1 - \epsilon) + \epsilon N}.$$

Here λ_m *is the maximum per-capita contact rate, and* ϵ *is a parameter describing the degree of social mixing within the population, with* $0 \leq \epsilon \leq 1$.

Using `Maple`, *modify the basic model (1), and investigate whether culling, or a vaccination program, is the better control strategy for this disease. (Ignore the ecological impact!) Use the parameter values from Section 8.7 and let* $\lambda_m = 5$, $\epsilon = 0.5$, $K = 10$.

Part III

Models of Heating and Cooling

Chapter 9

Formulating basic heat models

This chapter is concerned with modelling processes which involve heat conduction. We concentrate on formulating differential equations for the heat and temperature of systems, and emphasise the difference between them. In the absence of heat conduction we use the input–output principle, or the balance law introduced in Chapter 2, to determine the heat content of the object we are modelling. When conduction is involved, we also apply this input–output of heat energy, but to an arbitrary thin section. Maple *is used initially to establish the numerical solutions to the differential equations, whereas the analytic solutions are derived in the subsequent Chapters 10 and 11.*

9.1   Introduction

overview

We begin by introducing some problems which involve heating and cooling. It is of great importance to understand whether or not heat conduction is an important factor in any particular application, and to this end we consider a variety of problems, some of which require the consideration of heat conduction and others which do not.

Cooling of a cup of coffee

typical problem

Suppose we have a hot cup of coffee sitting on a bench top, too hot to drink. If the coffee is at a temperature of 60 °C, how long will it take to cool down to 40 °C? To begin to answer this, we introduce some basic physical concepts. In particular, we look at the distinction between temperature and heat, the two most important physical quantities in this problem.

temperature

First, consider *temperature*: this represents how hot the coffee is. If we can assume the temperature of the coffee is uniform throughout (we call this homogeneous) then the temperature will be a function of time alone. Temperature is measured in degrees Celsius °C, or in Kelvin K, where the temperature in °C is 273 less than the temperature in K, approximately. That is, °C + 273 = K.

heat

Second, consider the quantity *heat*. The temperature of the coffee drops because heat energy is transferred to the surrounding air. This is due to the fact that the surrounding air is at a lower temperature, and heat is always transferred from a region of high temperature to a region of lower temperature. Since heat is a form of energy it is measured in Joules, in units of the SI system.

compartmental diagram

Thus to answer the question about the cup of coffee we need to formulate a suitable equation for the temperature. But to do this we need to determine what is happening to the heat. So we construct a compartmental diagram for the heat input and output for the cup of coffee (see Figure 9.1.1).

Figure 9.1.1: Input–output diagram for the heat content in a cup of coffee.

formulating the equations

Equations which relate heat energy to temperature are derived in Section 9.2, using the concept of *specific heat*. We assume the temperature to be homogeneous throughout the compartment, and thus a function of time alone. We also need an equation which describes the rate of heat transfer to the surroundings. This too is introduced in Section 9.2 as is *Newton's law of cooling*.

A similar problem which we consider is that of determining the time taken for the body of water in a hot water tank to heat to a specified temperature. We again assume the temperature to be homogeneous, and thus a function of time. In this application we also need to account for the heat supplied to the system by the heating element.

water heater problem

Other heat conduction applications

Consider the problem of finding the rate at which heat can be transferred through a wall. This is important for determining the insulating properties of the wall, which are useful energy efficiency considerations.

wall insulation

The temperature on the outside of the wall will be different from that on the inside, which suggests we should also think of the temperature as a function of distance. There will be a temperature gradient in the wall and heat will flow in the direction of hot to cold (from the inside to the outside of a wall in winter) as a result of heat conduction.

temperature gradient

We also consider heat conduction through a hot water pipe insulating jacket. This is an interesting problem because we get counter-intuitive results. We see that, under certain circumstances, insulating a hot water pipe is less efficient (since more heat escapes) than not insulating it at all! Also, we consider the conduction effect of heat fins which are used to conduct heat away from a source, such as in computers or on a motorbike.

some other conduction problems

9.2   Some basic physical laws

Before formulating equations for the problems mentioned above, we need to understand some basic physical concepts of heat transport, as well as the distinction and relationship between heat and temperature.

overview

Heat and temperature

The result of applying heat, or heat energy, to an object is to raise its temperature. Similarly, as an object cools, its temperature drops as it loses heat to the surroundings. The greater the mass of the object, the more heat is required to raise its temperature, while if the temperature

heat and temperature

of a mass drops, then more heat energy is lost from a larger than a lesser mass, for an equivalent drop in temperature.

relating
heat to
tempera-
ture

When formulating a mathematical model for a problem involving heating or cooling, we need to take account of the amount of heat flowing into or out of a system. While it makes little sense to talk about a 'flow' of temperature, it is a change in temperature which we observe. Thus we need to be able to relate heat and temperature, and this can be done using the specific heat of a substance.

the word
equation

We assume the change in heat is directly proportional to the change in temperature and also the mass of the object. Thus we write

$$\left\{ \begin{array}{c} rate\ of \\ change\ of \\ heat\ content \end{array} \right\} = c \times \{mass\} \times \left\{ \begin{array}{c} rate\ of \\ change\ of \\ temperature \end{array} \right\} \qquad (1)$$

where c is a positive constant of proportionality, known as the specific heat of the material. (The definition of specific heat is, the amount of heat required to raise the temperature of 1 kg of a substance at a given temperature, by 1 °C.) In symbols, if we define Q as the rate of change of heat with time (measured in Watts), m as the mass of the material being heated or cooled (measured in kilograms), and U as the temperature[1], then

$$Q = cm\frac{dU}{dt}.$$

variation
in specific
heat

We have assumed c to be independent of the mass of the object and the temperature, but we know that metals absorb heat more easily than water. Thus a greater amount of heat is needed to raise the temperature of 1 kg of water by one degree than is needed to raise the temperature of 1 kg of metal by one degree. This implies that metals have a smaller specific heat than water. In Table 9.2.1, the specific heats of some common materials are given.

constant
specific
heat

The specific heat c is not actually constant over a large temperature range, however, if the temperature range is not too large use of (1), with constant specific heat c, provides reasonable predictions.

Newton's law of cooling

rate of
heat
loss/gain

A mechanism for heat to be lost from an object is that of exchanging heat energy with its surroundings. This takes place from the surface of

---

[1] U (not T) is used for temperature so that it is not confused with time t.

Table 9.2.1: Table of specific heats for some common substances. Units are standard SI units, i.e. $\mathrm{J\,kg^{-1}\,^\circ C^{-1}}$.

Substance	c	Substance	c
Aluminium	896	Asbestos	841
Copper	383	Brick	840
Stainless steel	461	Glass	800
Wood	2 385	Butter	2 300
Concrete	878	Lamb	3 430
Water (at 20 °C)	4 187	Potatoes	3 520

the object, and thus we would expect the rate of heat loss to increase with the exposed surface area of the object, and be directly proportional to this area. Further, if the difference in temperature between the surface of the object and the surroundings increases, then we would expect heat to be lost faster.

We therefore make the assumption that the rate of heat flow is directly proportional to the temperature difference between the surface and its immediate surroundings. Further, the existence of slowly moving air across the cooling object (a slight breeze), is assumed. Under these conditions we have Newton's law of cooling, which works equally well with heating problems. Newton's law of cooling states

Newton's law of cooling

$$\left\{ \begin{array}{c} rate\ heat \\ exchanged\ with \\ surroundings \end{array} \right\} = \pm hS\Delta U \qquad (2)$$

where ΔU is the temperature difference, S is the surface area from which heat is lost/gained (in units of $\mathrm{m^2}$), and h is a positive constant of proportionality. The constant h is called the *convective heat transfer coefficient*, or the *Newton cooling coefficient*. The units for h are Watts/$\mathrm{m^2}$/°C.

The correct sign for the temperature difference is determined in any specific problem by the direction of heat transfer, and whether the surface is gaining or losing heat. If S is the surface area of an object and if the temperature U of the object is greater than the temperature u_s of the surroundings, then we would write

correct sign

$$\left\{ \begin{array}{c} rate\ heat \\ exchanged\ with \\ surroundings \end{array} \right\} = hS\,(U - u_s) = hS\Delta U.$$

When the temperature of the surroundings is greater than the temperature of the object ($u_s > U$), then heat is exchanged in the opposite direction and so

$$\left\{ \begin{array}{c} rate\ heat \\ exchanged\ with \\ surroundings \end{array} \right\} = hS\left(u_s - U\right) = -hS\Delta U.$$

Returning to the cup of coffee

Applying the above expressions, we can now link the rate of change of heat in (1) directly to the rate of change of temperature, as in the following example.

Example 1: *Formulate a differential equation for the temperature of a cup of coffee over time.*

Solution: *To describe the conservation of heat energy we have, from the heat energy input–output diagram of Figure 9.1.1 in Section 9.1, that*

$$\left\{ \begin{array}{c} rate\ of \\ change\ of \\ heat\ content \end{array} \right\} = \left\{ \begin{array}{c} rate\ heat \\ lost\ to \\ surroundings \end{array} \right\}.$$

Using (2) and (1) we thus obtain the differential equation

$$cm\frac{dU}{dt} = -hS(U - u_s). \tag{3}$$

Note that a negative sign is chosen for the term on the RHS. The cup of coffee is cooling, and $(dU/dt) < 0$, and then since $\Delta U = U(t) - u_s$ is always positive we need to include a negative sign for consistency.

---

exponen-
tial decay
or growth

Once again we have an example of an equation describing exponential growth or decay, depending on whether the object is warm or cold compared with its environment. The heat loss from the hot cup of coffee left to stand follows an exponential decay process and, as we see in Section 10.1, it stabilises at the temperature of the surroundings u_s as expected.

wind-chill

We have assumed h to be a constant, however, this is a limitation of the model as the rate of heat lost from a surface will be affected by the air flow near it. This is sometimes called a *wind-chill* effect. Table 9.2.2 illustrates this by showing how the coefficient h increases with the velocity of the air-flow passing over a metal plate, and indicates that the effect is nontrivial.

Table 9.2.2: The value of the convective heat transfer coefficient h from a plate of length 0.5 m over which an airflow, given in metres per second (m/s), passes.

	h
Plate in still air	4.5
Air-flow at 2 m/s over plate	12
Air-flow at 35 m/s over plate	75

Summary of equations

To express a change of heat in terms of a change of temperature, we use the specific heat equation

$$\left\{ \begin{array}{c} rate\ of \\ change\ of \\ heat\ content \end{array} \right\} = cm\frac{dU}{dt}$$

where U is the temperature, m the mass of an object and c its specific heat. Newton's law of cooling is given by

$$\left\{ \begin{array}{c} rate\ heat \\ exchanged\ with \\ surroundings \end{array} \right\} = \pm hS\Delta U$$

where S is the surface area, ΔU the temperature difference and h a positive constant, namely the convective heat transfer coefficient or the Newton cooling coefficient.

Skills developed here:

- Distinguish between heat and temperature.
- Relate heat to temperature using the specific heat of a substance.
- Formulate a differential equation for the temperature of an object that is either heating or cooling.

9.3   Model for a hot water heater

overview   We derive a differential equation for the temperature of water being heated by an electric heating element. The problem applies principles similar to those in the example above, except that in this case heat is added to the system via the heating element. We write down a heat balance equation over a small time interval, and then let the time interval tend to zero to derive the differential equation. This process leads to a differential equation for the temperature as a function of time.

Problem description

description   Some domestic examples of water heaters include hot water systems, urns of problem   and kettles. Our problem specifically refers to an electrically heated hot water system, which is the standard for an average home. It usually contains 250 litres of water and is cylindrical with dimensions of height 1.444 m and diameter 0.564 m. Initially we assume the water to be at a temperature of 15 °C. The heating element when switched on supplies heat at a constant rate (per hour) of 3.6 kW. (The other commonly used elements supply 4.8 kW per hour.) In Figure 9.3.1 a schematic diagram illustrates the heat input and output for the system. We wish to determine how long it would take to heat the water to 60 °C.

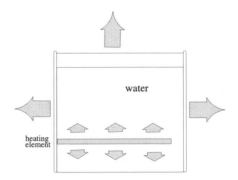

Figure 9.3.1: Schematic diagram of a hot water heater.

some   In order to study this problem we introduce some notation. We let $U(t)$
notation   denote the temperature of the water at time t. We denote by u_0 the initial temperature of the water, u_f as the final temperature, m as the mass of water in the heater and q as the rate of heat energy supplied. S in m^2 will denote the surface area from which heat may escape, which in the case of the hot water heater is the surface of the tank.

Other symbols will be introduced as required. Note that by assigning symbols to these quantities we increase the usefulness of our model, since we can later substitute various values for these symbols. We can thus use the model to explore the effect and sensitivity of change in these different physical quantities.

advantage of symbols

Model assumptions

We examine what simplifying assumptions can be made about the problem, and build the model using these.

some assumptions

- First, let us assume that the water in the tank is well stirred so that the temperature remains homogeneous throughout. Without this assumption the problem would be far more complex, since temperature would then be a function of both time and position and we would also have to take account of heat transfer (by conduction and convection) between different points in the tank. By assuming that the water is well stirred, we only have to consider the temperature of the water as a function of time.
- We also assume that heat is lost from the surfaces of the tank according to Newton's law of cooling.
- Finally, we assume that thermal constants, such as the specific heat and the Newton cooling coefficient, remain constant in our applications.

General compartmental model

We start formulating the model by writing down a word equation to describe the balance of heat energy in terms of rates of heat input and output. We must account for the heat produced by the heating element, and the heat lost from the surface of the water heater to the surroundings.

heat inputs and outputs

Example 1: *Give a suitable word equation describing the rate of change of heat.*

Solution: *We start by drawing an input–output diagram for the heat content of the system by applying the balance law. This is shown in Figure 9.3.2.*

Figure 9.3.2: Input–output diagram for the heat content in the domestic hot water system.

A heat balance for the system is described by the word equation

$$\left\{\begin{array}{c} \textit{rate of} \\ \textit{change} \\ \textit{of heat} \end{array}\right\} = \left\{\begin{array}{c} \textit{rate heat} \\ \textit{produced by} \\ \textit{heating element} \end{array}\right\} - \left\{\begin{array}{c} \textit{rate heat} \\ \textit{lost to} \\ \textit{surroundings} \end{array}\right\}. \tag{1}$$

Formulating the differential equation

rate of change of heat

As the temperature of the system rises, the net heat produced goes into raising the temperature of the system. To obtain a mathematical expression for the heat used to change the temperature, we use the fundamental equation relating heat to temperature through the specific heat from the first section of this chapter:

$$\left\{\begin{array}{c} \textit{rate of} \\ \textit{change of} \\ \textit{heat content} \end{array}\right\} = cm\frac{dU}{dt}$$

where c is the specific heat of the water, m is the mass of the water and $U(t)$ denotes its temperature at time t.

Example 2:   *Find expressions for the other terms in the word equation (1) and hence formulate a differential equation for the temperature.*

Solution: *It is assumed that the heating element produces heat at a constant rate per unit time, which we denote by q. Thus*

$$\left\{\begin{array}{c} \textit{rate heat} \\ \textit{produced by} \\ \textit{heating element} \end{array}\right\} = q. \tag{2}$$

For the rate of heat lost to the surroundings we use Newton's law of cooling, which states

$$\left\{\begin{array}{c} \textit{rate heat} \\ \textit{lost to} \\ \textit{surroundings} \end{array}\right\} = hS\left\{\begin{array}{c} \textit{temperature} \\ \textit{difference} \end{array}\right\} = hS\left(U(t) - u_s\right). \tag{3}$$

where S is the surface area of the heater and h is the Newton cooling coefficient. The temperature difference is the difference between the current water temperature $U(t)$ and the (constant) temperature of the surroundings u_s.

Now that we have suitable expressions for each of the quantities in equation (1) we can obtain a differential equation for the temperature. Substituting equations (2) and (3) into equation (1) we obtain the differential equation

$$cm\frac{dU}{dt} = q - hS\left(U(t) - u_s\right) \tag{4}$$

which describes the temperature variation of the water with time.

Numerical solution

Given an initial condition and suitable values for all the parameters, we can solve this differential equation numerically using `Maple`. (We also see how to obtain an analytic solution and make use of it in Section 10.2.)

example numerical solution

For a typical set of parameters we consider the standard 250 litre tank, of height 1.444 m and diameter 0.564 m. We take $m = 250$ kg, $c = 4200$ W kg^{-1} s^{-1}, $h = 12$ W m^{-2} °C^{-1}, $S = 3.06$ m^2, $q = 3600$ W and $u_s = 15$ °C (all in SI units). The plot of temperature against time is given in Figure 9.3.3.

typical parameter values

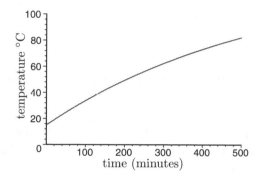

Figure 9.3.3: Plot of temperature against time (in minutes) using $m = 250$ kg, $c = 4200$ W kg^{-1} s^{-1}, $h = 12$ W m^{-2} °C^{-1}, $S = 3.06$ m^2, $q = 3600$ W and $u_s = 15$ °C. Note that the model no longer applies after the temperature reaches boiling point since it fails to account for the latent heat needed to boil the water.

In Figure 9.3.4 we repeat the solution with the heat input reduced considerably to only 500 W, and see that the temperature eventually approaches a constant value of approximately 30 °C. This constant value is an equilibrium solution to the differential equation. We investigate this further in Section 10.2.

an equilibrium solution

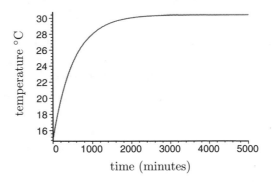

Figure 9.3.4: Plot of temperature against time (in minutes) using $m = 250$, $c = 4\,200$, $h = 12$, $S = 3.06$, $q = 500$ and $u_s = 15$ (all in SI units). The temperature approaches an equilibrium value.

9.4   Heat conduction and Fourier's law

overview   Before starting to solve any specific problems, we first lay some ground-work in the basic physics of heat conduction. We define *heat flux* to describe the flow of heat, and introduce *Fourier's law* which relates the heat flux to the temperature.

Heat and temperature

changing   In a given heating problem the temperature starts from some initial state
tempera-   and then, as a body is cooled or heated, the temperature changes. If
ture   the temperature is not the same at every point throughout the body
              then heat will also be conducted through the body from regions of high
              temperature to regions of lower temperature.

thermal   Eventually, the temperature will tend to some equilibrium state where
equilib-   there may still be a flow of heat through the body by conduction, but
rium   the temperature at any point within the body will not change with time.
              When the temperature is in this equilibrium state, none of the heat
              is used to increase the temperature, nor is any of the heat released to
              decrease the temperature. We can thus say, for *thermal equilibrium*

$$\left\{ \begin{array}{c} rate\ of \\ change\ of \\ heat\ content \end{array} \right\} = 0. \tag{1}$$

Heat conduction

Heat conduction can occur in a solid, liquid or gas, and involves the transfer of heat energy by the vibration of molecules. In this book we are concerned with heat *conduction*, however, there are other heat transport mechanisms such as *convection*, the transfer of heat in a liquid or gas by the fluid flow, and *radiation* which involves heat transfer by electromagnetic waves. *heat transport mechanisms*

Heat flux

Observation demonstrates that the rate at which heat is conducted through a body is directly proportional to the cross-sectional area through which the heat flows. It is thus reasonable to talk about a rate of heat flow per unit area through a cross-section. This quantity is called the heat flux. We formally define *heat flux $J(x)$* by *heat flux*

$$J(x) = \left\{ \begin{array}{c} \text{rate of} \\ \text{flow of heat} \\ \text{per unit time} \\ \text{per unit area} \end{array} \right\} \tag{2}$$

Heat flux is measured in SI units as Watts per square metre, where Watts are Joules per second.

Fourier's law of heat conduction

Common observation indicates that some substances conduct heat better than others. Heat flows more easily through certain metals than through substances such as brick and stone. *observations*

Fourier did some controlled experiments where he measured the heat conducted through thin plates of different materials. In his experiments he held the temperature constant, but at different values on either side of the plate. He noticed that the heat flux increased with the temperature difference, and decreased as the thickness of the plates increased. He thus came to the conclusion that the heat flux (rate of heat flow per unit area per unit time) is proportional to the temperature gradient, which is the gradient of the temperature expressed as a function of distance through the plate. *Fourier's heat conduction experiments*

We can express this result mathematically. If $J(x)$ denotes the heat flux at x, and $U(x)$ the temperature at x, then *Fourier's law* states *mathematical form of Fourier's law*

$$J(x) = -k\frac{dU(x)}{dx}. \tag{3}$$

The positive constant k is called the *conductivity*, which will be different for different materials.

why the minus sign?

The minus sign is necessary to ensure a positive rate for the conduction of heat. This means that when the temperature is decreasing (i.e. when the temperature gradient is negative) the heat flows in the positive direction. When the temperature is increasing (i.e. when the temperature gradient is positive) the heat flows in the negative direction. This reflects the fact that heat flows from regions of higher to regions of cooler temperatures, and not the other way around.

conduct-ivity

Table 9.4.1 gives values of the conductivity k for some different materials. Note that generally metals have higher conductivities than other materials, which means that heat flows more easily through them.

insulation

Indeed, it is sometimes desirable to use materials with low conductivities such as rock wool or polystyrene to provide good insulation, while at other times high conductivity is desired as in heating elements.

Table 9.4.1: Table of heat conductivities for some common materials (standard SI units).

Substance	k	Substance	k
Copper	386	Brick	0.38–0.52
Aluminium	204	Asbestos	0.113
Iron	73	Concrete	0.128
Stainless steel	14	Glass	0.81
Water (at 0 °C)	0.57	Wood	0.15
Lamb (at 5 °C)	0.42	Rock wool	0.04
Butter (at 5 °C)	0.20	Polystyrene	0.157

where Fourier's law breaks down

Over large temperature ranges the conductivity of a material k is not strictly constant. However, the assumption of constant conductivity over temperature ranges that do not vary dramatically still enables us to make sufficiently accurate predictions.

Summary

The important equations of heat transport are summarised as follows: For thermal equilibrium

$$\left\{ \begin{array}{c} rate\ of \\ change\ of \\ heat \end{array} \right\} = 0.$$

Fourier's law of heat conduction, where J is the heat flux and k the conductivity, is given by

$$J(x) = -k\frac{dU(x)}{dx}.$$

Skills developed here:

- Understand the physical meaning of thermal equilibrium, and relate it to a zero time-derivative for temperature.
- Describe the flow of heat in terms of heat flux.
- Understand the assumptions for, and application of, Fourier's law.

9.5 Heat conduction through a wall

We now formulate a differential equation for the simplest heat conduction problem. This involves the conduction of heat through a slab of some material in just one direction. We develop a differential equation for the equilibrium temperature reached as time increases.

Problem description

Imagine we have a plate with width L of some material. One side is hot and the other relatively cold, so that heat flows through the material in the x-direction towards the cooler side. A typical application is the wall of a building where the inside and outside are at different temperatures, as in Figure 9.5.1.

Figure 9.5.1: Coordinate system for the problem. We focus on the rate of heat going into and coming out of an imaginary section between the surfaces x and $x + \Delta x$.

General compartmental model

model approach

Our aim is to formulate a differential equation for the equilibrium temperature inside the material at any point x. We do this by first writing down a word equation describing the rate of heat input to and output from a small section of the material, assuming no heat is used to heat up the material. We then let the thickness of the section tend to zero.

a thin section as a compartment

We consider a thin section, or slice, of the material from x to $x + \Delta x$ as shown in Figure 9.5.1 as our compartment. This is a modified version of the approach we have taken before. The surface area of the face of this slice is A. We must account for all the inputs and outputs of heat for this section, which contribute to the rate of change of heat. We do this in the following example.

Example 1: *Obtain an input–output word equation for the arbitrary section of width Δx.*

Solution:

The only heat input is due to conduction into the section at x, and the only heat output is due to the heat conducted out of the section at $x + \Delta x$. The input–output diagram in Figure 9.5.2 illustrates this.

Figure 9.5.2: Input–output compartmental diagram describing the heat inputs and outputs to the arbitrary section of thickness Δx.

Hence, we have the word-equation

$$\left\{ \begin{array}{c} rate\ of \\ change\ of \\ heat\ in\ section \end{array} \right\} = \left\{ \begin{array}{c} rate\ heat \\ conducted \\ in\ at\ x \end{array} \right\} - \left\{ \begin{array}{c} rate\ heat \\ conducted \\ out\ at\ x + \Delta x \end{array} \right\}. \tag{1}$$

Formulating the differential equation

notation

First we introduce some notation. We let $U(x)$ denote the equilibrium temperature at the point x and let $J(x)$ be the flux of heat at x. Recall that this is the rate of flow of heat, per unit area per unit time, through the material at the point x.

heat input

The rate of heat entering the section is obtained by multiplying the heat flux $J(x)$ by the cross-sectional area (surface area of face of the section)

A. Thus

$$\left\{\begin{array}{c} \text{rate heat} \\ \text{conducted} \\ \text{in at } x \end{array}\right\} = J(x)A. \tag{2}$$

The rate of heat leaving the section is denoted by $J(x+\Delta x)$. Hence the rate of heat leaving the section, obtained by multiplying the flux by the area A, is

heat output

$$\left\{\begin{array}{c} \text{rate heat} \\ \text{conducted} \\ \text{out at } x + \Delta x \end{array}\right\} = J(x + \Delta x)A. \tag{3}$$

For equilibrium temperatures the LHS of the heat balance word equation (1) is zero. Substituting (2) and (3) back into (1), with the LHS set to zero, gives

putting it together

$$J(x)A - J(x + \Delta x)A = 0. \tag{4}$$

We are going to let $\Delta x \to 0$, but first we divide (4) by Δx. This leads to an expression involving the derivative with respect to x.

formulating the derivative

Example 2: *Obtain a differential equation for the temperature by taking the limit as $\Delta x \to 0$ in (4) and then applying Fourier's law, $J = -k\,dU/dx$.*

Solution: *We divide (4) by Δx and rearrange to obtain*

$$\frac{J(x + \Delta x) - J(x)}{\Delta x} = 0. \tag{5}$$

From the definition of the derivative[2] of J with respect to x, equation (5) becomes

$$\frac{dJ}{dx} = 0. \tag{6}$$

Since the heat flux through the material is due to conduction only we can apply Fourier's law to express the differential equation in terms of temperature. Substituting this into equation (6) gives

$$\frac{d}{dx}\left(-k\frac{dU}{dx}\right) = 0. \tag{7}$$

Since we assume the conductivity k to be a constant, (7) reduces to

differential equation

---

[2]From first principles the derivative of $J(x)$ with respect to x is

$$\frac{dJ}{dx} = \lim_{\Delta x \to 0} \frac{J(x + \Delta x) - J(x)}{\Delta x}.$$

$$\frac{d^2U}{dx^2} = 0. \tag{8}$$

This is a second-order differential equation for the equilibrium temperature $U(x)$. It is a trivial example of a differential equation but serves to illustrate how such equations arise. More complicated differential equations can occur when we include volumetric heat sources, as is the case when heat is produced by an electric current or chemical reaction.

Numerical solution

example numerical solution

Using `Maple` we can obtain a numerical solution, where we specify both the temperature and heat flux at $x = 0$ (see Figure 9.5.3). The `Maple` code to produce this is:

```
> restart: with(plots): with(DEtools):
> k := 1.0:
> de1 := diff(U(x),x$2) = 0;
> inits := U(0)=10, D(U)(0)=-1/k;
> plot1 := DEplot(de1, U, x=0..1, [[inits]] ):
> display(plot1);
```

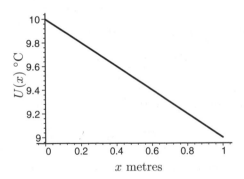

Figure 9.5.3: Numerical solution with $k = 1\,\mathrm{W\,m^{-1}\,{}^\circ C^{-1}}$, $U(0) = 10\,{}^\circ\mathrm{C}$ and $J(0) = 1\,\mathrm{W/m^2}$.

boundary value problems

However, more usually we would know either temperature or heat flux at both $x = 0$ and $x = l$. This changes the problem into a *boundary value problem* instead of an IVP. We consider such problems in greater detail later, and discuss methods for solving them in Chapter 11.

non-equilibrium temperatures

When the temperature is not at equilibrium, the temperatures are dependent on time as well as position. Furthermore, account must be taken of the amount of heat 'used up' in raising the temperature. This is examined in detail in Chapter 12 where we consider a problem similar to the present one, but derive a partial differential equation with two independent variables, x and t.

Modelling approach summary

The procedure adopted to derive models for heat conduction depends on a set of physical principles and is slightly different from what we have done before. A summary of the basic procedures for formulating differential equations describing heat transport problems is:

summary of basic procedure

- Consider all inputs and outputs of heat from a thin section of width Δx.
- Develop a word equation for the rate of change of heat.
- Divide by Δx and then take the limit as $\Delta x \to 0$ to derive a differential equation.
- Using Fourier's law, substitute to express heat flux J in terms of temperature U.

Skills developed here:

- Be able to reproduce the derivations of the equilibrium temperature equation.
- Reproduce the derivation of the second-order differential equation.
- Modify the word equation if there is an internal heat source, such as an electric current, providing heat at a given rate.

9.6 Radial heat conduction

Radial heat flow occurs in problems involving cylinders and spheres where the temperatures are the same at any given distance from the centre of the cylinder or sphere. We now formulate a differential equation for the equilibrium temperature in a cylinder. The procedure is similar to that used in the previous two sections but we must take account of the fact that the area through which heat is conducted now changes with the radial distance from the centre of the cylinder or sphere.

overview

Fourier's law for radial heat conduction

cylinders and spheres

Radial heat flow occurs in cylinders and spheres. In radial heat flow we need to account for the heat spreading out as it is transferred into or out of a cylindrical or spherical shell.

radial heat flux

If r denotes the radial distance of a point from the centre of a cylinder or sphere then we denote the heat flux at r by $J(r)$. The overall rate of flow of heat is

$$\left\{ \begin{array}{c} \text{rate of} \\ \text{flow of} \\ \text{heat} \end{array} \right\} = J(r)A(r)$$

where $A(r)$ denotes the area through which the heat flows. For a cylindrical surface of radius r and length l, $A(r) = 2\pi r l$, while for a spherical surface of radius r, $A(r) = 4\pi r^2$.

Fourier's law for radial heat conduction

For radial heat conduction Fourier's law takes the form

$$J(r) = -k\frac{dU(r)}{dr} \tag{1}$$

where $J(r)$ is the heat flux at r, $U(r)$ is the temperature at r and k is the conductivity. Thus Fourier's law states that the heat flux is directly proportional to the temperature gradient.

Model assumptions and approach

geometry of model

Let us consider a cylindrical object, which is hottest on the inside and coolest on the outer surface (see Figure 9.6.1). Define the outer radius of the cylinder at b and the inner radius at a, and then as a approaches zero the cylinder becomes solid, while for a close to b the cylinder models a hollow pipe. Take the length of the cylinder as ℓ, and let r denote the radial distance from its centreline.

some assumptions

We now make some assumptions on which to build the mathematical model.

- We assume that heat flows in the radial direction only; the temperature inside the cylinder will then depend on r and the time t only.
- If we also assume thermal equilibrium, then the equilibrium temperature will be a function of the radial distance r alone.

General compartmental model

compartmental diagram

Let us consider an imaginary small annular shell, inside the cylinder, from r to $r + \Delta r$. This is shown in Figure 9.6.1 and is the compartment.

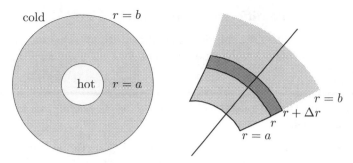

Figure 9.6.1: Radial heat flow in a cylinder. The first diagram illustrates the cross-section of a cylinder. Heat flows from the centre of the cylinder to the outside in all directions. The second diagram is of an arbitrary cylindrical shell, inside the cylindrical region, from r to $r + \Delta r$.

Then, applying the balance law, we consider any heat inputs into and outputs from this shell. Assuming no heat is generated inside the shell, heat input is through conduction only, as is the heat output, and thus we have the compartmental diagram Figure 9.6.2.

| heat conducted into shell → | heat content of Δr shell | heat conducted out of shell → |

Figure 9.6.2: Input–output diagram describing all the heat inputs and outputs for the arbitrary cylindrical shell.

From this we can formulate a word equation.   the word
                                                                               equation

Example 1: *Obtain a suitable word equation for a thin annular shell inside which no heat is generated.*

Solution:

Heat enters the shell at r and leaves at $r + \Delta r$ through the process of conduction. For this problem there is no heat generated internally in the shell. A word equation describing this balance of heat entering and leaving the cylindrical shell is thus

$$\left\{ \begin{array}{c} \text{rate of} \\ \text{change of} \\ \text{heat in shell} \end{array} \right\} = \left\{ \begin{array}{c} \text{rate heat} \\ \text{conducted} \\ \text{in at } r \end{array} \right\} - \left\{ \begin{array}{c} \text{rate heat} \\ \text{conducted} \\ \text{out at } r + \Delta r \end{array} \right\}. \qquad (2)$$

Formulating the differential equation

notation
We let $U(r)$ denote the equilibrium temperature at a distance r from the centreline. We also let $J(r)$ be the heat flux at a distance r from the origin.

Example 2: *Obtain a differential equation for the equilibrium temperature inside the annular shell.*

Solution: *The heat entering the cylindrical shell flows through an area* $A(r) = 2\pi r \ell$. *Similarly, the heat leaving the annular section flows through an area* $A(r + \Delta r) = 2\pi(r + \Delta r)\ell$. *In terms of the heat flux* $J(r)$ *we have*

$$\left\{\begin{array}{c} \text{rate heat} \\ \text{conducted} \\ \text{in at } r \end{array}\right\} = J(r)A(r),$$

$$\left\{\begin{array}{c} \text{rate heat} \\ \text{conducted} \\ \text{out at } r + \Delta r \end{array}\right\} = J(r + \Delta r)A(r + \Delta r). \tag{3}$$

For equilibrium temperatures the LHS of the word equation (2) is zero. Substituting (3) back into (2) gives

$$J(r)A(r) - J(r + \Delta r)A(r + \Delta r) = 0.$$

We can write this as

$$-\left[J(r + \Delta r)A(r + \Delta r) - J(r)A(r)\right] = 0. \tag{4}$$

Dividing by Δr *gives*

$$-\left[\frac{J(r + \Delta r)A(r + \Delta r) - J(r)A(r)}{\Delta r}\right] = 0 \tag{5}$$

and then taking the limit as $\Delta r \to 0$ *we have, from the definition of a derivative*[3],

$$-\frac{d}{dr}\left[J(r)A(r)\right] = 0. \tag{6}$$

Thus, regardless of the radius, this equation implies that the combination of the heat flux multiplied by the area is the same at any point inside the cylinder.

---

[3]Let $f(r) = J(r)A(r)$. Note that the derivative of f with respect to r is

$$\frac{df}{dr} = \lim_{\Delta r \to 0} \frac{f(r + \Delta r) - f(r)}{\Delta r} = \lim_{\Delta r \to 0} \frac{J(r + \Delta r)A(r + \Delta r) - J(r)A(r)}{\Delta r}.$$

Substituting for the area $A(r) = 2\pi r \ell$ and using Fourier's law, $(J = -k\,dU/dr)$ we obtain the second-order differential equation

$$-\frac{d}{dr}\left(-2k\pi\ell r \frac{dU}{dr}\right) = 0. \tag{7}$$

Since $2k\pi\ell$ is a constant, for the equilibrium temperature $U(r)$ this simplifies to

$$\frac{d}{dr}\left(r\frac{dU}{dr}\right) = 0. \tag{8}$$

Numerical solution

A numerical solution using `Maple` is given below (see Figure 9.6.3). For this solution, the temperature is specified at $r = 1\,\text{m}$ as $10\,°\text{C}$ and the heat flux at $r = 1\,\text{m}$ as $J(1) = 1\,\text{W m}^{-2}$. The `Maple` code used to produce the graph is:

example numerical solution

```
> restart: with(plots): with(DEtools):
> k :=1:
> de:= diff(r*diff(U(r),r),r) = 0;
> inits := U(1)=10, D(U)(1)=-1/k;
> plot1 := DEplot(de, U, r=1..5,[[inits]]):
> display(plot1);
```

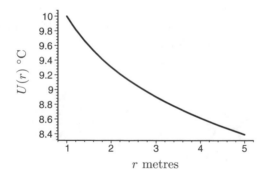

Figure 9.6.3: Numerical solution of cylindrical radial heat flow problem with $U(1) = 10\,°\text{C}$ and $J(1) = 1\,\text{W/m}^2$. The graph illustrates U as a function of r.

The solution indicates some exponential decay process of the temperature as the radius (r) increases. We return to the above differential equation in Section 11.2 and see how to derive its analytic solution. This provides us with the information to predict whether the diagram in Figure 9.6.3 represents typical behaviour, or whether it is simply one possible case.

Skills developed here:

- Derive a differential equation for the equilibrium temperatures in spherical heat flow.
- Derive the equilibrium temperatures for radial (cylindrical or spherical) heat flow with an internal generation of heat.

9.7   Heat fins

overview
We extend the approach adopted above to develop a mathematical model for a heat fin, which is a cooling device. Such a model describes the heat distribution along the fin, which will change with distance from the heat source.

Background

dissipating heat
Heat fins are used to enhance the dissipation of heat from machinery in which damage may be caused if heat is allowed to build up. Thus they are found on many types of machinery such as in motorcycle engines, refrigerators and computers. These fins work by increasing the surface area through which heat can be lost, and often have fans attached to them to force air across the fins, thus reducing the temperature more efficiently. There are many different forms which these fins can take, and some types are illustrated in Figure 9.7.1.

Figure 9.7.1: Some different types of heat fins.

a typical problem
A typical problem is to calculate the temperature distribution along a heat fin. From this, the rate of heat dissipation can be calculated, and the cooling efficiency for the heat fin determined. But first we need to formulate a differential equation for the changing temperature along the heat fin.

Model assumptions and approach

As our mathematical model we consider a rectangular heat fin of length ℓ, width w and thickness b (see Figure 9.7.2).

<small>flat rect-angular heat fin</small>

Figure 9.7.2: Schematic diagram for mathematical model.

We make some assumptions and build our mathematical model on these.

<small>some as-sumptions</small>

- We assume the thickness of the heat fin b to be small compared with its length ℓ.
- We assume that heat is lost from the surfaces of the heat fin to the surroundings (at temperature u_s) according to Newton's law of cooling.
- Because most of the heat loss occurs from the faces of the fin, both top and bottom, in our initial model we neglect the heat lost from the sides. Improving the model by including heat lost from the sides is not difficult and is explored in the exercises.
- A further assumption is to neglect the temperature variation over the cross-section of the fin. This assumption is reasonable provided the heat fin is thin (that is, $b \ll \ell$). In doing this, we can effectively assume the equilibrium temperature to be a function of x alone.

General compartmental model

Let us consider a small section of the heat fin from x to $x + \Delta x$, as shown in Figure 9.7.2. We first consider all the heat inputs into and outputs from this section, and later we let Δx tend to zero to establish the differential equation.

<small>applying the balance law</small>

Example 1: *Determine an input–output word equation for an arbitrary section of the heat fin.*

Solution: The thin section of the heat fin is shown in Figure 9.7.2 and we use the balance law to establish the inputs and outputs in Figure 9.7.3. Heat is conducted into the section at x and out from the section at $x + \Delta x$. We should also account for the heat lost from the top and bottom surfaces to the surroundings and for this we use Newton's law of cooling. Thus there is a single means by which heat enters the section and two mechanisms by which it leaves. (Recall that we have neglected any heat loss from the sides.)

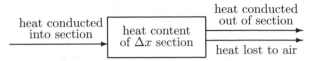

Figure 9.7.3: Input–output diagram for an arbitrary thin section of the heat fin.

The basic heat balance equation, for the section, is then

$$\left\{\begin{array}{c} \text{rate of} \\ \text{change of} \\ \text{heat in section} \end{array}\right\} = \left\{\begin{array}{c} \text{rate heat} \\ \text{conducted} \\ \text{in at } x \end{array}\right\} - \left\{\begin{array}{c} \text{rate heat} \\ \text{conducted} \\ \text{out at } x + \Delta x \end{array}\right\} - \left\{\begin{array}{c} \text{rate heat} \\ \text{lost to} \\ \text{surroundings} \end{array}\right\}. \tag{1}$$

For the equilibrium temperature the rate of change of heat is zero and so

$$\left\{\begin{array}{c} \text{rate heat} \\ \text{conducted} \\ \text{in at } x \end{array}\right\} - \left\{\begin{array}{c} \text{rate heat} \\ \text{conducted} \\ \text{out at } x + \Delta x \end{array}\right\} - \left\{\begin{array}{c} \text{rate of heat} \\ \text{lost to} \\ \text{surroundings} \end{array}\right\} = 0. \tag{2}$$

Formulating the differential equation

using the heat flux

The next step is to assign appropriate symbols to each of the terms in the word equation (2). We express the heat conduction quantities in terms of the heat flux J and later use Fourier's law of heat conduction to express these in terms of temperature. Once again we develop the model with symbols, so that we can examine the effect of changing some of these physical quantities, easily.

Example 2: *Express each of the quantities of (2) in terms of the heat flux J or the temperature U.*

Solution: *The area through which heat is being conducted is bw. In terms of the heat flux $J(x)$,*

$$\left\{\begin{array}{c} \text{rate heat} \\ \text{conducted} \\ \text{in at } x \end{array}\right\} = J(x)bw, \qquad \left\{\begin{array}{c} \text{rate heat} \\ \text{conducted} \\ \text{out at } x + \Delta x \end{array}\right\} = J(x + \Delta x)bw. \tag{3}$$

We use Newton's law of cooling for the heat loss from the surface of the section in the time interval. Recall that Newton's law of cooling states that the rate of heat loss is proportional to the surface area and the temperature difference between the object and its surroundings. The surface area of the section through which heat is lost to the surroundings is $2w\,\Delta x$. This corresponds to the top and bottom faces only, as we are neglecting heat lost from the sides.

To calculate the approximate heat loss by Newton's law of cooling we could use the temperature at x to give $hS(U(x) - u_s)$. (Recall that S is the surface area, u_s is the temperature of the surroundings and h is the Newton cooling coefficient.) But this would give a slight overestimate of the temperature since $U(x)$ decreases with x. Similarly, if we used the temperature at $x + \Delta x$ we would obtain a slight underestimate. The true value would use the temperature somewhere in between, say at x^*, where $x < x^* < x + \Delta x$. Thus we write

$$\left\{ \begin{array}{c} \text{rate heat} \\ \text{lost to} \\ \text{surroundings} \end{array} \right\} = 2hw\,\Delta x\,[U(x^*) - u_s]\,. \qquad (4)$$

(Although we do not know the exact value of x^*, it does not matter as eventually we let Δx tend to zero and then $x* \to x$.)

We now substitute the individual terms (3) and (4) into the heat balance equation (2) to obtain

$$J(x)bw - J(x + \Delta x)bw - 2hw\Delta x\,[U(x + \lambda\Delta x) - u_s] = 0, \qquad (5)$$

where $x^* = x + \lambda\Delta x$ and $0 \le \lambda \le 1$.

Taking the limit as $\Delta x \to 0$ to get the derivative, and then applying Fourier's law to give the equation in terms of temperature, we establish the differential equation required.

formulating the differential equation

Example 3: Let $\Delta x \to 0$ in equation (5) and obtain a differential equation for the temperature.

Solution: Dividing (5) by $bw\Delta x$ we are left with

$$-\left(\frac{J(x + \Delta x) - J(x)}{\Delta x} \right) - \frac{2h}{b}\,(U(x + \lambda\Delta x) - u_s) = 0. \qquad (6)$$

Then, taking the limit as $\Delta x \to 0$ and applying the definition of the derivative[4],

$$-\frac{dJ}{dx} - \frac{2h}{b}\,(U(x) - u_s) = 0. \qquad (7)$$

---

[4]By first principles, the derivative of $J(x)$ with respect to x is

$$\frac{dJ}{dx} = \lim_{\Delta x \to 0} \frac{J(x + \Delta x) - J(x)}{\Delta x}.$$

From (7), substituting for J and applying Fourier's law $(J(x) = -k\,dU/dx)$, we find

$$k\frac{d^2U}{dx^2} - \frac{2h}{b}\left(U(x) - u_s\right) = 0. \tag{8}$$

governing
equation This equation can be written in the compact form

$$\frac{d^2U}{dx^2} = \beta(U - u_s) \qquad \text{where} \quad \beta = \frac{2h}{kb} \tag{9}$$

establishing a second-order differential equation describing the heat flux in terms of temperature.

Numerical solution

example
numerical
solution Given the temperature and the heat flux at $x = 0$ we can easily find the temperature along the heat fin numerically, using Maple. Let us assume a given heat flux q_0 and a given temperature u_0 at $x = 0$. The code for solving this second-order equation is provided below. The graph it produces is given in Figure 9.7.4.

```
> restart: with(plots): with(DEtools):
> k := 386: h := 10: b := 0.001: us := 0: q0 := 4*10^4:
> de := diff(U(x),x$2) -2*h/(k*b)*(U(x)-us)=0;
> inits := U(0)=20, D(U)(0)=-q0/k;
> DEplot(de, U, x=0..0.1, [[inits]]);
```

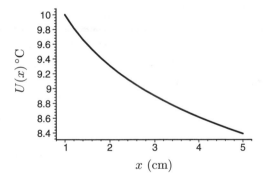

Figure 9.7.4: Numerical solution for a heat fin with conditions of known temperature $U(0) = 20\,°\text{C}$ and known heat flux $J(0) = 1\,\text{W/m}^{-2}$.

discussion
of results The temperature decreases along the heat fin as heat is lost to the surrounding air. In practice, we will not know the value of the temperature at $x = 0$ close to the heat source and instead we should impose some

condition on the end of the heat fin away from this source. It could be either a Newton cooling condition, or more simply, that no heat is lost from the end. We see in Section 11.5 how to find a solution analytically using this type of condition applied at the end of the fin, $x = \ell$.

Extensions and further problems

Heat fins come in many shapes and sizes. The derivation of the differential equation for heat fins with different cross-sections, such as circular, is almost identical to the above. In fact we end up with the same equation (9) but with a different value for β. *shapes and cross-sections differ*

Obtaining a differential equation for when the cross-section varies with distance from the origin, is more difficult but still follows the same procedure. Together with the previous section which deals with radial heat conduction, you have the required theory to deal with such problems. *cross-sections differ with length*

Other possible extensions include the incorporation of a heat production term. Another case is that where the temperature of the surroundings may be higher than the temperature of the fin, so that the fin gains heat rather than loses it. *other extensions*

Skills developed here:
- Derive the differential equation for a solid cylindrical heat fin (using ideas from this section and Section 9.6).
- Formulate differential equations for heat fins of rectangular, or other, cross-sections in the case of an equilibrium temperature.
- Formulate a differential equation for one-dimensional heat flow with a non-constant internal heat source.

9.8   Exercises for Chapter 9

1. **Simple heat calculations.** *The following questions require the use of appropriate equations from Chapter 9.*

(a) *If the heat flux through a wall is 30 (in SI units), how much heat is lost in 1 hour?*

(b) *How much heat does it take to heat a copper sphere of radius 1 cm from 5 °C to 25 °C?*

(c) *If the outside of a wall of a house loses heat to the surrounding at a rate of 20 W, and the temperature of the surroundings is 5 °C, what is the outside temperature of the wall. You may assume Newton's law of cooling.*

2. **Including an internal heat source.** *Heat flows inside a slab of material from left to right. Inside the material, an electric current generates heat at a constant rate Q_0 Watts m^{-3}.*

(a) *Write down a suitable word equation for the rate of change of heat in a section x to $x + \Delta x$ of the slab, where x is in the direction of heat flow.*

(b) *Hence deduce that the equilibrium temperature satisfies the differential equation of the form*

$$\frac{d^2 U}{dx^2} + \beta = 0,$$

where β is a constant you must determine.

3. **Formulating model extensions.** *Consider the formulation of the ordinary differential equation in Section 9.5 for the equilibrium temperature.*

(a) *Suppose, instead, heat flows in the opposite direction. Does the differential equation (8) remain the same? Justify your answer.*

(b) *Suppose, instead, the conductivity is not a constant but depends on temperature. Does the differential equation (8) remain the same? Justify your answer.*

4. Model formulation with circular cross-section. *A heat fin has circular cross-section of radius a. The heat fin loses heat to the surrounding air, which is at temperature u_a. Starting from a suitable word equation, show that the equilibrium temperature satisfies the differential equation of the form*

$$\frac{d^2U}{dx^2} - \alpha U = -\beta,$$

where α and β are constants you must determine.

5. Nuclear radiation. *In a shielding wall for a nuclear reactor, gamma-ray radiation causes the wall to heat up internally at a given rate $q(x)$ per unit time per unit volume, within the wall. Formulate the differential equation for the equilibrium temperature.*

6. Including an internal heat source. *Heat flows radially in a solid cylinder of length ℓ and radius a. Heat is also generated internally by an electric current at a constant rate Q_0 Watts per unit volume. The temperature on the inside is maintained at temperature u_a and heat is lost from the outer surface to the surroundings at temperature u_s, according to Newton's law of cooling.*

  (a) *Write down a word equation for the change in heat, for a cylindrical shell r to $r + \Delta r$ and over a time interval t to $t + \Delta t$.*

  (b) *What is the volume of the shell in (a)? Hence give an expression for the rate of heat generated by the electric current.*

  (c) *Obtain a differential equation for the equilibrium temperature. (Hint: not all of the information in the question need be used.)*

7. Equilibrium temperature inside a sphere. *Heat flows radially, by conduction, through a spherical shell of radius a. Formulate the differential equation for the equilibrium temperature inside the sphere.*

Figure 9.8.1: A cylindrical heat fin (see Question 8).

9 Formulating basic heat models

8.   *Cylindrical heat fin.* In the thin cylindrical heat fin (shown in Figure 9.8.1) of thickness ℓ, inner radius a and outer radius b, heat flows radially. Also, heat is lost from the surface of the fin to the atmosphere, at temperature u_s, according to Newton's law of cooling.

Deduce the differential equation for the equilibrium temperature assuming the temperature variation over the cross-section of the heat fin is negligible.

Chapter 10

Solving time dependent heat problems

In this chapter, we revisit the time dependent heating and cooling problems introduced in Chapter 9. These resulted in ODEs which we now solve using standard analytic methods. Given such analytic solutions, we can answer some general questions about the problems. We also consider some extensions of time dependent heat transport, without conduction, in two case studies: a model describing house temperatures, and a model for predicting the spontaneous combustion of fish and chip crumble.

10.1   *The cooling coffee problem revisited*

overview   In this section, we revisit the problem of determining the time it takes
for a hot cup of coffee to cool. The exponential growth/decay differential equation which arises can be solved analytically, using different
techniques, and this solution provides a far better general understanding
of the cooling process, than the particular numerical solution generated
in Chapter 9.

Review of the model

governing   In Section 9.1, we first introduced the problem of determining the tem-
equation   perature of a cup of coffee. In Section 9.2, we formulated the general
differential equation for the temperature $U(t)$ as a function of time, which
was

$$\frac{dU}{dt} = \frac{hS}{cm}(U - u_s),$$ (1)

where c is the specific heat of the coffee, m is the mass of the coffee, h is
the Newton cooling coefficient, S is the surface area through which heat
is lost and u_s is the constant temperature of the surroundings.

Analytic solution

solving the   We can write the differential equation as
problem
analyt-
ically
$$\frac{dU}{dt} = -\lambda(U - u_s),$$ (2)

where λ is a constant of proportionality and equivalent to hS/cm. The
solution to this differential equation can be found using the separable
technique, as in the following example.

Example 1:   *Solve the differential equation (2) subject to the initial condition*
$U(0) = u_0$.

Solution: *Separating the variables,*

$$\frac{1}{U - u_s}\frac{dU}{dt} = -\lambda.$$

Integrating both sides with respect to t, assuming $U > u_s$, gives

$$\ell n(U - u_s) = -\lambda t + K$$
$$U = e^{-\lambda t + K} + u_s.$$

Applying the initial condition $U(0) = u_0$ gives $e^K = u_0 - u_s$ and so

$$U(t) = (u_0 - u_s)e^{-\lambda t} + u_s. \tag{3}$$

Applying the coffee cooling model

We return to our original problem (see Section 9.1) to find how long it takes a cup of coffee to cool to $40\,°C$, and find the answer using the analytic solution we have just derived. A hot cup of coffee is at a temperature of $60\,°C$ and after 10 minutes the coffee has cooled to $50\,°C$. The room temperature is $20\,°C$. The following example shows how to find this time.

solution with Newton's law

Example 2: *Find how long it will take for the cup of coffee to cool to $40\,°C$.*

Solution: *We can find λ from the given information that $U = 50$ at time $t = 10$. Substituting this into the solution (3) we obtain*

$$50 = (60 - 20)e^{-10\lambda} + 20, \qquad and \quad \lambda \approx 0.0288.$$

Now, if U is $20\,°C$ then

$$40 = (60 - 20)e^{-0.029t} + 20 \qquad and \quad t \approx 24\,minutes.$$

Thus the coffee will reach the temperature of $40\,°C$ about 24 minutes after it was put into the room.

An alternative cooling law

The differential equation (2) applied Newton's law of cooling which assumes that air at room temperature is moving over and around the object. This is not necessarily the case in a room, or on a still day. A better model for cooling in still air is one in which the rate of the temperature increase/decrease is directly proportional to the (5/4)th power of the difference in temperature between the object and its surroundings. This is the law of *natural cooling* and is given by

cooling in still air

$$\left\{ \begin{array}{c} \text{rate of} \\ \text{change of} \\ \text{heat} \end{array} \right\} = \pm h_1 S(U - u_s)^{5/4}, \tag{4}$$

where the sign $\pm$ depends on whether the object is cooling or heating, h_1 is a positive constant of proportionality and S is the surface area. Note that $h_1 \neq h$, the Newton cooling coefficient in (1).

solution with natural cooling

For the problem of cooling a cup of coffee, the differential equation for the temperature is

$$\frac{dU}{dt} = -\lambda(U - u_s)^{5/4}, \qquad \lambda = \frac{h_1 S}{cm}. \tag{5}$$

(Recall that m is the mass of the coffee and c its specific heat.) This differential equation is also separable and can be solved using the same technique as before. (See Exercises Question 1.) Using its solution we can show that the model predicts it will take approximately 26 minutes for the coffee to cool from $60\,°\text{C}$ to $40\,°\text{C}$ in a room of temperature $20\,°\text{C}$, given that it takes 10.3 minutes to cool to $50\,°\text{C}$. This estimate for the time is slightly longer than that obtained using Newton's law of cooling. A comparison between these two cooling models is illustrated in Figure 10.1.1. The difference is not large, for this problem.

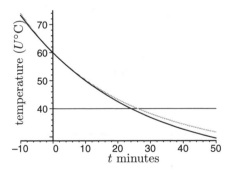

Figure 10.1.1: A comparison between Newton cooling (black line) and natural cooling (grey line). For comparison purposes, it has been assumed that the coffee takes 10.3 minutes to cool to $50\,°\text{C}$ for natural cooling.

interpretation of results

The physical meaning of this is that the natural cooling law gives a slower rate of cooling than Newton's law. While the factor $(U - u_s)^{5/4}$ is larger than $(U - u_s)$, the associated values for λ (0.0115 and 0.0288 respectively) effectively render the rate for natural cooling less than that for Newton cooling. This makes sense because we expect cooling in an environment with a slight breeze (Newton's law of cooling) to be more effective than cooling in perfectly still air (natural law of cooling).

Summary of equations

Newton's law of cooling (cooling in a breeze) is

$$\frac{dU}{dt} = \pm \frac{hS}{cm}(U - u_s).$$

The natural law of cooling (applicable when there is no breeze) is

$$\frac{dU}{dt} = \pm \frac{h_1 S}{cm}(U - u_s)^{5/4}.$$

Skills developed here:
- Solving a differential equation with a cooling law (e.g. Newton's law of cooling).
- Solving for a constant of proportionality and a constant of integration using a set of known results and an initial condition.
- Using a solution for the temperature to calculate the time of cooling (or time of heating).

10.2   The hot water heater problem revisited

We now return to the water heater problem introduced in Section 9.3. overview
Again, our aim is to determine and interpret the analytic solution to
indicate how such a solution provides a better general understanding of
the system, compared with any particular numerical solution.

Review of the model

brief
problem
description
Suppose we have a domestic hot water heater. We wish to determine a formula for how long it takes to heat the water to, say 60 °C, and what physical quantities this depends on. One interesting question that we can pose is:

> Does doubling the power output of the heating element halve the heating time?

governing
equation
We formulated the differential equation for the water temperature $U(t)$ in Section 9.3 as

$$cm\frac{dU}{dt} = q - hS\left(U - u_s\right), \qquad (1)$$

which describes the temperature variation of the water with time. Here q is the rate of heat supplied, m the mass of water, S the surface area through which heat is lost, c the specific heat of the water, h the Newton cooling coefficient, u_s the temperature of the surroundings and u_0 the initial temperature of the water. The final temperature will be denoted by u_f, and here $u_f = 60\,°C$.

Analytic solution

simpler
form
To solve the differential equation we first write it in the form

$$\frac{dU}{dt} = \beta - \alpha U, \qquad U(0) = u_0, \qquad (2)$$

where the constants α and β are defined as

$$\alpha = \frac{hS}{cm}, \qquad \beta = \frac{q + hSu_s}{cm}.$$

In this form, it is easier to see that the differential equation is a first-order linear equation.

Example 1: *Solve the differential equation (2) using the integrating factor technique.*

Solution: Since the differential equation is first-order linear, we can find an integrating factor,

$$R(t) = e^{\int_0^t \alpha\,dt} = e^{\alpha t}.$$

Hence, we can write the differential equation as

$$\frac{d}{dt}\left(Ue^{\alpha t}\right) = \beta e^{\alpha t}.$$

Integrating both sides with respect to t

$$e^{\alpha t}U = \frac{\beta}{\alpha}e^{\alpha t} + K,$$

$$\text{and} \quad U = \frac{\beta}{\alpha} + Ke^{-\alpha t},$$

where K is the constant of integration. Now applying the initial condition $U(0) = u_0$ gives $u_0 = \beta/\alpha + K$ and so $K = u_0 - \beta/\alpha$. Hence the solution is

$$U = u_0 e^{-\alpha t} + \frac{\beta}{\alpha}\left(1 - e^{-\alpha t}\right).$$

Substituting for α and β this gives

$$U(t) = \left(u_0 - u_s - \frac{q}{hS}\right)e^{-(hS/cm)t} + u_s + \frac{q}{hS}. \quad (3)$$

Note that this equation is also separable and could have been solved using that technique. (See Question 3 of the Exercises.)

Applying the hot water heater model

We apply the above theory to a typical hot water cylinder used in an average Australian home (of four individuals) to provide hot water. Then, by considering that energy used to heat the system has a cost, which most of us would like to minimise, we examine how best to operate the system. *using the solution*

We suppose that we have a standard home hot water tank, which holds 250 litres of water, and is cylindrical with a height of 1.444 m and a diameter of 0.564 m. The water is heated by a heating element immersed in the water which supplies heat at a constant rate of 3600 Watts ($q = 3600\,W = 3600\,J\,sec^{-1}$). We assume the mass of the water to be 250 kg (as a litre of water has the mass of approximately 1 kg) with the surface area of the tank approximately 3.06 m². We know the specific heat of water is $c = 4200\,J\,kg^{-1}\,°C^{-1}$ and the heat transfer coefficient is $h = 12\,W\,m^{-2}\,°C^{-1}$. Given that the temperature of the surroundings is $u_s = 15\,°C$, and that initially the water in the tank is the same temperature as its environment ($u_s = u_0$), we are interested in how long it would take the water to reach a temperature of 60 °C. *typical parameter values*

Example 2: *Find the time it would take to heat the tank of water to 60 °C.*

Solution: *Using solution equation (3) we need to solve for t. Again, we let*

$$\alpha = \frac{hS}{cm}, \quad \beta = \frac{q + hSu_s}{cm},$$

and then the solution is

$$U = \left(u_0 - \frac{\beta}{\alpha}\right)e^{-\alpha t} + \frac{\beta}{\alpha}.$$

Solving for $e^{-\alpha t}$ gives

$$e^{-\alpha t} = \frac{U - \beta/\alpha}{u_0 - \beta/\alpha}.$$

Hence we obtain

$$t = \frac{1}{\alpha}\ln\left|\frac{\beta/\alpha - u_0}{\beta/\alpha - U}\right|.$$

With the above parameter values substituted into this equation, the time required to heat the water from $u_0 = u_s = 15\,°\text{C}$ to $60\,°\text{C}$ is ≈ 4 hours and 50 minutes.

This result can be compared with the numerical solution in the Maple produced diagram of Figure 9.3.3 in Section 9.3.

insulation   It is possible to improve on the efficiency of the tank by insulating it well, and the standard used in Australia is approximately 3 cm of compacted fibreglass. Then, assuming some cost of power per Watt, we can compare the efficiency of the above tank with that for a tank which is insulated. (See Question 5 of the Exercises.)

efficiency questions   What is of major interest is how to run the most cost effective system. Clearly, the cost of heating the tank is proportional to the time it takes to heat the water, or the length of time for which the heater is on. Assuming all the constants as above, would it be cheaper to use a thermostat which switches the heater on each time the temperature drops below $50\,°\text{C}$, or to switch the heater off for 8 hours a day relying on the thermostat at other times? Would this result change if the temperature of the environment were to fall to $10\,°\text{C}$? (See Exercises, Question 4.)

doubling the power   We are also in a position to answer the question posed at the beginning of this section. This is done in the following example:

Example 3: *Does doubling the power output halve the heating time?*

Solution: We start with (1)

$$cm\frac{dU}{dt} = q - hS\left(U(t) - u_s\right)$$

and its solution as an expression for t is

$$t = \frac{cm}{hS}\ln\left|\frac{\frac{q}{hS} + u_s - u_0}{\frac{q}{hS} + u_s - U}\right|.$$

It is clear that if q is doubled, its effect on t is not to halve it. The effect of q is not quite so simple. While it will certainly reduce the natural logarithm function with the effect of decreasing the time, in general it will not halve it. Figure 10.2.1 illustrates the behaviour of t as q increases or decreases.

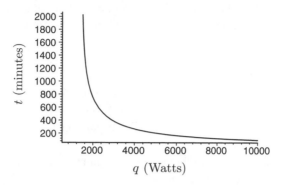

Figure 10.2.1: A plot of q in Watts as a function of t in minutes. U is fixed at $u_f = 60\,^{\circ}\mathrm{C}$ and other parameters are as for Example 3.

The equilibrium temperature

Physically the differential equation (1) represents a balance between heat produced by the heating element and heat lost from the surface. This suggests the existence of an equilibrium temperature, or thermal equilibrium. To find the equilibrium temperature we set $dU/dt = 0$ and then solve for U. Thus

$$q - hS\left(U - u_s\right) = 0.$$

It follows that the equilibrium temperature U is

$$U = u_s + \frac{q}{hS}\,. \tag{4}$$

Note that, since q, h and S are all positive constants, the equilibrium temperature is always greater than the temperature of the surroundings. This makes sense intuitively.

Now consider the solution to differential equation (1) which is given by (3). If we take the limit of this solution as t becomes large, we see that

$$U \to u_s + \frac{q}{hS} \qquad \text{as } t \to \infty.$$

We see that this limiting temperature, called the steady state temperature, is just the equilibrium temperature of (4).

physical
interpret-
ation

The physical significance of this is that, as the steady state temperature is
approached, heat is no longer used up to change the temperature. Instead
a balance is attained where the heat input is equal to the heat escaping
from the surface. Thus, in mathematical terms from the differential
equation describing the heat balance, equation (1), the derivative term
is zero, which is the condition for an equilibrium temperature.

Skills developed here:

- Formulate a differential equation for a hot water heater with
  insulation of varying degrees of effectiveness.
- Formulate a model to describe a cooling device, such as a
  refrigerator.
- Be able to ascertain how each parameter in the model affects
  the heating/cooling times.

10.3   Case Study: It's hot and stuffy in the attic

We have considered some simple applications of heat transport, but how
do these relate to larger systems? In the case study which follows, we
adapt these ideas to consider the changing temperatures in a building.
This is essential for good insulation characteristics.

To understand heat transfer in complicated structures, we need a model
for the heat transfer between two regions at different temperatures. We
assume

$$\left\{ \begin{array}{c} rate \\ heat \\ transferred \end{array} \right\} = \frac{S \times \Delta U}{R}$$

where S is the surface area between the two regions, ΔU is the temperature difference between them and R is the resistance to heat transport.
For walls of houses R represents the total *heat resistance* due to the
resistance of the material making up the wall, which might include insulation materials. The value of R for a given wall can be determined
experimentally, or theoretically (as we see later in Chapter 11).

The following was taken (and substantially modified) from Sansgiry and
Edwards (1996).

In different parts of the world dwellings and public buildings take on many different and imaginative forms. Mud, corrugated iron, fibro (asbestos) sheeting and glass are just some of a huge variety of materials used. One aspect considered in their construction is the climate. Certain materials used, or the aspect and design chosen, may effectively use the environment to enhance the heating/cooling mechanisms to suit a particular climate. Others, lacking this forethought, may be stuffy or cold.

To understand how the materials used for heating/cooling processes work, we consider a model for the temperature of a simple brick house with a tiled roof, shown in the Figure 10.3.1. The building has two compartments: the interior at temperature $U_1(t)$, and the roof cavity (between the ceiling and the roof) at temperature $U_2(t)$. The temperature of the surrounding air is assumed to be $u_s = 35\,°\mathrm{C}$ corresponding to a house in a hot climate.

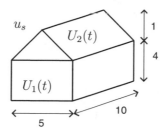

Figure 10.3.1: Schematic diagram for a model of a house with two compartments: an interior and a roof cavity. Here $V_1 = 200\,\mathrm{m}^3$ and $V_2 = 25\,\mathrm{m}^3$.

A system of differential equations, dependent on heat conduction, can be derived by considering the heat inputs and outputs in each of the two compartments. The resulting governing equations are

$$M_1\frac{dU_1}{dt} = \frac{S_{10}}{R_{10}}(u_s - U_1) + \frac{S_{12}}{R_{12}}(U_2 - U_1),$$
$$M_2\frac{dU_2}{dt} = \frac{S_{20}}{R_{20}}(u_s - U_2) + \frac{S_{12}}{R_{12}}(U_1 - U_2), \tag{1}$$

where S is surface area, R total heat resistance due to the building material and M_1 and M_2 are parameters representing the overall thermal mass (air plus solid materials) of the two regions. For example, $M_1 = \rho c V_1 + \rho_b c_b V_b$, where ρ is the density and c the specific heat of air, V_1 the volume of region 1, ρ_b is the brick wall density, c_b the brick wall specific heat and V_b is the volume of the brick wall. (For the results generated here, these values are calculated as $M_1 = 0.6498018 \times 10^8$ and $M_2 = 0.5184672 \times 10^7$, both in SI units.) Also

S_{10}, S_{12} and S_{20} denote the surface areas of the various boundaries of the regions (or compartments) and R_{10}, R_{12} and R_{20} are the corresponding thermal resistances. Typical values for these heat resistances are $R_{10} = 0.3$ for walls without insulation, $R_{12} = 0.1$ for ceilings (without insulation) and $R_{20} = 0.2$ for the roof. With wall and ceiling insulation these R values increase, with typically $R_{10} = 2$ and $R_{12} = 3$. Clearly, which building and insulation materials are chosen, has a substantial impact on the resulting house temperatures.

Using Maple to find a numerical solution to the system provides the results illustrated in Figure 10.3.2.

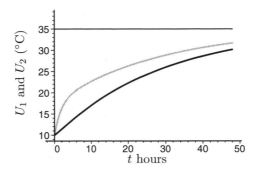

Figure 10.3.2: Plot of temperatures against time for the model without radiation from the roof tiles. The black line is U_1, the interior temperature, the grey line is U_2, the roof cavity temperature and the horizontal line is the temperature of the outside environment.

A valid criticism of the above model is that over long periods of time the roof cavity temperature tends to the outside temperature. In reality, in warm climates the temperature of the roof cavity is often much hotter than the outside temperature, particularly on very hot days, as in the case under consideration. One reason for this is that the roof tiles heat up to a temperature substantially greater than that of the surroundings, due to their exposure to the sun, and then radiate this heat into the roof cavity. To incorporate this aspect into the model we include the transfer of heat by radiation.

Let us assume the temperature of the roof tiles to be $80\,^\circ$C. A model for the heat transfer by radiation is given by the Stefan–Boltzmann law,

$$\left\{ \begin{array}{c} \text{rate} \\ \text{heat} \\ \text{transferred} \end{array} \right\} = \epsilon \sigma S \left(T_r^4 - T_2^4 \right)$$

where $\sigma = 5.6697 \times 10^{-8}$ is the Stefan–Boltzmann constant, the parameter ϵ is the emittance (typically $\epsilon \simeq 0.8$), T_r is the absolute temperature of the roof tiles (measured in degrees Kelvin) and T_2 is the absolute temperature of the roof cavity (also in degrees Kelvin). S is the surface area of the roof, as before.

The quantities $u_s + 273$, $U_1(t) + 273$, $U_2(t) + 273$ and $u_r + 273$ now represent the absolute temperatures in degrees Kelvin (K), where u_r is the temperature of the roof tiles.

Incorporating this extra heat source into our model results in a modification of the second equation; note that the temperature is now expressed in degrees Kelvin. Setting $T_r = (u_r+273)\,\mathrm{K}$ and $T_2 = (U_2+273)\,\mathrm{K}$ the resulting governing equations are now

$$M_1\frac{dU_1}{dt} = \frac{S_{10}}{R_{10}}(u_s - U_1) + \frac{S_{12}}{R_{12}}(U_2 - U_1),$$

$$M_2\frac{dU_2}{dt} = \frac{S_{20}}{R_{20}}(u_s - U_2) + \frac{S_{12}}{R_{12}}(U_1 - U_2)$$
$$+ \epsilon\sigma S_{12}\left((u_r + 273)^4 - (U_2 + 273)^4\right). \tag{2}$$

Temperature predictions for this new system are given in Figure 10.3.3. Note that the temperature in the roof cavity $U_2(t)$ is now greater than the outside temperature. This has also caused the temperature of the interior $U_1(t)$ to increase compared with the first model, as would be expected.

Figure 10.3.3: Inclusion of radiation from roof tiles. Plot of roof cavity temperature U_2 (grey) and interior temperature U_1 (black) against time in hours. The roof cavity temperature exceeds the temperature of the surroundings, $u_s = 35\,^\circ\mathrm{C}$ (horizontal line).

Clearly, over a number of days, the outside temperature would change and also vary cyclically with a 24-hour period, and this time dependence could be incorporated into the model. However, it would still be appropriate to use this model to explore the effects of different insulation strategies. For example, would it be more cost efficient and effective to increase the insulation in the walls or in the ceiling? Or would different roofing materials improve, and possibly solve, the problem of a hot stuffy attic?

10.4 *Spontaneous combustion*

overview   Spontaneous combustion may occur in a variety of situations when heat
produced by a chemical reaction cannot escape from a system fast enough.
Thus heat may build up and result in ignition. In this section, the basic mathematical theory describing this process is developed, with both
an analytical and numerical analysis. The complete theory (including
conduction) is complex, beyond the scope of this book; thus, what is
presented here is the simplest case, that of a *stirred chemical reactor*
where there is no conduction of heat.

Background

cause of   Spontaneous combustion occurs when the temperature inside a body in-
spontan-   creases at a rate faster than heat is able to escape from its surface. The
eous   higher the temperature, the more heat is produced by the chemical reac-
combus-   tion (usually oxidation), so it is important when designing systems which
tion   involve heat production to ensure that heat can escape sufficiently easily.

the   The important mechanisms are the rate of heat production and the rate
physical   of heat conduction away from this heat source. The reaction causing
process   the heat production may be one where some organic material intersects
with the atmosphere or oxygen, such as woollen fibres in bales. It is
characterised by its *exothermicity*, that is, the amount of heat produced
by the reaction, and by the *rate of the reaction*, which determines the
rate at which the heat is produced. The process by which heat is able to
escape is typically conduction, and thus the surrounding materials and
their ability to conduct heat also have a substantial effect on the system.
Before formulating the differential equations governing this process, we
need to introduce some further physical principles.

Arrhenius law for heat

heat gain   The rate of the reaction (associated with the rate of heat gain) can be
described by an exponential function $Ae^{-E/(RT)}$ where E is the *activation
energy*, with R the universal gas constant[1], given by $R = 8.314\,\mathrm{J\,K^{-1}\,mol^{-1}}$,
and where T is the absolute temperature, measured in degrees Kelvin (K)
from absolute zero. (T will be used here to distinguish between temperature measured in K and temperature U measured in °C.) The parameter
A is a further constant dependent on the type of the reaction. It is called
the *pre-exponential factor* and has units of $\mathrm{s^{-1}}$, where s is the SI unit of

---

[1]The universal gas constant also occurs in the famous law $PV = nRT$ relating
pressure, volume and temperature.

time, seconds. The rate of gain of heat (per unit volume) is given by

$$\left\{\begin{array}{c} rate \\ heat\ generated \\ by\ reaction \\ per\ unit\ volume \end{array}\right\} = \rho Q A e^{-E/(RT)} \tag{1}$$

where ρ is the density (mass per unit volume) and Q is the 'heat of reaction', measured in Joules.

Clearly, this function increases monotonically as the temperature rises, with $\lim_{T \to \infty} A^{-E/RT} = A$. From a graph of this reaction rate, we note that there is a domain of temperatures for which a small increase in T will result in a large increase in the reaction rate (see Figure 10.4.2 for the scaled equation, to come). This is associated with a substantial increase in the heat production.

properties of Arrhenius law

Model assumptions and approach

We now make some simplifying assumptions on which to build the mathematical model.

some assumptions

- We neglect heat conduction by assuming a uniform temperature throughout the reacting material.
- The rate of heat created is given by the Arrhenius law.
- Heat is lost from the surface of the reacting material according to Newton's law of cooling.

General compartmental model

We can apply the balance law to this process, where the compartment is the container of material (drums, wool-bales, etc.). The heat input is via some exothermic reaction and the output is through the mechanism of heat conduction away from the surface of this compartment. (We have assumed above that the temperature inside the compartment is uniform.)

the balance law

The process is illustrated in the input–output diagram of Figure 10.4.1.

compartmental diagram

Figure 10.4.1: Input–output compartmental diagram for the heat production and dissipation process.

In words, the heat balance equation is

the word equation

$$\left\{\begin{array}{c} rate\ of \\ change\ of \\ heat\ content \end{array}\right\} = \left\{\begin{array}{c} rate \\ heat\ generated \\ by\ reaction \end{array}\right\} - \left\{\begin{array}{c} rate \\ heat\ lost \\ to\ surroundings \end{array}\right\}. \tag{2}$$

Formulating the differential equation

heat loss   The rate of heat loss is proportional to the surface area (S) of the body
of material. It also depends on the type of material, for which there is
an associated heat transfer coefficient h. Thus mathematically, the heat
loss is proportional to the temperature difference between the material
and its surroundings, and can be represented as

$$\left\{\begin{array}{c} \text{rate of} \\ \text{heat} \\ \text{loss} \end{array}\right\} = hS(T - T_a), \tag{3}$$

where h is the Newton cooling coefficient, S is the surface area from which
heat is lost, and T_a is the temperature of the surroundings, namely the
ambient temperature (measured in degrees Kelvin).

putting it   Combining the expressions for heat loss (3) and gain (1) into an equation
together   describing the rate of temperature change (2), and using the specific heat
to evaluate the rate of change of heat in terms of the rate of change of
temperature, we obtain the differential equation

$$\rho V c \frac{dT}{dt} = \rho V Q A e^{-E/(RT)} - hS(T - T_a), \tag{4}$$

where ρ is the reacting substance density (moles per unit volume), Q is a
measure of heat gain associated with the reaction (Joules per mole) and
c is the specific heat of the reacting substance. V and S are the volume
and surface area of the material, respectively, and T_a is the ambient
temperature (measured in degrees Kelvin from absolute zero).

Numerical solution

scaling the   In equation (4) there are numerous parameters, which make the analysis
tempera-   cumbersome. We therefore simplify by scaling the temperature. We
ture   define a dimensionless temperature θ (since E/R has the dimension of
temperature)

$$\theta = \frac{T}{E/R}. \tag{5}$$

Then, with this change of variable, the system can be rewritten as

$$\sigma \frac{d\theta}{dt} = \lambda e^{-1/\theta} - (\theta - \theta_a) \tag{6}$$

with

$$\theta_a = \frac{T_a}{E/R}, \qquad \lambda = \frac{\rho A V Q}{hS(E/R)}, \qquad \sigma = \frac{\rho V c}{hS}. \qquad (7)$$

The constant θ_a denotes the dimensionless ambient temperature. The ·interpretation of parameters· parameter λ is known as the reaction efficiency parameter, since high values of λ correspond to high levels of heat generation from the chemical reaction, compared with the heat loss through Newton cooling. Both λ and θ_a are dimensionless parameters. Parameter σ is a measure of the speed of the reaction. Equation (6) contains only three parameters[2] θ_a, λ, and σ.

Using **Maple** we can solve the differential equation (6) numerically, for ·numerical solution· different values of the reaction efficiency parameter λ. The results are illustrated in Figure 10.4.2. In each case, the solution tends to a steady-state, however, as we vary λ from $\lambda = 2.84$ through to $\lambda = 2.85$ the amount by which the steady-state jumps increases dramatically.

With reference to equations (7), increasing λ can be considered as increasing the volume to surface area ratio. The system finds it more difficult to ·physical interpretation· release heat to the surroundings, and so the reaction increases in intensity. The sudden jump in equilibrium temperature, apparent in the right hand diagram of Figure 10.4.2, corresponds to the spontaneous ignition of the material.

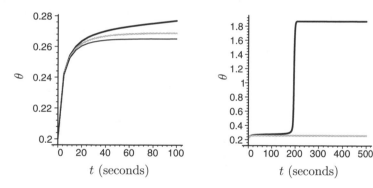

Figure 10.4.2: Numerical results for dimensionless temperature $\theta(t)$ against t, on solving equation (6). The parameter λ is gradually increased: $\lambda = 2.83$ (thin black), $\lambda = 2.84$ (grey) and $\lambda = 2.85$ (thick black). In the second figure the numerical solution is run over a longer time. There is sudden jump in steady-states between $\lambda = 2.84$ (grey) and $\lambda = 2.85$ (thick black).

---

[2]We could reduce this to only two parameters by writing $t = \sigma\tau$ and using the chain rule $\sigma d\theta/dt$ becomes $d\theta/d\tau$. The equation would then be completely dimensionless.

Analytic solution

finding the equilibrium

Ideally, we would like to determine exactly which values of λ correspond to the sudden jump in equilibrium temperature. To find this value, we set $d\theta/dt = 0$, for equilibrium, and solve the resulting equation graphically. Thus for equilibrium,

$$\lambda e^{-1/\theta} = \theta - \theta_a, \qquad \lambda = \frac{\rho V AQ}{hS(E/R)} = \frac{\rho RV AQ}{hSE}. \qquad (8)$$

stable and unstable equilibrium points

Figure 10.4.3 illustrates the two curves for heat loss and heat gain given by

$$y = e^{-1/\theta}, \qquad y = \frac{1}{\lambda}(\theta - \theta_a),$$

on the same system of axes. Clearly there can be at most three points of intersection where the heat loss equals the heat gain, that is $(d\theta/dt) = 0$ or $(dT/dt) = 0$. From (6) if the heat loss is greater than the heat gain, then $d\theta/dt < 0$, and if the heat loss is less than the heat gain then $d\theta/dt > 0$.

interpretation

With reference to the solid straight line for heat loss in the figure (Figure 10.4.3), if the temperature is low so that $\theta < \theta_1$, then the heat gain is greater than the heat loss and the temperature increases with $\theta \to \theta_1$. If the temperature is such that $\theta_1 < \theta < \theta_2$, then the heat loss is greater than the heat gain and the temperature will decrease so that $\theta \to \theta_1$. Thus we consider the intersection at θ_1 as a stable equilibrium for the system. Arguing in a similar manner, we see that at θ_2 there is an unstable equilibrium, and at θ_3 another stable point.

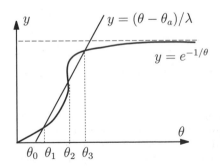

Figure 10.4.3: Sketch of the graphical solution of equation (6). There are, at most, 3 equilibrium solutions (θ_1, θ_2 and θ_3), although for some values of λ there may be only one solution.

If the initial condition is such that the system approaches the equilibrium at θ_3, we have the conditions for spontaneous ignition. However, if the initial conditions are different, ignition can be avoided. If we start with an initial temperature such that $\theta(0) = \theta_a$, then the temperature tends to the closest equilibrium. However, if λ is sufficiently large (i.e. the slope of the line, $1/\lambda$, is sufficiently small), for only one equilibrium point (corresponding to θ_3) to exist, then ignition will occur.

ignition conditions

As we increase λ, which decreases the slope of the line $1/\lambda(\theta - \theta_a)$, we reach critical values where the number of equilibrium values change: from one through two to three, or from three through two to one. The values of λ for which this change occurs are called bifurcation points.

equations for critical λ

Example 1: Given $\theta_a = 0.2$ find the critical value of the parameter λ where ignition occurs.

Solution: *The critical value occurs where the slope of the line in Figure 10.4.3 is tangential to the curve (which may occur in two places). Hence we solve the simultaneous equations*

$$e^{-1/\theta_c} = \frac{1}{\lambda_c}(\theta_c - 0.2), \qquad \frac{1}{\theta_c^2}e^{-1/\theta_c} = 1/\lambda_c,$$

where the first equation describes the point(s) of intersection, and the second equates the first derivatives of each, corresponding to the line and curve having the same slope. With a little algebra, we can deduce

$$\lambda_c = \theta_c^2 e^{1/\theta_c}, \qquad \theta_c^2 - \theta_c + 0.2 = 0.$$

Solving the quadratic equation gives two roots

$$\theta_{c_1} = \frac{1}{2}\left(1 - \sqrt{1 - 4 \times 0.2}\right) \simeq 0.277,$$

$$\theta_{c_2} = \frac{1}{2}\left(1 + \sqrt{1 - 4 \times 0.2}\right) \simeq 0.724.$$

After substituting these back into the expression for λ_c we get critical values of λ corresponding to a bifurcation of the equilibrium temperatures,

$$\lambda_c \simeq 2.847, \qquad \lambda_c \simeq 2.085.$$

As we have seen above, it is the larger value $\lambda_c = 2.847$ which corresponds to spontaneous ignition. This value is consistent with the results of Figure 10.4.2.

Alternatively, we can hold λ fixed and think of varying θ_a. This corresponds to the slope of the line in Figure 10.4.3 remaining constant, but

critical temperature

the intersection point it makes with the θ-axis moving to the right as θ_a increases. Again, we move from one equilibrium temperature (θ_1) to three equilibrium temperatures (θ_1, θ_2, θ_3) and back to just one equilibrium temperature (θ_3), for the larger values of θ_a. It is the second bifurcation here that corresponds to thermal ignition, since, owing to the stability of the equilibrium points, the temperature to which the system evolves jumps suddenly from θ_1 to θ_3. The value of θ_a where this occurs is the critical ambient temperature, θ_{ac}. (See Exercises, Question 10 for a numerical example of determining the critical ambient temperature.)

Further interpretation

the ambient temperature is crucial

Clearly the critical ambient temperature is crucial to any strategy for avoiding spontaneous combustion. The analysis above indicates that both the storage conditions and temperature of the material, as well as the ambient temperature, need to be carefully monitored in processes where spontaneous ignition may occur.

avoiding combustion

Spontaneous combustion can be avoided completely if the ambient temperature is sufficiently low so that there is only one stable equilibrium point on the lower branch of the heat gain curve. This occurs for low temperatures, below the ambient temperature θ_{ac}. However, such temperatures are not always practicable in the 'real world' and the regular monitoring of all conditions may be required.

Limitations and extensions

neglected factors

There are some important factors of which our model has not taken account.

- As the reaction proceeds fuel is used up. Here we have neglected the consumption of fuel.
- We have assumed the reactants are mixed (a stirred chemical reactor). In practice, perfect mixing will not occur, and the temperature is different at different points throughout the reacting material, which allows heat conduction to occur. This factor is especially important in solid materials.

including fuel consumption

Both these assumptions can be relaxed, but this leads to more complicated models. If we relax the first assumption, the model will comprise two simultaneous ODEs, one for the temperature and one for the concentration. Equilibrium values can be found by solving the two appropriate algebraic equations simultaneously, and bifurcations can be established as a parameter is varied. This is discussed briefly in Fulford and Broadbridge (2000).

If we relax the second assumption (no conduction), then our model results in a PDE for the temperature (see Section 12.1 for how partial derivatives arise). For further details of how to deal with this, see Fulford and Broadbridge (2000); Jones (1993). *including conduction*

In Section 10.5, we discuss a case study involving the spontaneous ignition of fish and chip crumble. Another practical example involving spontaneous ignition of moist milk powder is given in Rivers et al. (1996). In Lignola and Maio (1990), inclusion of reactant consumption in a stirred reactor, shows that some interesting oscillatory behaviour can occur in these processes. *further reading*

Skills developed here:

- Use `Maple` to find a numerical solution for the temperature in a stirred chemical reactor.
- For a given dimensionless ambient temperature find the critical value of the reaction efficiency.
- For a given reaction efficiency parameter find the critical ambient temperature such that spontaneous ignition will occur.

10.5 Case Study: Fish and chips explode

There have been many recorded cases of spontaneous combustion, for example, the sudden combustion of a barn filled with hay, the spontaneous ignition of wool bales in transit or storage, and a case of spontaneous ignition in a fish and chips shop in Napier (New Zealand), which is the subject of the case study which follows. Such events are of great interest and importance to insurance companies, which are called upon to pay for the damage. The information for the following case derives from such a report, written for a New Zealand insurance company.

This discussion is adapted from a report by Smedley and Wake (1987).

Fish and chips shops are not usually associated with explosives, however, in 1986 in Napier, New Zealand, one such shop was severely damaged when a drum of waste from the cooking scraps ignited spontaneously.

Fish and chips are cooked in large sieves submerged in vats of animal fat which are heated to about 180 °C. The hot crumble, or waste, which detaches during cooking, is allowed to cool and drain above the vat before being tipped into drums. The concern was that over a period of time the pile of crumble would (and the case in New Zealand proved that it could) spontaneously ignite through some heat producing mechanism within the waste.

The authors approached the problem from the point of view of constructing a mathematical model applicable to general problems of this nature, but which would be sufficiently simple for non-mathematicians to apply in situ. That is, they needed to provide adequate conditions to test for on site in order to prevent such 'explosive' events.

Spontaneous combustion is possible if the temperature inside the pile of crumble increases at a faster rate than heat is able to escape from the surface, where the crumble meets the air and drum wall. That is, if the exothermicity and the reaction rate are sufficiently great, such that the heat cannot dissipate, but instead builds up to a point of ignition. These processes need to be examined closely.

The exothermic reaction, which is the reaction producing the heat, is that of the organic material consisting of the fish and chip residue reacting with the atmosphere or oxygen. Since spontaneous combustion has occurred, this indicates that the exothermicity of this reaction is sufficient to cause ignition. In the case of oxidation for fish and chip residue, the activation energy was measured to be $E/R = 30\,000$. (In comparison, for wool $E/R = 9\,036$, or for cotton $E/R = 18\,000$.) As mentioned above, heat is lost through the interface between the crumble and the air as well as through the crumble-drum wall interface. Thus a change in this area will affect this heat loss, and there is the possibility of changing the dimensions of the drums in order to exercise a degree of control with respect to the heat loss. For the fish and chips shop in question, the drums used were cylinders with a diameter and height of 38 cm.

In order to avoid spontaneous combustion, which is the aim, consider the relationship between these heat loss and heat gain processes, illustrated in Figure 10.4.3 of Section 10.4. The diagram illustrates the existence of a critical temperature (T_c) above which spontaneous combustion will always occur, and also a critical initial temperature (T_{0c}), dependent on the ambient temperature, below which combustion can be avoided. In order to predict the dynamics of such a system, particularly to avoid the event of spontaneous combustion, both the storage conditions and temperatures, as well as the ambient temperatures, need to be monitored closely in the waste disposal process.

We return to the formulation of strategies to adopt *in situ* which will avoid a repeated disaster in the fish and chips shop and keep the insurance company happy. The authors used a modified version of the model in Section 10.4. It was based on a heat balance, but was more complex, including details of the combustion process, and an approximation of the Arrhenius term (see Jones (1993)). Using the predictions of this more sophisticated model they were able to provide a simple look-up table of values from which the proprietor could read the critical ambient temperature and the associated critical initial temperature. Such figures could be derived for a variety of convenient drum sizes.

In practice the safety procedure is simple. A quick reference to the look-up table, and some intermittent monitoring of the temperatures will provide the required conditions for an uneventful evening of business down at the fish and chips shop.

10.6 Exercises for Chapter 10

1. Coffee cooling. Imagine you have a cup of coffee cooling according to the natural law of cooling, so that the temperature $U(t)$ satisfies

$$\frac{dU}{dt} = -\lambda(U - u_s)^{5/4},$$

where λ is a positive constant and $u_s = 20\,°C$ is the temperature of the surroundings.

(a) Assuming $U(0) = u_0$, separate the variables to obtain

$$\lambda t = 4\left[(U - u_s)^{-1/4} - (u_0 - u_s)^{-1/4}\right].$$

(b) If the cup of coffee cools from an initial temperature $u_0 = 60\,°C$ to $50\,°C$ in 10.3 minutes, obtain λ and hence find the time it takes for the cup of coffee to cool to $40\,°C$.

(c) Using `Maple`, compare the solution of this natural cooling model with the solution for Newton's law of cooling by plotting them on the same system of axes, as in Figure 10.1.1. (Assume that with Newton's law of cooling the coffee cools from $u_0 = 60\,°C$ to $50\,°C$ in 10 minutes.)

2. Beer warming. *A cold beer is at a temperature of* 10 °C. *After* 10 *minutes the beer has warmed to a temperature of* 15 °C. *If the room temperature is* 30 °C, *how long will it take the beer to warm to* 20 °C?

3. Separating variables. *Solve the equation developed to describe the hot water heater, (equation (2) from Section 10.2) using the separable technique, and show the solution to be equivalent to that provided in the text.*

4. The hot water heater. *Consider the problem of a water heater with parameters and model as in Section 10.2:*

$$\frac{dU}{dt} = \beta - \alpha U, \qquad U(0) = u_0, \tag{1}$$

where the constants α *and* β *are defined as*

$$\alpha = \frac{hS}{cm}, \qquad \beta = \frac{q + hSu_s}{cm}.$$

(Parameters $q = 3600\,\mathrm{W}$, $S = 2.7\,\mathrm{m}^2$ *and* $u_s = 15\,°\mathrm{C}$ *with constants* $h = 12\,\mathrm{W\,m^{-2}\,°C^{-1}}$ *and* $c = 4200\,\mathrm{J\,kg^{-1}\,°C^{-1}}$.)

(a) *Suppose the temperature of the water in the hot water tank is* 60 °C, *the heater is turned off so that no heat is supplied and the water is allowed to cool. What will the temperature of the water be after* 8 *hours?*

(b) *With heat once again supplied* ($q = 3600\,\mathrm{W}$), *how long would it take to heat the water back up to* 60 °C, *after it has cooled for* 8 *hours in (a)?*

(c) *Suppose again the water is* 60 °C, *that the heater is turned off so that no heat is supplied and the water is allowed to cool. How long will it take for the water to reach* 50 °C?

(d) *With the heater turned on again, how long will it take to heat the tank of water from* 50 °C *to* 60 °C?

(e) *What is of paramount interest is how to run the most cost efficient system and clearly the cost of heating the tank is directly related to the time for which the heater is on. Assuming all constants as above, would it be cheaper to use a thermostat which switches the heater on each time the temperature drops below* 50 °C *and off each time the temperature reaches* 60 °C, *or to switch the heater off for* 8 *hours each day relying on the thermostat at other times? Use the model and the results found above to substantiate your answer. Although actual costs are not needed to calculate the most economical solution, typical values in Australia are* 8.22 c *per kW hr.*

(f) *Would your choice of the most economical system change if the temperature of the environment were to drop to 10 °C?*

5. Insulating the hot water heater. *Assume the model and parameters of
Question 4. Suppose with good insulation no heat is lost to the environment.
Adapt the model to accommodate this change and then, with a surrounding
temperature of 15 °C, and the water in the tank at this temperature, establish
how long it would take to heat the tank to 60 °C.*

6. Horse temperatures. *A vet is taking the temperature of a sick horse.
Initially, the temperature of the thermometer is 27.8 °C. Three minutes after
insertion the reading is 32.2 °C and three minutes later it is 34.4 °C. The horse
then has a violent convulsion which destroys the thermometer completely so
that no final reading can be taken. You may assume that the rate of change
of temperature of the thermometer is directly proportional to the difference
between the temperature of the thermometer and that of the horse.*

(a) *Let $U(t)$ be temperature at time t, with $U(0) = u_0$ the initial temperature of the thermometer and u_h the temperature of the horse. Model
the system with a differential equation and show that its solution is*

$$U(t) = u_h + (u_0 - u_h)e^{\lambda t},$$

where λ is the constant of proportionality.

(b) *Since $u_0 - u_h$ is a positive constant, show that*

$$e^{3\lambda} = \frac{U(3) - u_h}{u_0 - u_h} = \frac{U(6) - u_h}{U(3) - u_h}.$$

(c) *Now find the temperature of the horse and also the value of the constant
λ.*

7. Modelling house temperatures. *Develop a model for the temperature
inside a building comprising two rooms with a flat ceiling (no roof cavity),
which is described below.*

*One room has a heater which supplies heat at a constant rate q_0. You may
assume that heat is lost to the surroundings, from both rooms, according to
Newton's law of cooling and that the heat transfer from one room to the other
is proportional to the temperature difference between the two rooms. You may
also assume the air inside the building is well mixed so that heat conduction
can be ignored. Any radiation can also be ignored.*

*Give either a compartmental diagram or suitable word equations to help
explain your formulation and define any symbols introduced. You are not
required to solve any equations.*

8. Stirred chemical reactor. *For a stirred chemical reactor the heat balance equation may be given in terms of the scaled temperature (θ) as*

$$\sigma \frac{d\theta}{dt} = \lambda e^{-1/\theta} - (\theta - \theta_a),$$

where θ_a, λ and σ are positive constants.

(a) *What type of heat processes have been modelled by the two terms on the RHS of the differential equation?*

(b) *Let θ_1, θ_2 and θ_3 be the three possible equilibrium temperatures of the heat balance equation. Draw three separate graphs with temperature θ (along the vertical axis) and time (along the horizontal axis): one for each of the cases of 1, 2 or 3 equilibrium points. Choose the three illustrations such that spontaneous combustion is a possible outcome in each.*

Indicate clearly on the graphs how the temperature (θ) changes over time using a variety of initial conditions. Include in the graph the equilibrium temperatures (constant with time). Explain clearly why you have the temperature increasing or decreasing in each case.

(c) *When would you have spontaneous combustion, regardless of the initial temperature? Refer to your diagrams above to answer this.*

(d) *How would you change the model if heat loss followed natural cooling, that is, cooling in still air? (Give the unscaled differential equation. No analysis is required.)*

9. Spontaneous ignition and critical values. *Consider the equation for the scaled absolute temperature in an exothermic oxidation reaction*

$$\sigma \frac{d\theta}{dt} = \lambda e^{-1/\theta} - (\theta - \theta_a).$$

For $\theta_a = 0.3$, determine the appropriate critical value of λ where spontaneous ignition occurs.

10. Critical ambient temperature. *Consider the equation for the scaled absolute temperature in an exothermic oxidation reaction, with $\lambda = 2$,*

$$\sigma \frac{d\theta}{dt} = 2e^{-1/\theta} - (\theta - \theta_a).$$

(a) *Deduce that the critical ambient temperature θ_{ac} where bifurcations occur, satisfies the equations*

$$\theta_{ac} = \theta_c - \theta_c^2, \qquad \theta_c^2 = 2e^{-1/\theta_c},$$

where θ_c is the critical value of θ for which bifurcations occur.

(b) *By solving the second equation graphically, find the two values of θ_c and hence determine the approximate critical ambient temperature θ_{ac} above which spontaneous ignition occurs.*

11. Liquid explosive. *Dinitroxydiethyl-nitramine (DINA) is a liquid explosive, with which Merzhanov et al. (1961) carried out experiments in a cylindrical vessel of diameter 6.3 cm and length 6.3 cm. The mass used was 259 g. Given the values*

$$E = 45\,000 \,\mathrm{cal\,mol}^{-1}, \qquad A = 3.8 \times 10^{10}\,\mathrm{s}^{-1},$$
$$Q = 950\,\mathrm{cal/mol}, \qquad h = 1.5 \times 10^{-2}\,\mathrm{cal\,cm}^{-2}\,{}^{\circ}\mathrm{C}^{-1}\,\mathrm{s}^{-1},$$

calculate the critical ambient temperature for ignition. (The experimentally measured critical ambient temperature was 152 °C.)

Note that it is not necessary to convert the units into SI units here, as the units given are consistent. Calories (cals) and moles (mol) are units which are commonly used in chemistry.

Chapter 11

Solving heat conduction problems

In the previous chapter we formulated differential equations for some heat conduction problems. We now see how to solve these differential equations, and apply suitable boundary conditions to obtain the temperature distribution.

11.1 Boundary condition problems

overview
In the previous chapter we discussed the formulation of differential equations. To solve these, we first need to solve the equations, and then evaluate the arbitrary constants using some known conditions. The problems we look at involve specifying conditions on the boundaries of a region. This is, technically, slightly different from applying initial conditions, although in practice the process is similar. We concentrate here on how to apply boundary conditions to the general solution of a differential equation. Such problems are known as boundary value problems.

Typical problems

heat loss through a wall
The first problem we consider is the overall rate of heat loss through the wall of a building. We solve the appropriate differential equation to find the equilibrium temperature at any point inside the wall and then use Fourier's law to calculate the heat flux through the wall.

insulating a pipe
The second problem analyses the rate of heat lost through an insulating jacket surrounding a water pipe. This problem is similar to the previous one, but involves radial heat flow rather than heat flow in one direction.

cooling a computer chip
In our final problem, we examine the cooling of a computer chip by mounted heat fins. A problem similar to this is the cooling of a motorcycle engine, also using heat fins.

common features
Each of these problems involve first solving a differential equation for the equilibrium temperature. After finding the general solution, we then apply suitable boundary conditions to find the arbitrary constants, a process which requires the solution of simultaneous equations.

Boundary conditions

number of arbitrary constants
When we solve a second-order differential equation we obtain a solution with two arbitrary constants. In general, the number of arbitrary constants is the same as the order of the differential equation. (General principles used in solving second-order equations can be found in Appendix A.2 and Appendix A.5.)

example differential equation
For example, consider the differential equation

$$\frac{d^2U}{dx^2} - 4U = 0.$$

Since the differential equation is linear and its coefficients are constant we look for a solution of the form $U(x) = e^{\lambda x}$. Substitution of this into the differential equation gives

$$\lambda^2 e^{\lambda x} - 4e^{\lambda x} = 0$$

and hence we obtain $\lambda^2 - 4 = 0$. Thus we find two possible values for λ, $\lambda = 2$ or $\lambda = -2$. This yields two solutions e^{2x} and e^{-2x}.

The general solution of the differential equation is obtained by taking a *general solution* linear combination of these two solutions

$$U(x) = C_1 e^{2x} + C_2 e^{-2x}. \tag{1}$$

This solution has two arbitrary constants C_1 and C_2 whose values are determined by applying boundary conditions.

Boundary conditions are conditions applied on either boundary of a region. Suppose the region is $0 \le x \le 1$. One pair of boundary conditions might be,

$$U(0) = 0, \qquad U(1) = 3.$$

example boundary conditions

Another example is

$$U(0) = 1, \qquad U'(1) = 0.$$

This second example involves the derivative of U as one of the boundary conditions.

In solving differential equations where the independent variable is time, we have so far encountered initial conditions. The difference between initial and boundary conditions is the value of the independent variable they are applied to. Initial conditions are always applied at the same value (usually $t = 0$). Boundary conditions, however, are applied at two different values (e.g. $x = 0$ and $x = 1$).

initial versus boundary conditions

Applying boundary conditions

The process of applying boundary conditions is exactly the same as applying initial conditions. Applying the conditions leads to, in general, two simultaneous equations for the two arbitrary constants.

overview

As an example, consider the following differential equation with boundary conditions,

example

$$\frac{d^2 U}{dx^2} - 4U = 0, \qquad U(0) = 0, \quad U(1) = 3.$$

The solution to this differential equation is given by (1). Substituting $x = 0$ and then $x = 1$ into this solution, and then applying the boundary conditions, produces the pair of simultaneous equations

$$C_1 + C_2 = 0, \qquad C_1 e^2 + C_2 e^{-2} = 3.$$

solving for the arbitrary constants Solving simultaneously for C_1 and C_2 gives $C_1 = 3/(e^2 - e^{-2})$ and $C_2 = -3/(e^2 - e^{-2})$. We now substitute these values for C_1 and C_2 back into the general solution (1). Hence the solution U, which satisfies both boundary conditions, is

$$U(x) = 3\frac{e^{2x} - e^{-2x}}{e^2 - e^{-2}}$$

for x between 0 and 1.

hyperbolic form of solution This can also be expressed in terms of the hyperbolic function $\sinh(2x) = (e^{2x} - e^{-2x})/2$ as

$$U(x) = 3\frac{\sinh(2x)}{\sinh(2)}.$$

(For details on hyperbolic functions see Appendix B.4.)

practical problems In the following sections we consider some practical problems. These involve solving differential equations derived in the previous chapter, and then applying suitable boundary conditions.

Skills developed here:

- Solve simultaneous equations to find the constants of integration.
- Differentiate between, and apply, initial conditions and boundary conditions.
- Understand the principles involved in solving simple second-order differential equations, with boundary conditions.

11.2   Heat loss through a wall

overview We solve a differential equation and apply boundary conditions to find the equilibrium temperature inside a brick wall. The boundary condition on the outside of the wall follows from Newton's law of cooling, while on the inside the wall is kept at a constant temperature.

Problem description

The design of energy efficient buildings has become more important as energy costs have risen, and there is a greater awareness of the responsible use of resources. As was seen in the earlier case study (Section 10.3), which examined the temperatures in a building, a crucial concern in such design is that of calculating the rate of heat loss from a building wall. We take a closer look at this heat loss process. problem applica- tion

Consider a building with walls of thickness $\ell = 20\,\mathrm{cm}$. The inside of the wall is maintained at a comfortable temperature $u_i = 20\,°\mathrm{C}$. The temperature is maintained by a thermostat on the wall. The outside of the wall is exposed to air at a temperature $u_o = 5\,°\mathrm{C}$, however, the outside of the wall is not at $5\,°\mathrm{C}$, but loses heat to the atmosphere (see Figure 11.2.1). (Subscripts i and o denote variables on the inside and outside of the wall, respectively.) inside and outside tempera- tures

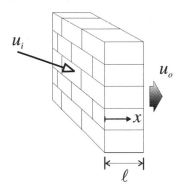

Figure 11.2.1: Heat flow through the wall of a house, $0 \le x \le \ell$.

We are interested in determining the rate of heat flow through the wall. One possible question we might ask is: a question

> If the thickness of the wall is doubled what happens to the rate of flow of heat through the wall?

Model assumptions and approach

We let the x-axis measure distance from the inside of the wall, outwards. Closer to the inside the temperature will be closer to $20\,°\mathrm{C}$. Hence we may think of the temperature as a function of two variables x and t. For the equilibrium temperature, however, the temperature will be a function of x only. We make some assumptions and then build the model on these. equilib- rium tempera- ture a function of x

- We assume heat flows directly from the inside to the outside through the wall. Thus we neglect any complicated heat flow at the edges, top and bottom. For most of the wall the heat flow will be in the x-direction.

- We also assume the wall to have no windows.
- Since the inside of the wall has a thermostat, once thermal equilibrium is established the temperature will be constant. We assume it stays constant at $u_i = 20\,°\mathrm{C}$.
- For thermal equilibrium the temperature on the outside of the wall will also be constant but it will not be the same as the temperature of the surroundings, $u_o = 5\,°\mathrm{C}$, or of the inside $u_i = 20\,°\mathrm{C}$. It will be somewhere between these temperatures. We assume the outside of the wall satisfies Newton's law of cooling. The temperature of the outside can be determined from the solution for the temperature inside the wall.

The differential equation

equation at thermal equilibrium
In the previous chapter we formulated a differential equation for the equilibrium temperature of some material through which heat was conducted in the x-direction. The differential equation (from Section 9.5, equation (8)) is

$$\frac{d^2U}{dx^2} = 0. \tag{1}$$

This is a second-order differential equation so we expect two arbitrary constants in the solution.

interpretation of the equation
The equation arose from a heat balance in thermal equilibrium. Note that the differential equation does not depend on the conductivity of the material. The conductivity will be incorporated through the boundary conditions.

Analytic solution

solving the differential equation
It is a simple matter to solve this differential equation. We can integrate both sides twice. By integrating with respect to x once we obtain

$$\frac{dU}{dx} = C_1, \tag{2}$$

where C_1 is an arbitrary constant and then integrating again gives

$$U(x) = C_1 x + C_2. \tag{3}$$

Here C_1 and C_2 are the two arbitrary constants. In order to solve for these constants we need two boundary conditions.

The boundary conditions

We require information at each boundary, $x = 0$ and $x = \ell$.   boundaries

In modelling this problem we assumed that the temperature of the room is regulated by a thermostat so that the temperature of the inside of the wall is always at a given temperature u_i. Thus the boundary condition at $x = 0$ may be written as   boundary condition at $x = 0$

$$U(0) = u_i. \tag{4}$$

We also assumed that the outside of the wall loses heat to the surroundings according to Newton's law of cooling, which states that the rate of heat loss is proportional to the surface area of the wall face and the temperature difference between the wall and the surroundings. We can equate this to the rate of heat conducted through the wall at $x = \ell$, given in terms of the heat flux as $J(\ell)A$, where A is the area through which the heat passes. Then, dividing by the surface area,   boundary condition at $x = \ell$

$$J(\ell) = h(U(\ell) - u_o), \tag{5}$$

where h is the convective heat transfer coefficient.

Since the wall is hotter than the surroundings and the x-axis points towards the surroundings the sign is correct giving a positive heat flux for $U(\ell) > u_o$.   check the sign

We now apply both boundary conditions (4) and (5) to determine the arbitrary constants C_1 and C_2. This gives us two equations for the two unknowns C_1 and C_2.   applying the boundary conditions

---

Example 1:   *Apply the boundary conditions (4) and (5) to determine the arbitrary constants C_1 and C_2 of equation (3). Hence obtain an expression for the temperature profile in the wall.*

Solution: *The boundary condition $U(0) = u_i$ states that $U = u_i$ at $x = 0$. To apply this we substitute $x = 0$ into the general solution (3) and obtain the equation*

$$C_1 \times 0 + C_2 = u_i. \tag{6}$$

From this we can solve directly for C_2 giving

$$C_2 = u_i. \tag{7}$$

To apply the Newton cooling boundary condition (5) we first calculate $J(x)$ using Fourier's law $J = -k(dU/dx)$, where k is the conductivity. Using equation (3), we obtain

$$J(x) = -k\frac{dU(x)}{dx} = -kC_1.$$

At $x = \ell$ equation (5) becomes

$$-kC_1 = h\left(C_1\ell + C_2 - u_o\right). \tag{8}$$

Substituting (7) into (8) and dividing through by h we obtain

$$-\frac{k}{h}C_1 = C_1\ell + u_i - u_o.$$

Rearranging gives

$$-\left(\ell + \frac{k}{h}\right)C_1 = u_i - u_o,$$

which implies that

$$C_1 = \frac{-(u_i - u_o)}{\ell + \frac{k}{h}}. \tag{9}$$

Substituting the above expressions for the arbitrary constants (7) and (9) back into the general solution (3) gives the solution for the equilibrium temperature

$$U(x) = u_i - \left(\frac{u_i - u_o}{\ell + \frac{k}{h}}\right)x. \tag{10}$$

---

sketch of solution

This is a simple linear function which is illustrated in Figure 11.2.2. The temperature $U(x)$ decreases with x. But note that the temperature of the outside of the wall is at a higher temperature to that of the surroundings, u_o.

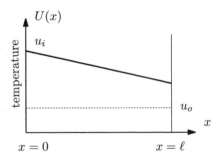

Figure 11.2.2: Equilibrium temperature in a wall where the inside of the wall is at temperature u_i and on the outside wall heat is lost to the surroundings, which are at temperature u_o.

limit checks

As a quick check on the solution, note that as $h \to 0$, $U(\ell) \to u_i$. Physically, $h \to 0$ corresponds to not allowing any heat to escape so that the temperature on the outside would be the same as the temperature on the inside. Alternatively, consider the limit as $h \to \infty$. In this case $U(\ell) \to u_o$. This corresponds physically to the situation where the temperature of the outside wall is the same as the temperature of the outside surroundings.

Rate of heat loss from a wall

The rate of heat loss from the wall can be calculated by using Fourier's law. This gives the heat flux (the rate of flow of heat per unit time per unit area). By Fourier's law the heat flux at $x = \ell$, $J(\ell)$, can be evaluated from

$$J(x) = -k\frac{dU(x)}{dx}.$$

using Fourier's law

Example 2: *Find an expression for the heat flux through the wall.*

Solution: *Using the solution for the temperature (10) we obtain*

$$J(\ell) = -k \times -\frac{u_i - u_o}{\ell + \frac{k}{h}} = \frac{u_i - u_o}{\frac{\ell}{k} + \frac{1}{h}}. \tag{11}$$

As a quick check we see that this gives a positive value since $u_i > u_o$.

Let us carry out some calculations with typical numerical values. Suppose the inside temperature is $u_i = 20\,^\circ\text{C}$, whereas the outside temperature is $5\,^\circ\text{C}$. Consider a brick wall of thickness $\ell = 15\,\text{cm} = 0.15\,\text{m}$. The conductivity of brick is $k = 0.5$ SI units, and the convective heat transfer coefficient $h = 10$ SI units. With these values we calculate $\ell/k = 0.3$ and $1/h = 0.1$, so the the heat flux through the wall is

some typical values

$$J = \frac{15}{0.3 + 0.1} = 37.5\,\text{W/m}^2. \tag{12}$$

Thus we can obtain an estimate for the total rate of heat loss from the walls of a brick house by multiplying this value by the total area of the walls.

Note that J is directly proportional to the temperature difference but not directly proportional to the conductivity k or to the thickness of the wall ℓ. Thus doubling the wall thickness will not halve the heat loss from the wall, and nor would doubling the conductivity (except in cases where $1/h$ is much smaller than ℓ/k). However, for decreased k and h there is a reduction in the heat flux J.

doubling the wall thickness

Heat resistance R-values

R-values   There is a neat interpretation to equation (11), which we can write as

$$J = \frac{u_i - u_o}{R}, \qquad \text{where} \quad R = \frac{\ell}{k} + \frac{1}{h}. \tag{13}$$

The parameter R is interpreted as a lumped thermal resistance. We can think of R as made up of two separate resistances:

- $R_1 = \ell/k$, the resistance to heat flow through the material, and
- $R_2 = 1/h$, the surface resistance to heat flow to the air.

Note that the resistance to heat flow through the material, ℓ/k, increases with the width of the material ℓ and decreases with the conductivity k, as we might expect. Companies which provide insulation batts often talk about the R-value of their batts. For example, $R2.0$ batts have a thermal resistance of $\ell/k = 2.0\,\mathrm{W\,°C^{-1}}$.

heat loss   Using this notion of thermal resistances we can calculate the rate of heat
through a   loss for more complicated structures, such as an insulated wall of a house.
wall   Here, the total resistance is

$$R = \frac{\ell_1}{k_1} + \frac{\ell}{k} + \frac{1}{h},$$

where k_1 and ℓ_1 are the conductivity and width of the insulation batts, respectively, and k and ℓ are those values corresponding to brick. With $R2.0$ batts ($\ell_1/k_1 = 2.0$), a brick wall ($\ell/k = 0.3$) and $h = 10$ units this gives a total heat flux of

$$J = \frac{u_i - u_o}{R} = \frac{15}{2.0 + 0.3 + 0.1} = 6.25\,\mathrm{W/m^2}.$$

This is much lower than the previously calculated flux through a single brick wall.

heat loss   For a pane of glass, the total thermal resistance is made up of the re-
through a   sistance to heat passing through the glass ℓ/k together with the surface
window   resistances for both the inside and outside surfaces (since there would
               not be a thermostat on a pane of glass). This gives

$$J = \frac{(u_i - u_0)}{\frac{1}{h} + \frac{\ell}{k} + \frac{1}{h}}.$$

For a 5-mm thickness of glass, $\ell = 5 \times 10^{-3}\,\mathrm{m}$. The conductivity of glass is $k = 0.8$ SI units and we take $h = 10$ SI units. Substituting for these data values gives $\ell/k = 0.0062$ and $1/h = 0.1$. We obtain a value for the heat flux through the glass pane as

$$J = \frac{15}{0.1 + 0.0062 + 0.1} \simeq 73 \, \text{W/m}^2.$$

For a window, the ratio ℓ/k measuring the resistance to heat flow through the material $(0.1/0.0062 \simeq 16)$ is much greater than it is for a brick wall $(0.1/0.3 \simeq 0.33)$. This indicates why heat loss through windows is generally more significant than through walls, especially when the walls are insulated. Note that if we had not included the two surface resistance terms (the two $1/h$ terms), then we would have calculated the heat flux erroneously, as $15/0.0062 \simeq 2.5 \times 10^3 \, \text{W/m}^2$. An error of order 10^2! *interpretation*

Skills developed here:

- From a word description, write in mathematical form the boundary conditions corresponding to

    - prescribed temperature on a boundary,

    - prescribed heat flux on a boundary, and

    - Newton heat exchange with surroundings, on a boundary.

- Be able to apply the above boundary conditions to find the temperature for heat flow in one direction.

- Obtain the heat flux from the temperature.

- Be able to interpret the heat flux in terms of thermal resistances.

11.3 Case Study: Double glazing: what's it worth?

Is double glazing what it's cracked up to be? We saw in Section 11.2 that there is a simple relation between the heat flux and the temperature difference for heat flow through a slab of any material. It is

$$J = \frac{u_1 - u_2}{R},$$

where R is the heat resistance (commonly called the R-value). We consider here a case study which makes use of this result in an examination of the heat transfer process in windows. In particular we are interested in the extent to which double glazing reduces this heat loss and improves efficiency.

The following case study is adapted from Mesterton-Gibbons (1989) and information brochures on double glazing from manufacturing companies.

Advertising promotes double glazing as a very effective way of preventing heat loss from your home, however, how effective is it? The windows are expensive to buy and install so we need to know to what extent the benefits outweigh the cost.

A wall of any building transfers heat in one direction if the temperatures on either side are different. In the case of glass window panes there is a heat transfer outwards in winter and inwards in summer, neither of which may be desirable. Double glazing companies manufacture windows typically consisting of two 3 mm panes of glass separated by a 12 mm gap. Using the theory of heat flux we can get a very good idea of just how much heat loss is prevented by installing these double glazed windows.

Let u_1 be the temperature inside a room and u_2 the temperature outside. We are concerned with winter heat loss so $u_1 > u_2$. Let ℓ_g be the thickness of the glass in each pane and ℓ_a the width of the air cavity between the sheets of glass. We take u_a and u_b to be the temperatures on either side of the air cavity, as illustrated in Figure 11.3.1.

Figure 11.3.1: Schematic diagram of double glazing.

Let J (the heat flux) be the amount of heat transfered through a unit area (per unit time) of a plane perpendicular to the direction of the x-axis, at x. We are not interested here in the initial heat transfer to the glass or to the cavity, but rather in the behaviour of the system after it has settled down to a state of thermal equilibrium, where J is a constant.

The heat flux J is given by

$$J = \frac{u_1 - u_2}{R}, \tag{1}$$

where R is the total heat resistance and $R = 2R_g + R_a$. This can be derived from the relation

$$J = \frac{u_1 - u_a}{R_g} = \frac{u_a - u_b}{R_a} = \frac{u_b - u_2}{R_g},$$

which expresses the equality of heat flux through each part of the system resulting from thermal equilibrium, or constant J. R above is the sum of the three individual heat resistances, due to the two panes of glass R_g and the air cavity R_a. We later relate the heat resistances R_g and R_a to the glass and air conductivities and thicknesses.

We need to compare this flux with that in the case of a single paned window, that is, one which is not double glazed ($\ell_a = 0$), but where the pane now has thickness $2\ell_g$. The heat flux for the single paned window, J_s, is

$$J_s = \frac{(u_1 - u_2)}{2R_g}.$$

(We have taken the single pane to have a thickness of $2\ell_g$ so that the only difference between J and J_s is due to the air gap.) We define a relative heat loss Δ by

$$\Delta = \frac{J_s - J}{J_s} = \frac{2R_g + R_a}{2R_g}. \tag{2}$$

If we assume the temperature of the window is exactly the temperature of the air, then $R_g = \ell_g/k_g$ where ℓ_g is the width of a single pane of glass and k_g is its conductivity. Similarly, $R_a = \ell_a/k_a$ where ℓ_a is the distance between the panes and k_a is the conductivity of the air. This essentially follows the calculations of Mesterton-Gibbons (1989). Although this assumption is not very accurate, we shall see where it leads, and then extend the model with this assumption relaxed.

Using approximate values of the conductivities $k_g = 0.8\,\mathrm{J\,m^{-1}\,s^{-1}\,°C^{-1}}$ and $k_a = 0.05\,\mathrm{J\,m^{-1}\,s^{-1}\,°C^{-1}}$, we have $k_g/k_a \simeq 16$. Also, we set $\ell_a/\ell_g = r$. Substituting these values into (2) we obtain the relative heat loss Δ_0 as

$$\Delta_0 = \frac{16r}{16r + 2} = \frac{r}{r + 1/8}.$$

The graph of this function can be seen in Figure 11.3.2. It is a monotonically increasing function with $\Delta = 0$ when $r = 0$ and $\lim_{r \to \infty}(r/(r + 1/8)) = 1$.

Thus for $r = 1$ the reduction in heat loss is 88%. For $r = 4$ the reduction has increased substantially to 97% but for $r = 8$ the reduction is not much better at 98%. To conclude then, if $\ell_a \approx 4\ell_g$ the heat loss can be reduced by 97% and not much is gained by increasing the gap (ℓ_a) between the glass panes. This is indeed a substantial reduction in heat loss, and these specifications are exactly those manufactured as standard by the companies considered.

However, the above calculations assumed the surface temperature of the glass to be the same as the air temperature. This is a unrealistic assumption for a typical pane of glass and can result in a huge overestimate of the heat flux. (Possibly by a factor of 3000% for a 15 °C temperature difference!) In an improved model, we do not take the temperature of the glass surface to be the same as that of the air, but instead replace this with a Newton cooling condition. This results in heat resistances being increased by an amount $1/h$ for each glass–air contact, where h is the Newton cooling coefficient. If we now include the Newton cooling terms for only the external glass–air contacts (but neglect the ones for the internal air cavity) our expression for R_g becomes $R_g = \ell_g/k_g + 1/h$ and, with $k_g/k_a = 16$, $\ell_a/\ell_g = r$, the relative heat loss Δ_1 is

$$\Delta_1 = \frac{16r}{2\sigma + 16r + 2}, \qquad \text{where} \quad \sigma = \frac{k_g}{h\ell_g}.$$

If we also allow for Newton cooling from the glass to the air cavity and from the air cavity to the glass then R_a becomes $R_a = \ell_a/k_a + 2/h$ and we obtain, from (2), Δ_2

$$\Delta_2 = \frac{2\sigma + 16r}{4\sigma + 16r + 2}, \qquad \text{where} \quad \sigma = \frac{k_g}{h\ell_g}.$$

A comparison of the results from these three models, Δ_0, Δ_1 and Δ_2, is illustrated in Figure 11.3.2.

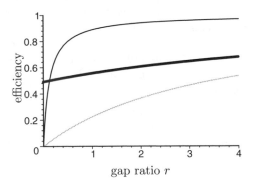

Figure 11.3.2: Relative heat loss efficiency calculations Δ_0 (thin black), Δ_1 (grey) and Δ_2 (thick black) plotted against r, the ratio of air gap to glass width. We have set $h = 10 \, \text{W} \, \text{m}^{-2} \, °\text{C}^{-1}$ and $\ell_g = 0.003 \, \text{m} = 3 \, \text{mm}$.

The reduction in heat loss here is not quite the 97% of the first model, however, there is still a substantial improvement, particularly as window panes are a *major cause of heat loss from homes during winter. Furthermore, in summer, the transmission of heat inwards is reduced to a significant degree, and double glazing provides a substantial reduction in sound transmission as well.*

It should be mentioned here that k_a applies to dry, still air, and so this model is not applicable where air may contain moisture (which increases k_a and also the heat loss). For this reason double glazing requires a fully sealed cavity without moisture. Wall cavities, floors and ceilings use other materials for insulation, as it is not feasible to seal the cavities involved. In these cases, where there is moving air, heat can be convected (transport by a moving medium) whereas in the above double glazing example only the conduction of heat was considered (transport through a medium at rest).

Clearly, from the results of the above models, a substantial reduction in heat loss can be achieved through windows which are double glazed. Perhaps they are nearly as good as the manufacturers claim and worth the cost. Those without the economic resources, or in rental accommodation, have been known to use plastic-wrap (transparent sheet of plastic) attached against the inside of single window panes to create a crude type of double glazing in the colder months—with remarkably rewarding effects.

11.4   Insulating a water pipe

In this section we adapt the theory developed in Section 9.6, concerning the radial conduction of heat, to consider the insulation of a cylindrical pipe. After examining the physical processes we arrive at an unusual result—that sometimes we can be worse off with insulation than with no insulation at all! *overview*

Problem description

In cold climates a significant heat loss can occur from water pipes exposed to cold air. In extreme conditions, the heat loss may be such that the temperature of the water in the pipes drops below $0\,°\mathrm{C}$ and freezes. As the water freezes it expands and can burst the pipes. To restrict this heat loss and prevent such problems, a jacket of insulating material is placed around the pipe, or the pipe is lagged. *problem application*

It is therefore useful to calculate the rate of heat loss from a pipe of given length. One of the things we should investigate is: *a question*

> *How does this rate of heat loss change as the thickness of the insulating material increases?*

Model assumptions and approach

model
geometry

For a mathematical model we examine radial heat conduction inside the insulation jacket. We let the inner radius of the jacket be a and the outer radius be b. Hence the thickness is $(b - a)$ and the temperatures u_a and u_b are those on the inside and outside of the insulated pipe respectively. We denote by r the radial distance from the centre of the pipe. The geometry is shown in Figure 11.4.1.

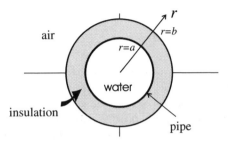

Figure 11.4.1: Geometry of the pipe insulation problem.

some as-
sumptions

We make some simplifying assumptions and then build our model on these.

- Since metal is such a good conductor of heat we assume the pipe metal to be at the same temperature as the insulation jacket, and to have good contact with it. We assume the inside of the insulation jacket is always at that temperature.
- On the outside of the insulation jacket the air temperature is u_b. We assume that heat is lost from this surface according to Newton's law of cooling.

The differential equation

equation
at thermal
equilib-
rium

In a previous chapter we formulated the differential equation for radial heat conduction in a cylinder or cylindrical shell. The equilibrium temperature satisfied the differential equation (Section 9.6, equation (8))

$$\frac{d}{dr}\left(r\frac{dU}{dr}\right) = 0. \tag{1}$$

This differential equation is of second-order, but is not a constant coefficient differential equation. Using the product rule we can write this differential equation in the form

$$r\frac{d^2U}{dr^2} + \frac{dU}{dr} = 0.$$

This shows that the differential equation has coefficients which are not constant with respect to r. However, the original form is more convenient here for finding its solution.

Analytic solution

To obtain the solution to this differential equation we integrate both sides twice with respect to r.

integrating twice to solve

Example 1: *Solve the differential equation (1).*

Solution: *Integrating once gives*

$$r\frac{dU}{dr} = C_1,$$

where C_1 is an arbitrary constant of integration. Hence

$$\frac{dU}{dr} = \frac{C_1}{r}.$$

Integrating again with respect to r $(r > 0)$ gives

$$U(r) = C_1 \ln(r) + C_2, \qquad (2)$$

where C_2 is the second arbitrary constant of integration.

We have two arbitrary constants which are required for the general solution of a second-order differential equation. The equilibrium temperature depends on the logarithmic of the radial distance from the centreline, rather than the linear distance as for the problem solved in Section 11.2.

form of solution

The boundary conditions

We need two boundary conditions in order to solve for the two arbitrary constants. For this problem the boundary conditions are at the edges of the insulation at $r = a$ and $r = b$.

boundaries

At $r = a$ we specified the temperature as u_a. In mathematical form this is written as

boundary condition at $r = a$

$$U(a) = u_a. \qquad (3)$$

At $r = b$ the surface of the insulation loses heat to the surroundings. Newton's law of cooling states the rate of loss of heat to the surroundings is proportional to the surface area and the temperature difference. Since the heat flux is the rate of heat loss divided by the surface area, we can write

$$J(b) = h(U(b) - u_b), \tag{4}$$

where u_b is the temperature of the surrounding air. Since the temperature of the insulation is greater than the temperature of the air this gives a positive heat flux for $U(b) > u_b$.

Example 2: *Apply the boundary conditions (3) and (4) and hence find an expression for the equilibrium temperature $U(x)$.*

Solution: *Applying the first boundary condition (3) at $r = a$, we obtain the equation*

$$C_1 \ln(a) + C_2 = u_a. \tag{5}$$

To apply Newton's law of cooling at the boundary (4) we first calculate the heat flux. Using the general solution (2) and Fourier's law

$$J(r) = -k\frac{dU(r)}{dr} = -k\frac{C_1}{r} \tag{6}$$

and hence

$$J(b) = -k\frac{C_1}{b}. \tag{7}$$

The Newton cooling condition at $r = b$ now reduces to the equation

$$-k\frac{C_1}{b} = h(C_1 \ln(b) + C_2 - u_b). \tag{8}$$

Equations (5) and (8) give two simultaneous equations for C_1 and C_2:

$$\ln(a)C_1 + C_2 = u_a, \tag{9}$$

$$\left(\ln(b) + \frac{k}{bh} \right) C_1 + C_2 = u_b. \tag{10}$$

Solving these equations we get

$$C_1 = \frac{-(u_a - u_b)}{\ln(b/a) + \frac{k}{bh}} \quad \text{and} \quad C_2 = u_a + \frac{(u_a - u_b)}{\ln(b/a) + \frac{k}{bh}} \ln(a),$$

and substituting these values for C_1 and C_2 back into the general solution (2) we have

$$U(r) = u_a - \frac{u_a - u_b}{\ln(b/a) + \frac{k}{bh}} \ln(r/a). \tag{11}$$

Rate of heat loss from a pipe

We calculate the total rate of heat loss from the pipe. For this we use Fourier's law of heat conduction to find the heat flux at $r = b$.

rate of
heat loss

Example 3: *Calculate the total rate of heat loss at $r = b$ of an insulated pipe of length l.*

Solution: *Let Q denote the total rate of heat loss from a pipe of length ℓ. We can calculate this quantity by multiplying the heat flux at $r = b$ by the surface area of the insulation jacket (at $r = b$). The total surface area of the insulation jacket is $2\pi b\ell$. Hence*

$$Q = 2\pi b\ell J(b).$$

Using Fourier's law, $J = -k\, dU/dr$, the rate of heat loss per unit length becomes

$$Q = -2k\pi b\ell \frac{dU(b)}{dr}$$

where we evaluate dU/dr at $r = b$.

Using the solution obtained for $U(r)$ from equation (11), we obtain

$$Q = -2k\pi b\ell \left.\frac{d}{dr}\left(u_a - \frac{u_a - u_b}{\ln(b/a) + \frac{k}{bh}} \ln\left(\frac{r}{a}\right)\right)\right|_{r=b}$$

$$= 2k\pi b\ell \frac{u_a - u_b}{\ln(b/a) + \frac{k}{bh}} \left.\frac{d}{dr} \ln\left(\frac{r}{a}\right)\right|_{r=b}.$$

Evaluating the derivative and then substituting for $r = b$ gives

$$Q = 2k\pi b\ell \frac{u_a - u_b}{\ln(b/a) + \frac{k}{bh}} \times \frac{1}{b}. \tag{12}$$

Hence

$$Q = \frac{2k\pi\ell(u_a - u_b)}{\ln(b/a) + \frac{k}{bh}}. \tag{13}$$

As a quick check note that $u_a > u_b$ and $b/a > 1$, so we get a positive rate of heat loss per unit length, as would be expected.

a quick
check

Results and interpretation

interpreting the formulae

In order to predict how the rate of heat loss varies as the insulation thickness is increased, we consider the above model with some appropriate parameter values.

some example values

Take $\ell = 1\,\mathrm{m}$, $a = 15\,\mathrm{mm}$, $u_a = 60\,^\circ\mathrm{C}$, $u_b = 15\,^\circ\mathrm{C}$ and $h = 5\,\mathrm{W\,m^{-2}\,^\circ C^{-1}}$. We vary the outer radius b. The computed values of Q are given in Table 11.4.1 where we look at two different values of conductivity $k = 0.05$ and $k = 0.17$ SI units. The first value $k = 0.05$ corresponds to polyurethane and the value $k = 0.17$ corresponds to asbestos felt.

Table 11.4.1: Table of values of heat loss from an insulated water pipe for different thicknesses (in mm) of insulation and for two different conductivities (in SI units).

Insulation thickness $(b - a)$	Rate of heat loss $k = 0.05$	Rate of heat loss $k = 0.17$
0	21.2	21.2
1	20.5	22.0
2	19.8	22.6
5	18.0	24.2
10	15.5	25.7

unusual behaviour

Note from Table 11.4.1 that the rate of heat loss for conductivity $k = 0.05$ decreases as the thickness of the insulation increases, as we might expect. However, for the larger conductivity of $k = 0.17$, we see that the rate of heat loss increases as the thickness of the insulation is increased. In this case it appears that we are better off not insulating the pipe at all!

graph the heat loss

Figure 11.4.2 gives a graph of the rate of heat loss from the insulated water pipe as a function of the outer radius of the insulation b. Note that there is a maximum rate of heat loss at the value $b = k/h$.

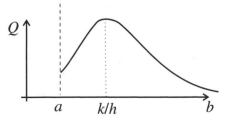

Figure 11.4.2: Diagram of the rate of heat loss from an insulated water pipe versus the outer radius of the insulation. There is a maximum at $b = k/h$. It illustrates the general behaviour for any values of k, h and b.

This suggests that there is a critical value for the outer radius b. For the case cited above, where the conductivity is $k = 0.05$, the critical value of the outer radius is $10\,\text{mm}$, but for $k = 0.17$ the critical radius is $34\,\text{mm}$. In this latter case one would need an outer radius of greater than $34\,\text{mm}$ before the insulation would have the effect of reducing the heat loss from the pipe. If this critical value is less than the inner radius, then placing insulation around the pipe has the desired effect of reducing the heat loss from it. If the critical value is greater than the inner radius, however, then it may be better not to insulate the pipe at all.

critical radius

The physical explanation for this is that there is a balance between two competing effects. Increasing the thickness of the insulation means that there is more insulation through which the heat need flow. However, increasing the thickness of the insulation also increases the surface area through which heat is lost.

reason for the surprising behaviour

Solid cylinders and spheres

A different problem is that of a solid cylinder (or sphere). For this type of problem we usually do not have a boundary condition given at $r = 0$. Note that the general solution for a cylinder (2) cannot be evaluated at $r = 0$ due to the $\ln(r)$ term, which tends to $-\infty$. And for a solid sphere, there is a term r^{-1} which becomes infinite at $r = 0$.

solid cylinders and spheres

For these problems a 'hidden', or implicit, boundary condition is applied. Physically we require the equilibrium temperature inside the cylinder to be finite. The only way this can be true is if the arbitrary constant coefficient in the $\ln(r)$ term is identically zero. This condition sets the value of one of the arbitrary constants and hence acts like a boundary condition, but in an implicit way.

implicit boundary condition

> **Skills developed here:**
> - Solve for the temperature in a cylindrical shell, with any combination of prescribed temperature, heat flux or Newton cooling boundary conditions.
> - Solve for temperature in a spherical shell with any combination of prescribed temperature, heat flux or Newton cooling boundary conditions.
> - Extend to a cylinder or sphere with internal heat production.
> - Solve for a solid cylinder or sphere with an implicit boundary condition, using that the temperature is finite at the centre ($r = 0$) to set one of the arbitrary constants to zero.

11.5   Cooling a computer chip

overview
We now consider a single heat fin from a set of heat fins attached to a computer chip. We solve the appropriate differential equation to find an analytic solution, and then examine this to understand what it tells us about the general behaviour of the temperature in the chip.

Problem description

CPU chips generate heat
In a computer, modern processing chips can generate a lot of heat. If the heat is not dissipated adequately the temperature of the chip can increase significantly, possibly causing the chip to cease functioning properly or to become permanently damaged. The Cyrix 6x86-P166 chip generates heat at a rate of approximately 24 W and Cyrix advise that the chip should not reach a temperature of over 70 °C (see Cyrix (1998)).

using heat fins
One method of dissipating heat is to mount heat fins on the chip with a cooling fan to blow air between the fins. For the Cyrix chip, 12 heat fins are used with a fan mounted above them. Each heat fin has dimensions of width $w = 5.1$ cm, thickness $b = 1$ mm and length $\ell = 1.5$ cm. A schematic diagram of the heat fin assembly and fan for a Cyrix 6x86 CPU chip is shown in Figure 11.5.1.

a quick calculation
The Cyrix 6x86 chip, for example, generates heat at a rate of $q = 24$ Watts, Cyrix (1998). If the only way the heat can dissipate is from the

Figure 11.5.1: Heat dissipation assembly for a Cyrix CPU computer chip. It consists of a cooling fan and 12 heat fins attached to the chip.

surface of the chip, then we can model the temperature of the chip by using Newton's law of cooling.

Example 1: *Use Newton's law of cooling to calculate the equilibrium temperature of the chip if there are no cooling fins and the dimensions of the chip are 5.1 cm × 5.3 cm.*

Solution: *Newton's law of cooling states that the rate of heat loss Q is*

$$Q = hS\Delta U, \tag{1}$$

where ΔU is the temperature difference between the chip and the ambient temperature. For the Cyrix chip, with surface area $S = 5.1\,\text{cm} \times 5.3\,\text{cm} = 2.7 \times 10^{-3}\,\text{m}^2$ and using a value of $h = 5\,\text{W}\,\text{m}^{-2}\,°\text{C}^{-1}$ we obtain

$$\Delta U \simeq \frac{24}{5 \times 2.7 \times 10^{-3}} \simeq 1800\,°\text{C} \tag{2}$$

for the temperature difference between the chip face and the surrounding air.

Of course, the chip would cease to function long before this temperature were reached. Clearly we need to dissipate the heat more efficiently. interpretation

We are interested in calculating the temperature of the chip with the heat fins attached. We are also interested in answering the question: a question

>   How does this temperature vary if we increase or decrease the length of the heat fins?

Model assumptions and approach

We develop a model for a single heat fin where the geometry of the model is shown in Figure 11.5.2. single heat fin model

We make some simplifying assumptions on which to build the mathematical model. some assumptions

- We assume the heat loss is distributed evenly over the 12 heat fins.

Figure 11.5.2: Diagram of model used for a single heat fin. Heat enters at $x = 0$ (left end) with flux q generated from the computer chip. We assume the heat flux from $x = \ell$ (right end) is zero.

- We assume the heat loss from the sides of the fin to be negligible compared with the heat loss from the top and bottom.
- We also neglect any heat loss from the end of the heat fin.

The differential equation

heat flux for a single heat fin

We can calculate the heat flux entering a single heat fin. This heat flux, denoted by q, is obtained by taking the rate of heat loss for a single fin $24/12 = 2\,\mathrm{W}$ and then dividing by the area of the cross-section, bw. This gives a heat flux of $q = 3.92 \times 10^4\,\mathrm{W/m^2}$.

equation at thermal equilibrium

In Chapter 9 (Section 9.7), we formulated an appropriate differential equation for the equilibrium temperature distribution, $U(x)$, along a single heat fin. This differential equation is

$$\frac{d^2U}{dx^2} = \beta(U - u_s), \qquad \text{where} \quad \beta = \frac{2h}{kb}. \tag{3}$$

Here $U(x)$ is the temperature, x the distance along the heat fin, k the conductivity, h the convective heat transfer coefficient and u_s the temperature of the surrounding air. This differential equation is an inhomogeneous, constant coefficient, second-order differential equation.

relative temperature

For simplicity we set the temperature of the heat fin relative to the temperature of the surrounding air. On this temperature scale, $u_s = 0$ and the differential equation (3) becomes

$$\frac{d^2U}{dx^2} = \beta U \qquad \text{where} \quad \beta = \frac{2h}{kb} \tag{4}$$

which is now homogeneous. This is not a major restriction as it is simple, with a change of variable, to convert a problem with non-zero u_s to one with $u_s = 0$. Equation(4) represents a *heat balance*. The LHS term comes from heat conduction and the RHS comes from heat lost to the surrounding air.

Analytic solution

First we derive a general solution to the differential equation. Then we find the associated constants and hence the particular solution, using the boundary conditions.

overview

Example 2: *Find the general solution to the differential equation (4).*

Solution: *The general solution can be found by substituting a trial solution of the form $U(x) = e^{\lambda x}$ into the differential equation (4). This gives rise to the characteristic equation $\lambda^2 = \beta$. Since β is a positive constant the roots of the characteristic equation are $\pm\sqrt{\beta}$. A general solution of the differential equation (4) is then*

$$U(x) = C_1 e^{\sqrt{\beta}x} + C_2 e^{-\sqrt{\beta}x} \qquad (5)$$

where C_1 and C_2 are arbitrary constants.

It is also possible to express this solution in terms of the hyperbolic functions[1] $\cosh(\sqrt{\beta}x)$ and $\sinh(\sqrt{\beta}x)$. Often this form can lead to simpler algebra when applying certain types of boundary conditions. (Further details of hyperbolic functions can be found in Appendix B.4.)

solution in hyperbolic form

The boundary conditions

We require two boundary conditions, one at the end $x = 0$, the origin, and one at the end $x = \ell$, to give us two simultaneous equations for the two arbitrary constants C_1 and C_2. At $x = 0$ we know the value of the heat flux to be q. We write this boundary condition as

two boundary conditions

$$J(0) = q. \qquad (6)$$

---

[1]The hyperbolic functions cosh and sinh are defined by

$$\cosh(z) = \frac{1}{2}\left(e^z + e^{-z}\right), \quad \sinh(z) = \frac{1}{2}\left(e^z - e^{-z}\right).$$

At the end $x = \ell$ we have assumed we can neglect the heat loss. This is expressed mathematically as

$$J(\ell) = 0. \tag{7}$$

Example 3: *Apply the boundary conditions to obtain an expression for the temperature along the heat fin.*

Solution: *Both boundary conditions involve the heat flux J. From Fourier's law $(J(x) = -k\,dU/dx)$ and the general solution (5), we obtain*

$$J(x) = -k\sqrt{\beta}\left(C_1 e^{\sqrt{\beta}x} - C_2 e^{-\sqrt{\beta}x}\right). \tag{8}$$

Applying the first boundary condition $J(0) = q$ and evaluating this at $x = 0$ gives the equation

$$C_1 - C_2 = \frac{-q}{k\sqrt{\beta}}. \tag{9}$$

To apply the second boundary condition, $J(\ell) = 0$, we substitute $x = \ell$ into the expression for $J(x)$ to get

$$C_1 e^{\sqrt{\beta}\ell} - C_2 e^{-\sqrt{\beta}\ell} = 0. \tag{10}$$

We solve equations (9) and (10) simultaneously giving an expression for C_2. Substituting this into (9) gives

$$\left(1 - e^{2\sqrt{\beta}\ell}\right)C_1 = \frac{-q}{k\sqrt{\beta}},$$

which allows us to solve for C_1 as

$$C_1 = \frac{q}{k\sqrt{\beta}}\left(\frac{1}{e^{2\sqrt{\beta}\ell} - 1}\right). \tag{11}$$

Substituting this back into our expression for C_2 we obtain

$$C_2 = \frac{q}{k\sqrt{\beta}}\left(\frac{e^{2\sqrt{\beta}\ell}}{e^{2\sqrt{\beta}\ell} - 1}\right). \tag{12}$$

With these expressions for C_1 and C_2 we have, from the general solution (5), an expression for the temperature along the heat fin, namely

$$U(x) = \frac{q}{k\sqrt{\beta}}\left(\frac{e^{\sqrt{\beta}x} + e^{2\sqrt{\beta}\ell}e^{-\sqrt{\beta}x}}{e^{2\sqrt{\beta}\ell} - 1}\right). \tag{13}$$

From this general solution for the temperature in equation (13), we can immediately write down an expression for the temperature of the computer CPU chip, which is the quantity of interest. This is the temperature at $x = 0$ and is given by

$$U(0) = \frac{q}{k\sqrt{\beta}} \left(\frac{e^{2\sqrt{\beta}\ell} + 1}{e^{2\sqrt{\beta}\ell} - 1} \right). \tag{14}$$

temperature of the chip

Results and interpretation

Using some appropriate parameter values for copper heat fins we take $k = 386$ and $h = 32$ SI units. The value of the convective heat transfer coefficient h is higher than the one we used in the calculation in equation (2) to incorporate the effect of the fan.[2] We also have $b = 1\,\mathrm{mm} = 10^{-3}\,\mathrm{m}$, $w = 5.1 \times 10^{-2}\,\mathrm{m}$ and $\ell = 1.5\,\mathrm{cm} = 1.5 \times 10^{-2}\,\mathrm{m}$. We then calculate $\beta = 174$ (in SI units). We have previously calculated $q = 3.92 \times 10^4\,\mathrm{W/m^2}$ for each heat fin. Substituting these values into (14) we obtain

some parameter values

$$U(0) = 41.4\,^{\circ}\mathrm{C}.$$

This is the temperature rise above the ambient temperature, or the surrounding temperature, as our model assumed $u_s = 0$.

In Figure 11.5.3, we plot the temperature along the length of the heat fin. The temperature is highest at the chip surface and decreases along the length of the heat fin, as expected.

temperature variation along the heat fin

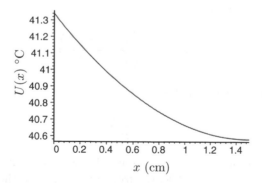

Figure 11.5.3: Graph of temperature in °C along the length of the heat fin.

It is more interesting to sketch a graph of the temperature of the com-

chip temperature versus length of heat fin

---

[2]To calculate h a formula from Holman (1981) was used, assuming an air speed of $1\,\mathrm{m\,s^{-1}}$.

puter chip for different lengths of the heat fin. This graph, illustrated in
Figure 11.5.4, provides some insight into how long to make the heat fin
for the best results. The temperature of the chip decreases as we increase
the length of the heat fin. Physically this makes sense, since increasing
the length of the heat fin increases the area through which heat is lost.
However, $U(0)$ appears to approach a limiting value as ℓ (the length of
the fin) increases.

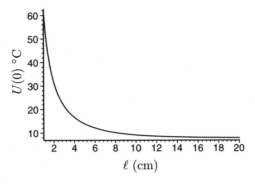

Figure 11.5.4: The temperature $U(0)$ in °C of the computer chip as a function of the
length l of the heat fin.

investigat-
ing large
ℓ

From (14) we can calculate the observed limiting temperature as we let
ℓ become very large. This will provide a lower bound for the coolest
temperature which the chip can reach. To find this limit we let $\ell \to \infty$
in (14). Dividing both top and bottom by $e^{2\sqrt{\beta}\ell}$, to avoid infinite limits
in the numerator and denominator, yields

$$U(0) = \frac{q}{k\sqrt{\beta}} \left(\frac{1 + e^{-2\sqrt{\beta}\ell}}{1 - e^{-2\sqrt{\beta}\ell}} \right). \tag{15}$$

Hence, the limiting value as $\ell \to \infty$ is

$$U_{\min} = \frac{q}{k\sqrt{\beta}}. \tag{16}$$

For the same parameter values as above, this gives a minimum tempera-
ture for U of $U_{\min} = 8.1\,°\mathrm{C}$. Clearly, from Figure 11.5.4, the temperature
is close to 8.1°C for a fin length of much less than $\ell = \infty$, and thus not
much is gained by increasing the fin length beyond a certain point (say
10 cm). However, there is a substantial gain by increasing the length
from 2 to 4 cm.

> **Skills developed here:**
> - Obtaining the general solution to the homogeneous heat fin equations.
> - Applying boundary conditions to this general solution, including other combinations of boundary conditions, such as one or both boundaries at prescribed temperatures.
> - Interpreting the solution for the temperature of a computer chip to find a theoretical minimum temperature.

11.6   Exercises for Chapter 11

1. Solving a boundary value problem. *Consider the differential equation*

$$\frac{d^2U}{dx^2} = 1$$

which satisfies the boundary conditions

$$U(0) = 1, \qquad U(2) = 0.$$

(a) *Find the general solution.*

(b) *Apply the boundary conditions to find the solution.*

(c) *Find the solution if the boundary condition at $x = 1$ is replaced with the boundary condition (at $x = 1$)*

$$\frac{dU}{dx} = 0.$$

2. Formulating a boundary value problem. *Write down, in mathematical form, boundary conditions for the following. Express them in terms of temperature U or heat flux J.*

(a) *A wall of a furnace gains heat from the end $x = 0$ at a fixed rate of $300\,\mathrm{W}$, while the other end is maintained at temperature $30\,^{\circ}\mathrm{C}$.*

(b) The outside of a wall of a house loses heat according to Newton's law of cooling, to the surrounding air at temperature $10\,^{\circ}\mathrm{C}$, while the inside gains heat according to Newton's law of cooling from the inside of the house which is at a temperature of $25\,^{\circ}\mathrm{C}$.

(c) A slab of material has its right end held at temperature $80\,^{\circ}\mathrm{C}$ and the left end gaining heat according to Newton's law of cooling.

3. Heat flux through a wall. In an internal wall of thickness d, the temperature at $x = 0$ is maintained (by a thermostat) at temperature u_1 and the other side is maintained at temperature u_2.

(a) Find an expression for the temperature at any point inside the wall.

(b) Hence find an expression for the heat flux through the wall.

4. Heat flux through a window. Consider the flow of heat through a window pane of width ℓ. Assume that heat is gained from the inside and lost from the outside of the window according to Newton's law of cooling. The temperature of the air on the inside is u_i and the temperature on the outside is u_o. The Newton cooling coefficients are h_i and h_o for the inside surface and outside surface of the window, respectively. You are given that the equilibrium temperature, inside the window, satisfies

$$\frac{d^2 U}{dx^2} = 0.$$

(a) What is the general solution for the equilibrium temperature?

(b) What are the boundary conditions for this problem?

(c) Find the equilibrium temperature as a function of x.

(d) Hence deduce the heat flux from the outer surface $x = \ell$ is given by the expression

$$\frac{k(u_i - u_o)}{\ell + \frac{k}{h_i} + \frac{k}{h_o}}.$$

(e) Interpret this formula in terms of heat resistances.

5. Heat flux and thermal resistance. *Suppose we have two different materials with widths d_1 and d_2, conductivities k_1 and k_2, and which are joined together at $x = 0$. The equilibrium temperatures $U_1(x)$ and $U_2(x)$ both satisfy the basic equilibrium heat equations*

$$\frac{d^2 U_1}{dx^2} = 0, \qquad \frac{d^2 U_2}{dx^2} = 0.$$

The temperature on the inside $(x = -d_1)$ is held (by a thermostat) at a temperature u_i and the temperature on the outside $(x = d_2)$ is held (by a thermostat) at a temperature u_o.

(a) *Find expressions for the two temperatures $U_1(x)$ and $U_2(x)$. You may assume the temperature and heat flux are continuous where the materials join at $x = 0$.*

(b) *Show the heat flux is given by*

$$J = \frac{u_i - u_0}{\frac{d_1}{k_1} + \frac{d_2}{k_2}}.$$

(c) *Interpret this in terms of thermal resistances.*

6. Internal floor heating. *Houses are often heated via a heating source located within a concrete slab floor (e.g. by attaching an electrical resistance wire to the reinforcing bars embedded within the concrete). Suppose such a heat source is located in a slab of thickness d and generates heat at a constant rate q (per unit volume). The equilibrium temperature $U(x)$ satisfies the differential equation*

$$k\frac{d^2 U}{dx^2} + q = 0$$

where k is the thermal conductivity. If each side of the slab is maintained at the same temperature u_0 find an expression for $U(x)$, the temperature inside the slab.

7. Nuclear reactor. *In the shielding wall of a nuclear reactor the rate of heat generated by gamma rays interacting with the wall, the internal heat generation (per unit volume) $q(x)$, may be modelled by*

$$q(x) = q_0 e^{-ax}$$

where q_0 and a are positive constants. The equilibrium temperature inside the wall satisfies the differential equation

$$k\frac{d^2 U}{dx^2} + q(x) = 0.$$

(a) Sketch $q(x)$ as a function of x.

(b) Determine the general solution for the equilibrium temperature.

(c) Given that the inside of the wall $x = 0$ is maintained at a constant temperature of u_1 and the outside of the wall is maintained at a constant temperature of u_2, find an expression for the equilibrium temperature.

8. Combining heat resistances. *To prove that the individual heat resistances are additive, consider*

$$J = \frac{u_1 - u_a}{R_g} = \frac{u_a - u_b}{R_a} = \frac{u_b - u_2}{R_g}.$$

Find expressions for u_a and u_b and substitute them back into one of the equations for J. This will establish equation (1) of the double glazing case study (Section 11.3).

9. Double glazing. *Following the case study on double glazing (Section 11.3), assume that the inside room temperature is $20\,°C$, the outside environment temperature is $0\,°C$, the thickness of each of the two panes of glass is $d = 4\,mm$, the conductivity of glass is $k_g = 0.8\,J\,m^{-1}\,s^{-1}\,°C^{-1}$ and the conductivity of air is $k_a = 0.05\,J\,m^{-1}\,s^{-1}\,°C^{-1}$. Calculate the heat flux for a gap width of twice the thickness of a single pane of glass. Compare this to the heat flux through a single pane of glass with width $2d$.*

10. Comparing glazing models. *Using Maple, draw graphs of the three delta functions in the case study on double glazing (Section 11.3). Look at limits over time, and with the standard specifications of manufacturers how these three compare.*

11. Cooling lizards. *Lizards at rest are known to generate heat internally at a rate of $q = 0.5\,W/kg$. We might model the lizard as a cylinder of radius a and length ℓ. The equilibrium temperature satisfies the differential equation*

$$\frac{k}{r}\frac{d}{dr}\left(r\frac{dU}{dr}\right) = \rho q,$$

where ρ is the (average) density of lizard tissue, k is the conductivity, and r measures radial distance from the centre of the lizards body (a cylinder).

(a) Starting from a suitable word equation, outline a derivation of this differential equation.

(b) Find the general solution.

(c) For a solid cylinder (our lizard) we require the equilibrium temperature to be finite at $r = 0$. What does this tell you about the value of one of the arbitrary constants?

(d) Suppose heat is lost to the surroundings according to Newton's law of cooling, where the surroundings are at temperature u_0. Find an expression for the equilibrium temperature inside the lizard (cylinder). Where is the equilibrium temperature a maximum?

12. Equilibrium temperature in a spherical shell. The equilibrium temperature inside a spherical shell, of inner radius a and outer radius b satisfies the differential equation

$$\frac{d}{dr}\left(r^2 \frac{dU}{dr}\right) = 0.$$

(a) Find the general solution of the differential equation.

(b) Find the equilibrium temperature if the inner surface $r = a$ is maintained at temperature u_1 and the outer surface $r = b$ is maintained at temperature u_2.

13. Hyperbolic functions. Verify that

$$U(x) = D_2 \cosh(\sqrt{\beta}x) + D_1 \sinh(\sqrt{\beta}x)$$

is a general solution of the differential equation

$$\frac{d^2U}{dx^2} = \beta U.$$

Now find the solution which satisfies the boundary conditions applied at $x = 0$,

$$U(0) = 1, \qquad \frac{dU(x)}{dx} = 0 \quad \text{at } x = 0.$$

For details on hyperbolic functions see Appendix B.4.

14. Cooling an engine. Heat fins are often attached to engines to cool them when in operation. Find an expression for the temperature inside a heat fin where one end of the heat fin is attached to a metal engine, and is maintained at temperature u_1. You may neglect any heat loss from the other end of the heat fin.

The equilibrium temperature satisfies the differential equation

$$\frac{d^2U}{dx^2} = \beta U,$$

where β is a constant and the temperature of the surroundings is zero.

15. Rectangular heat fin. *In Section 11.5 it was shown that the relative temperature in a rectangular heat fin satisfies the differential equation*

$$\frac{d^2 U}{dx^2} = a^2 U, \qquad a^2 = \frac{2h}{kb},$$

where b is the thickness of the heat fin, k is the conductivity, h the Newton cooling coefficient, and we have taken the temperature of the surrounding air to be zero (see Section 11.5, equation (3)).

(a) *Write down the boundary conditions assuming that the temperature at the end attached to the heat source is u_1 and the heat lost from the other end is given by Newton's law of cooling.*

(b) *Hence obtain an expression for the equilibrium temperature. (Hint: you will find the algebra easier if you use the hyperbolic form of the general solution*

$$U(x) = C_1 \cosh(ax) + C_2 \sinh(ax).$$

where C_1 and C_2 are arbitrary constants.)

16. Formulating boundary conditions. *In the following, write down boundary conditions involving the temperature U or the heat flux J. (It may be useful to draw a diagram).*

(a) *In a spherical shell defined by $r = a$ to $r = b$, heat flows radially. The temperature on the outside is u_2 and the inside of the shell receives heat at a rate of $10\,\mathrm{W}$.*

(b) *In a wall of length L the temperature at one end $x = L$ is u_1 and the temperature at the other end is u_2.*

(c) *In a cylindrical heat fin, of small radius a, heat is conducted along the length of the wire. The heat fin is attached to a surface whose temperature is maintained at $100\,^\circ\mathrm{C}$. The other end of the heat fin is assumed to lose no heat.*

(d) *A wall located between $x = 0$ and $x = \ell$ has temperature at $x = \ell$ fixed at u_1. The other side of the wall gains heat from the surroundings, at temperature u_2, according to Newton's law of heat exchange.*

(e) *The material inside a cylindrical insulating layer, of inner radius 1 and outer radius 2, gains heat from the surroundings at temperature $3u_1$, according to Newton's law of heat exchange. The temperature at the inner radius is maintained at u_1.*

Part IV

Models with Partial
Differential Equations

Chapter 12

Introduction to partial differential equations

This chapter provides a brief introduction to the derivation of partial differential equations (PDEs). It is a natural extension of the previous models of heat and lake pollution, and indicates how they might be improved and generalised. A case study is included to illustrate how such equations are applied in practice.

12.1   The heat equation

overview   In previous chapters we have mentioned the dependence of certain functions on two independent variables: time and position. We now see how to derive a PDE for the temperature when the system is not in thermal equilibrium, that is when there are changes in both of these independent variables.

Introduction

tempera-
ture
depends
on both x
and t

So far we have assumed thermal equilibrium so that, for all calculations, the heat flux is dependent on x (position) alone. Let us now consider what happens when the material is not in thermal equilibrium. We have to take account of the heat used in raising (or lowering) the temperature. The temperature will now depend on both position x and time t. Hence we let $U(x,t)$ be the temperature, and denote the heat flux as $J(x,t)$, at position x and time t.

Formulating a PDE

the word
equation

We return to the problem of heat flow through a wall, where the wall is now no longer in thermal equilibrium. From Section 9.5, the word equation describing the rate of change of heat content in a small section x to $x + \Delta x$ of the wall is

$$\left\{\begin{array}{c} rate\ of \\ change\ of \\ heat\ in\ section \end{array}\right\} = \left\{\begin{array}{c} rate\ heat \\ conducted \\ in\ at\ x \end{array}\right\} - \left\{\begin{array}{c} rate\ heat \\ conducted \\ out\ at\ x + \Delta x \end{array}\right\}. \tag{1}$$

Since the temperature is no longer in thermal equilibrium we must regard the temperature and heat flux as functions of two variables, $U(x,t)$ and $J(x,t)$, where x measures distance and t measures time.

formulat-
ing the
PDE

The rate of change of heat is calculated from the rate at which heat is 'converted' into an increase in temperature. For the LHS of this equation we use equation (1), from Section 9.2, which relates heat to temperature as

$$\left\{\begin{array}{c} rate\ of \\ change\ of \\ heat\ in\ section \end{array}\right\} = c \times \{mass\} \times \frac{\partial U}{\partial t}, \tag{2}$$

with c the specific heat. Note the partial derivative with respect to time, since U depends on both x and t. (For more details on partial derivatives see Appendix B.2.) The mass of the section can be expressed in terms of the wall density ρ, as $\rho A \Delta x$, where A is the usual surface area.

The heat conduction terms can be expressed in terms of the heat flux

$$\left\{\begin{array}{c} \text{rate heat} \\ \text{conducted} \\ \text{in at } x \end{array}\right\} = AJ(x,t), \qquad \left\{\begin{array}{c} \text{rate heat} \\ \text{conducted} \\ \text{out at } x + \Delta x \end{array}\right\} = AJ(x + \Delta x, t). \quad (3)$$

Then, from equations (1) and (2),

$$c\rho \frac{\partial U}{\partial t} \Delta x = -J(x + \Delta x, t) + J(x, t). \quad (4)$$

Example 1: For equation (4) take the limit as $\Delta x \to 0$ and then use Fourier's law to obtain a differential equation for the temperature.

Solution: We first divide through by Δx to obtain

$$c\rho \frac{\partial U}{\partial t} = -\left[\frac{J(x + \Delta x, t) - J(x, t)}{\Delta x}\right].$$

Recall the definition of a partial derivative,

$$\frac{\partial J}{\partial x} = \lim_{\Delta x \to 0} \frac{J(x + \Delta x, t) - J(x, t)}{\Delta x},$$

where we are looking at the change in J with respect to the x variable, while the t variable remains constant. If we now take the limit $\Delta x \to 0$, we obtain

$$c\rho \frac{\partial U}{\partial t} = -\frac{\partial J}{\partial x}. \quad (5)$$

Finally, we apply Fourier's law to express the equation in terms of temperature. For quantities which vary with both space x and time t, Fourier's law must be written as

$$J(x, t) = -k \frac{\partial U}{\partial x}. \quad (6)$$

Substituting Fourier's law (6) back into (5) gives

$$c\rho \frac{\partial U}{\partial t} = -\frac{\partial}{\partial x}\left(-k \frac{\partial U}{\partial x}\right).$$

For constant k this simplifies to

$$\frac{\partial U}{\partial t} = \alpha \frac{\partial^2 U}{\partial x^2}, \qquad \text{where} \quad \alpha = \frac{k}{c\rho}. \quad (7)$$

the heat
equation

This second-order partial differential equation is called the *heat equation*, and the parameter α is known as the thermal diffusivity. Note that for equilibrium temperatures, that is for $\partial U/\partial t = 0$, (7) reduces to

$$\frac{d^2 U}{dx^2} = 0,$$

which was the ordinary differential equation obtained in Section 9.5.

Other applications of the heat equation

diffusion

This heat equation (7) is widely known, arising in many other applications. One area of application is finance, where another partial differential equation called the Black–Scholes equation for share option prices, can be transformed into the heat equation by a change of variables. See Wilmott (1998). Variants of this equation are also used in population dynamics with spatial dependence, as in Murray (1990).

lake
pollution

Another application is in the diffusion of chemicals. Partial differential equations would provide a valuable extension to the model we developed in Chapter 2 on lake pollution. There we assumed that pollution concentration within the lake was uniformly distributed, while in reality it is a function of both time and distance from its source.

Skills developed here:

- Formulate a PDE to describe the diffusion of a pollutant from some source in a lake, such as a river.
- Distinguish cases where the use of an ODE is appropriate, and where a PDE is appropriate.

12.2   *Oscillating soil temperatures*

overview

We consider an interesting and fundamental property of heat processes, namely the oscillatory nature of temperatures of large masses, such as the ocean and the soil. It will provide a theoretical basis for the case study concerning the detection of land mines.

Problem description

It is well-known that soil acts as an excellent thermal insulator for varia- wine
tions in temperature. Since wine demands an environment with as close cellars and
to a constant temperature as possible, wine cellars are constructed un- weather
derground making use of this thermal property. The thermal mass of the
ocean, which has a substantial impact on the milder climatic temperatures experienced in coastal regions, is another well-known case of this
phenomenon.

The theory we develop concerns the soil mass and its temperatures over soil tem-
time. We intend to develop a model which reflects the oscillatory nature peratures
of this thermal mass.

Model assumptions and approach

Let us consider the problem of determining the temperature in the soil model
due to periodic heating by day and cooling by night. We consider an geometry
infinite soil depth, with the surface $x = 0$ corresponding to the soil
surface and the x–axis pointing into the ground.

We let $U(x,t)$ denote the relative temperature variation in the ground, notation
at some time t, and at a distance x into the ground. We measure the
temperature relative to some mean temperature, in this case the mean
surface temperature. From Section 12.1 the governing differential equation is the standard non-equilibrium heat conduction equation,

$$\frac{\partial U}{\partial t} = \alpha \frac{\partial^2 U}{\partial x^2}, \tag{1}$$

where α is the thermal diffusivity for soil. (Recall that $\alpha = k/c\rho$, with k
the conductivity, ρ the density and c the specific heat.)

Boundary conditions for the PDE

On the surface of the ground, $x = 0$, and we have to specify the heat flux boundary
$Q(t)$ (for which we develop a model shortly). Due to heating by the sun condition
during the day and then cooling at night, we require a suitable function on ground
which oscillates with time. surface

The change in temperature is driven by a diurnal cycle. Thus we model diurnal
the surface heat flux $Q(t)$ with a simple trigonometric function, $\cos(\omega t)$, cycle
where ω is determined from the period being (1 day), that is,

$$Q(t) = q\cos(\omega t), \qquad \omega = \frac{2\pi}{60 \times 60 \times 24} \simeq 7.3 \times 10^{-5} \text{ seconds.} \tag{2}$$

Hence, using Fourier's law of heat conduction, we can write the boundary condition on the ground surface $(x = 0)$ as

$$-k\frac{\partial U}{\partial x} = q\cos(\omega t) \quad \text{at } x = 0. \tag{3}$$

(We could alternatively have considered a more complicated boundary condition by allowing Newton cooling/heating from the ground surface, but we shall not do so here.)

boundary condition at infinite depth   The second boundary condition states that the temperature variation goes to zero at a long distance from the surface, that is,

$$-k\frac{\partial U}{\partial x} \to 0 \quad \text{as } x \to \infty. \tag{4}$$

Complex valued temperatures

assuming a form of solution   To solve the PDE (1), with the two boundary conditions (3) and (4), we assume a suitable form of the solution. Bearing in mind the PDE (1) has constant coefficients, we would assume exponential forms for both the x-dependence and the t-dependence. However, this would not match with the cosine term in the boundary condition (3). Similarly, assuming a form $U(x,t) = F(x)\cos(\omega t)$, while matching in the boundary condition (3) will not help to solve the PDE (1).

simplifying with complex numbers   A useful trick is to use complex numbers. This technique allows us to simplify the problem we have to solve. We write the problem in terms of complex exponentials creating a problem which is easier to solve. Since our original problem is the real part of the complex problem, the solution to our problem is the real part of the complex problem solution. We represent the trigonometric functions as complex exponentials, using Euler's identity

$$e^{iz} = \cos(z) + i\sin(z), \qquad \text{where} \qquad i = \sqrt{-1}.$$

(See Appendix B.3 for an introduction to complex numbers.) We define a complex valued temperature $\hat{U}$, where U is the real part of $\hat{U}$,

$$U(x,t) = \text{Real}\left(\hat{U}(x,t)\right). \tag{5}$$

complex form of boundary condition   The problem we now solve is

$$\frac{\partial \hat{U}}{\partial t} = \alpha\frac{\partial^2 \hat{U}}{\partial x^2}, \quad -k\frac{\partial \hat{U}}{\partial x} = qe^{i\omega t} \text{ at } x = 0, \quad -k\frac{\partial \hat{U}}{\partial x} = 0 \text{ as } x \to \infty. \tag{6}$$

We then take the real part of the solution, which corresponds to the solution of the original problem (1).

Analytic solution

Assume the solution for the complex valued temperature can be written in the form   *substituting into the PDE*

$$\hat{U}(x,t) = F(x)e^{i\omega t}. \tag{7}$$

Now substituting (7) into equation (6) we obtain

$$i\omega F(x)e^{i\omega t} = \alpha \frac{d^2 F(x)}{dx^2} e^{i\omega t}.$$

Note that the derivative is an ordinary derivative, which is the case since F is a function of x alone. Dividing each equation by $e^{i\omega t}$, we obtain an ODE for $F(x)$,

$$\alpha \frac{d^2 F}{dx^2} = i\omega F. \tag{8}$$

We now find its general solution.

Example 1:   *Find the general solution to the differential equation (8).*

Solution: *The equation is a constant coefficient differential equation. Hence let $F(x) = e^{\lambda x}$. Substituting this solution form into equation (8),*

$$\lambda^2 = \frac{i\omega}{\alpha}, \qquad so \qquad \lambda = \pm\sqrt{\frac{\omega}{\alpha}} \times i^{1/2}.$$

From Appendix B.3 we see that

$$i^{1/2} = \pm\frac{(1+i)}{\sqrt{2}}.$$

So the solution is (substituting back into $F(x) = e^{\lambda x}$ and taking a linear combination of the solutions)

$$F(x) = C_1 e^{\sqrt{\frac{\omega}{2\alpha}}(1+i)x} + C_2 e^{-\sqrt{\frac{\omega}{2\alpha}}(1+i)x},$$

where C_1 and C_2 are arbitrary complex valued constants. Defining

$$b = \sqrt{\frac{\omega}{2\alpha}}(1+i) \tag{9}$$

we can write

$$F(x) = C_1 e^{bx} + C_2 e^{-bx}. \tag{10}$$

---

applying
the
boundary
conditions

Using (7) and (10) the complex valued temperature is given by

$$\hat{U}(x,t) = \left(C_1 e^{bx} + C_2 e^{-bx}\right) e^{i\omega t}, \qquad \text{where} \quad b = \sqrt{\frac{\omega}{2\alpha}}\,(1+i). \qquad (11)$$

The arbitrary constants C_1 and C_2 are found by applying the boundary conditions in (6), as shown in the following example.

Example 2:   *Apply the boundary conditions to find values for C_1 and C_2 and hence obtain the complex valued temperature.*

Solution: *Applying the boundary condition at $x = 0$, $\hat{U} = q e^{i\omega t}$ gives*

$$-kb\,(C_1 - C_2)\,e^{i\omega t} = q e^{i\omega t}, \qquad (12)$$

and hence

$$C_1 - C_2 = \frac{-q}{kb}.$$

Applying the other boundary condition, as $x \to \infty$, noting that $e^{-bx} \to 0$ and $e^{bx} \to \infty$ (see Exercises, Question 4), the condition can only be satisfied if

$$C_1 = 0.$$

Hence we obtain $C_2 = q/kb$.

Substituting for C_1 and C_2 back into (11), the complex valued temperature $\hat{U}(x,t)$ is now given by

$$\hat{U}(x,t) = \frac{q}{kb} e^{-bx} e^{i\omega t}. \qquad (13)$$

---

final real
tempera-
ture
solution

The actual temperature is obtained by taking the real part of this complex valued solution. Using the expression for b from (9), and Euler's identity $e^{i\theta} = \cos(\theta) + i\sin(\theta)$ (whence $(1+i)/\sqrt{2} = e^{i\pi/4}$), allows us to write (13) in the form

$$\hat{U}(x,t) = \left(\frac{q}{k}\sqrt{\frac{\alpha}{\omega}}\right) e^{-i\pi/4} e^{-\sqrt{\omega/2\alpha}(1+i)x} e^{i\omega t}$$

$$= \left(\frac{q}{k}\sqrt{\frac{\alpha}{\omega}} e^{-\sqrt{\omega/2\alpha}\,x}\right) e^{i(\omega t - \sqrt{\omega/2\alpha}\,x - \pi/4)}.$$

Taking the real part of $\hat{U}(x,t)$ (using Euler's identity) yields the expression for the real-valued temperature

$$U(x,t) = \left(\frac{q}{k}\sqrt{\frac{\alpha}{\omega}} e^{-\sqrt{\omega/\alpha}\,x}\right) \cos\left(\omega t - \sqrt{\frac{\omega}{\alpha}}\,x - \pi/4\right). \qquad (14)$$

This can also be done using Maple, with the built-in functions Re and evalc.

```
> restart;
> alias(I=I, i=(-1)^(1/2));
> assume(omega>0, alpha>0, k>0);
> b := (omega/alpha)^(1/2)(1+i)/2^(1/2);
> U := -q/(k*b)*exp(-b*x)*exp(i*omega*t);
> evalc(Re(U));
```

We can summarise the solution technique described above for oscillatory problems involving linear PDEs:

summary of technique

- Introduce a complex valued function $\hat{U}(x,t) = U(x,t)+iV(t)$ where $U(x,t)$ is the solution we seek, and assume its form to be $\hat{U} = F(x)e^{i\omega t}$.
- Substitute this into the PDE, where all the $e^{i\omega t}$ terms cancel leaving a constant coefficient, ODE for $F(x)$.
- Substitute for $F(x) = e^{\lambda x}$ and hence obtain the general solution for $\hat{U}$.
- Apply the boundary conditions to find the arbitrary constants.
- Take the real part of $\hat{U}(x,t)$ to obtain $U(x,t)$ (using Euler's identity).

Interpreting the solution

A general oscillation can be expressed in the form

form of general oscillation

$$A\cos\left(\omega(t - \phi)\right),$$

where A is the *amplitude* of the oscillation and ϕ is the *phase-shift*. The amplitude measures the maximum extent of the oscillation and the phase-shift measures the time difference between the oscillatory behaviour of the above function and that of $A\cos(\omega t)$. The first term in brackets in (14), which is the coefficient of the cosine term, represents the amplitude of the oscillation. This term decreases exponentially with depth x. The phase-shift is proportional to the depth. This means the oscillation lags behind what is happening at the ground surface, by an amount which increases with depth. Furthermore, the temperature on the surface also lags behind the heat flux applied to the surface by an amount $\pi/(4\omega)$ (which is equivalent to 1/8 of a day). Figure 12.2.1 provides a typical sketch of the temperature profiles at different times, which illustrates this decrease in amplitude, and an increase in phase-shift, with depth.

In practice, for soils, growth in the exponential term is sufficiently large for the amplitude to die off quickly. This implies that soil is a very good

practical considerations

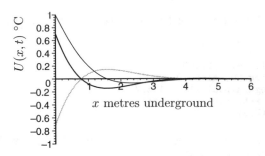

Figure 12.2.1: **Maple** graph of the ground temperature at different times resulting from an oscillating heat flux applied at the surface. (Solution to equation (14) with parameters $k = 1$, $\alpha = 1$ and $q = 1$.) The temperature is plotted at times $\omega t = 0$ (black), $\omega t = \pi/4$ (thin black) and $\omega t = \pi$ (grey) radians, with $\omega = 1$.

insulator from temperature variations caused by daily fluctuations. As was mentioned at the beginning of the section, wine requires a close to constant temperature for storage. Thus wine is stored in cellars, which do not have to be very deep to reduce the temperature variation to less than $1\,°\mathrm{C}$.

Skills developed here:

- Write down appropriate governing equations for a heat conduction problem with an oscillatory driving temperature.
- For a problem with oscillating temperatures, convert to a complex valued problem to simplify the analysis.
- Convert back from the complex valued temperatures to real valued temperatures.

12.3   Case Study: Detecting land mines

In the following case study we use the above theory of oscillatory temperatures of a soil mass to detect non-soil objects buried beneath the earth's surface. The main idea is that such objects change the oscillatory pattern sufficiently to be detected and, furthermore, different buried objects such as land mines or rocks produce changes sufficiently different to allow us to distinguish between them.

This case study is based on Miller (1995), Davies (1994) and discussions with Tony Richings.

Land mines are a serious problem. Decades after a war land mines still maim and cripple civilians with an enormous impact on their communities. Large areas on the Falkland islands are fenced off and unsafe for use as a result of a war fought more than 15 years ago. In November 1998, Hurricane Mitch together with floods and mud slides in Nicaragua and Honduras exposed long buried land mines, carrying them to previously clear regions and creating further distress. Many countries around the world are affected by mines: El Salvador, Ethiopia, Peru, Argentina, Afghanistan, Angola, Mozambique, Iraq, Iran, Kuwait, Korea and Cambodia, to name some. Cambodia is probably one of the worst affected countries in that the mines are widespread and regularly moved by monsoon floods. Although the war there ended some six years ago, there were in 1999 an average of 350 accidents a month due to land mines, which represents a substantial proportion of the total Cambodian population of six million. The safe detection and removal of land mines thus assumes crucial importance.

A common method for spotting land mines is to use metal detectors, however, land mines are now often made of plastic to avoid detection. Another approach, being investigated by the Defence Science and Technology Organisation in Australia, is to use thermal infrared imaging. This works because the temperature of the ground just above a buried land mine is different from the temperature of the surrounding ground. The mine produces a 'hot spot' on the surface during the day and a 'cold spot' at night.

To gain a better understanding of this process, mathematical models were developed by CSIRO (Miller (1995)), which could be used to investigate such effects as the depth and size of the land mine, the thermal properties of land mine and soil, as well as the weather conditions. In fact two models were produced: a full three dimensional model used for quantitative predictions and a one dimensional model for initial investigations. We consider a slightly simplified version of the one dimensional model.

Consider a land mine buried at a depth d of $d = 5\,\text{mm} = 0.005\,\text{m}$. The ground is heated by the sun during the day and cools at night, and we make use of this oscillatory heat flux $Q(t)$ to model the heating and cooling processes.

We neglect any horizontal heat flow in the ground above the land mine and thus consider a one dimensional model where the heat flow is only in the vertical direction. This is illustrated in Figure 12.3.1 below. We also let $U_1(x, t)$ denote the temperature of the soil and $U_2(x, t)$ denote the temperature of the land mine, where t is time and x denotes distance into the ground from the surface. The surface of the soil is taken at $x = -d$ and the interface between the soil and mine at $x = 0$.

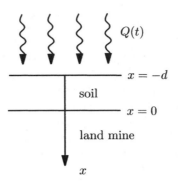

Figure 12.3.1: The 1-D model for thermal detection of a land mine. The soil surface is heated by day and cools at night; this is represented by an oscillatory heat flux $Q(t)$ applied to the soil surface.

The temperature at any point will be oscillating with time due to the diurnal cycle of heating and cooling of the sun, therefore the temperatures will not reach equilibrium but depend continuously on time and depth. The heat equation is used to model these temperatures of the soil above the land mine, as well as the land mine (with the configuration illustrated in Figure 12.3.1) resulting in the equation system

$$\frac{\partial U_1}{\partial t} = \alpha_1 \frac{\partial^2 U_1}{\partial x^2}, \qquad -d < x < 0, \tag{1}$$

$$\frac{\partial U_2}{\partial t} = \alpha_2 \frac{\partial^2 U_2}{\partial x^2}, \qquad 0 < x < \infty, \tag{2}$$

where

$$\alpha_1 = \frac{k_1}{\rho_1 c_1}, \qquad \alpha_2 = \frac{k_2}{\rho_2 c_2}. \tag{3}$$

The parameter values k_1 and k_2 are the conductivities of the soil and land mine respectively, ρ_1 and ρ_2 the densities of the soil and land mine and c_1 and c_2 the specific heats of the soil and mine.

The associated boundary conditions are as follows. On the surface of the soil ($x = -d$) the heat flux has a specified value $Q(t) = q \cos(\omega t)$, where q is the amplitude of the oscillatory heat flux. By Fourier's law this gives the boundary condition at $x = -d$

$$-k_1 \frac{\partial U_1}{\partial x} = q \cos(\omega t) \quad \text{at } x = -d. \tag{4}$$

Where the soil and the top of the land mine join ($x = 0$) we impose a condition for the continuity of temperature and heat flux,

$$U_1(0,t) = U_2(0,t), \qquad -k_1 \frac{\partial U_1}{\partial x} = -k_2 \frac{\partial U_2}{\partial x} \quad \text{at } x = 0. \tag{5}$$

A further boundary condition, specified for $x \to \infty$, states that the rate of flow of heat, or heat flux, is zero. Hence

$$-k_2 \frac{\partial U_2}{\partial x} = 0, \quad \text{at } x = \infty. \tag{6}$$

The governing equations can be solved to obtain a solution for the soil surface temperature (at $x = -d$), given by the real part of the following complex function:

$$\hat{U}_1(-d, t) = \frac{q}{k_1 b_1} \left(\frac{(1+r) + (1-r)e^{-2b_1 d}}{(1+r) - (1-r)e^{-2b_1 d}} \right) e^{i\omega t}, \tag{7}$$

where

$$r = \frac{k_2 b_2}{k_1 b_1}, \qquad b_1 = \sqrt{\frac{\omega}{\alpha_1}} \frac{(1+i)}{\sqrt{2}}, \qquad b_2 = \sqrt{\frac{\omega}{\alpha_2}} \frac{(1+i)}{\sqrt{2}}.$$

Note that, in the limit as $d \to \infty$, this expression for the surface temperature reduces to the (complex valued) temperature for the surface temperature for a semi-infinite soil. (Semi-infinite soil here refers to deep soil with no buried land mine, as would be found in the vicinity of the land mine.) This is the same as setting $r = 1$, so that the thermal parameters of the land mine are the same as those of the surrounding soil.

One thing we notice about the solution (7) is that the temperature is proportional to q, the amplitude of the periodic heat flux term. In practice this quantity might be difficult to measure directly. However, it can be determined by measuring the temperature in soil where there are no buried objects.

We might then be interested in comparing the relative temperature of the soil surface above the land mine with the temperature in an infinite homogeneous soil. This is shown in Figure 12.3.2 for a MK5 land mine, buried under 5 mm of soil. We have taken

$$\alpha_1 = 0.6 \times 10^{-6}, \quad \alpha_2 = 0.1 \times 10^{-6}, \quad k_1 = 1.2, \quad k_2 = 0.26.$$

The temperature on the soil surface above the land mine (solid line) at approximately noon, $t = 0$ days, is greater than the surface temperature for soil without any buried mine (dotted line). Similarly, at night ($t = 0.5$ days) the temperature is lower than that for the soil without a land mine. (Actually, there is a small phase-lag in the soil temperatures without the land mine which can be observed by examining the graph carefully.) Thus, where the land mine is buried, we see a 'hot spot' on the soil surface during the day and a 'cold spot' at night.

A similar calculation for a buried rock ($\alpha_2 = 1.1 \times 10^{-6}$, $k_2 = 2.5$) shows the opposite, that is, a 'cold spot' during the day and a 'hot spot' at night. This makes thermal detection attractive from the point of view of discriminating land mines from other buried objects.

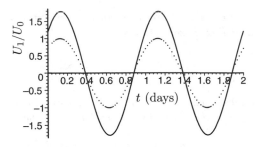

Figure 12.3.2: Normalised temperature of soil surface versus time in days t. The solid line is the normalised surface temperature for an MK5 land mine buried 5 mm below the surface. The dotted line is the normalised temperature of the soil surface when there is no buried object. Both temperatures have been expressed as multiples of the maximum surface temperature, $U_0 = q/(kb)$, for a soil with no buried land mine.

Our model predictions are not as accurate as we might have hoped: the predicted temperature variations are far too large. This is because the model is only one-dimensional and fails to account for heat conduction in the horizontal direction. When the model is extended to include this conduction the temperature variations become far smaller.

Hopefully, such models will be relevant in the clearing of land mines in affected countries. There is a current political drive to ban land mines completely and most countries, including Australia, have signed such an agreement at the Ottawa Convention in 1997. However, the USA, China and Vietnam refused. It is an interesting (and possibly unrelated) fact that while land mines are manufactured in several countries, the USA, China and Vietnam are the principal manufacturers of land mines today.

12.4   Lake pollution revisited

overview

We return to the problem of modelling the levels of pollution in a lake from Section 2.5. There are a number of processes by which a mass of pollutant may travel through a lake. Previously we considered a lake in which the pollution was well mixed and was transported with the water flow. This means of transport is advection, which is equivalent to the convection of heat. Below, we relax the 'well mixed' assumption, allowing the pollution levels to be different at different locations within the lake, although again advection is responsible for the movement of pollution. Modelling this process leads to a PDE for the pollution concentration, which we derive and solve analytically using a technique known as the method of characteristics. We also indicate how `Maple` can be used to solve the equation, and generate solutions for given boundary conditions. In the final section, the much slower diffusive process of pollution transport is introduced, and a PDE to model this developed.

Model assumptions and approach

We assume pollution to travel solely by water flow through the lake, that is, by advection. We assume a unidirectional flow of water from its entrance into, to its exit from, the lake. Thus if x is in the direction of water flow, and y and z are the other axes in 3D space, then, for any fixed x, there is no change in pollution concentration in the y and z directions. Once again, we consider the volume V to be constant, as well as the flow of water into, and out of, the lake, and the pollution source will be from a feeder river at one end. However, unlike in our previous lake pollution model from Section 2.5, in this case pollution will not be assumed to be of uniform concentration throughout the lake. Alternatively, we model pollution levels at any location x within the lake, and at any time t. To do this we make an assumption about the shape of the lake: it is taken as rectangular! As justification for this simplification, in many lakes there is a main channel of flow from entrance to exit, where advection is the dominant means of pollution movement through the lake. This channel is often, roughly, rectangular in shape. As the formulation progresses below, you will see how the non-rectangular shape of real lakes might be dealt with, as well as changes in pollution concentration in the y and z directions.

assumptions

The concentration C of pollutant depends on both position x (horizontal distance from its entry point) in the lake, and time t. Hence we let $C(x,t)$ be the concentration, and examine how it varies throughout the lake with time. The pollution mass flux (across a unit of area per unit of time) will be denoted $J(x,t)$, at position x and time t. We consider the movement of pollution in one direction only: the direction of changing x.

dependence on x and t

Consider a lake, estimated by a rectangular shape of length $(b-a)$, and constant cross-sectional area A, as illustrated in Figure 12.4.1. Thus $a \leq x \leq b$. $F(x,t)$ is taken as the horizontal water flow (across area A per unit of time), so that $F(a,t)$ is the horizontal water flow across area A at the entrance to the lake, and $F(b,t)$ the horizontal water-pollution flow across area A leaving the lake.

simplified lake

Formulating the PDE

The input–output principle, or balance law, provides us with a basis on which to build the equations. Applied to the total mass of pollutant in

the word equation

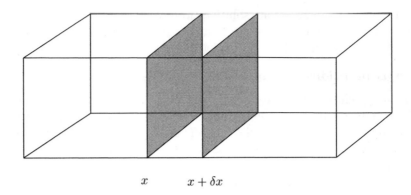

Figure 12.4.1: Diagram of a 'rectangular' lake, from $x = 0$ to $x = b$.

a section of width δx,

$$\left\{ \begin{array}{c} \textit{rate of change} \\ \textit{of pollutant} \\ \textit{mass} \end{array} \right\} = \left\{ \begin{array}{c} \textit{rate pollutant} \\ \textit{mass enters} \\ \textit{at } x \end{array} \right\} - \left\{ \begin{array}{c} \textit{rate pollutant} \\ \textit{mass leaves} \\ \textit{at } x + \delta x \end{array} \right\}. \qquad (1)$$

formulat-
ing the
PDE

If we let $C(x,t)$ denote the pollutant concentration, and let $J(x,t)$ denote the pollutant mass flux at position x and time t, and if A is the area of the cross-section (assumed constant), then equation (1) becomes

$$\frac{\partial}{\partial t}(CA\delta x) = J(x,t)A - J(x+\delta x, t)A. \qquad (2)$$

Note that each of the three terms in (2) has units of mass divided by time. Furthermore, the rate of change with time is given by a partial derivative since the quantities involved depend on both time t and position x. We divide throughout by $A\delta x$ to obtain

$$\frac{\partial C}{\partial t} = -\left(\frac{J(x+\delta x, t) - J(x,t)}{\delta x} \right).$$

Letting $\delta x \to 0$ yields the equation

$$\frac{\partial C}{\partial t} = -\frac{\partial J}{\partial x}. \qquad (3)$$

expression
for the
flux

We have assumed that the pollutant is transported solely by the motion of the fluid: that is, via advection. We need an expression for the pollutant mass flux due to the motion of the water–pollutant mixture. F was defined as the flow rate (litres per second) of the mixture, which we assumed to be constant. So the rate of flow of pollutant mass (at the point x) is FC. The pollutant mass flux (rate of flow of mass per unit area per unit time) is thus given by

$$J(x,t) = \frac{FC(x,t)}{A}. \tag{4}$$

Substituting this into equation (3) yields a PDE for the pollutant concentration,

$$\frac{\partial C}{\partial t} + \frac{F}{A}\frac{\partial C}{\partial x} = 0. \tag{5}$$

On the boundaries of the lake we place conditions, which allow the constants of integration to be established, just as initial conditions were required for the solution of differential equations in previous sections. (For further details on boundary conditions see Section 11.1.) We assume that pollution is entering the lake at $x = a$, such that

> the boundary conditions

$$C(a,t) = \frac{g(t)}{F}, \tag{6}$$

where $g(t)$ is a time dependent function describing the total amount of pollution entering the lake at a. A second condition is placed on the pollution concentration at location x, when $t = 0$: that is, the initial concentration at each location x within the lake.

$$C(x,0) = P(x). \tag{7}$$

Analytical solution

We now need to solve the PDE (5), together with its boundary conditions, equations (6) and (7). We examine the simple case where the system has settled to some steady state, or equilibrium, in the exercises (see Question 9), and deal with the general case below.

> two cases to solve

We are interested in establishing the changing level of pollution over time, at any location in the lake, before any steady-state solution is reached. For this we introduce the *method of characteristics*, in its simplest form. It is a method commonly used for the solution of PDEs, but the idea is not extended in this text. (The reader is referred to any text on PDEs for further details on this, and other, solution methods.)

> general case

Example 1: *Suppose pollution is entering a lake from a river at one end. Modelling the spread of pollution with (5), establish the level of pollution at any time and position (horizontal distance from the pollution source) in the lake.*

Solution: *We need to solve,*

$$\frac{\partial C}{\partial t} + \frac{F}{A}\frac{\partial C}{\partial x} = 0$$

$$C(a, t) = \frac{g(t)}{F} \quad (possibly \quad 0)$$

$$C(x, 0) = P(x). \tag{8}$$

Rearranging the PDE above

$$\frac{\partial C}{\partial t} = -\frac{F}{A}\frac{\partial C}{\partial x}.$$

This implies that the rate of change of C with respect to t is a constant $(-F/A)$ times the rate of change of C with respect to x. Set constant $(F/A) = v$. Then,

$$\frac{\partial C}{\partial x} \Big/ \frac{\partial C}{\partial t} = -\frac{A}{F} = -\frac{1}{v} = -\frac{dt}{dx}.$$

So v is a velocity, and solving the simple differential equation on the RHS, $t = x/v - K_1$ (with K_1 an arbitrary constant), or $x = vt + K$ where $K = vK_1$. What we now show is that pollution concentration C is constant along these lines (curves) defined by x and t.

From the definition for the derivative of a function of two variables $C(x, t)$, we have that

$$\frac{dC(x, t)}{dt} = \frac{\partial C}{\partial x}\frac{\partial x}{\partial t} + \frac{\partial C}{\partial t}\frac{\partial t}{\partial t}$$

$$= \frac{\partial C}{\partial x}\frac{\partial x}{\partial t} + \frac{\partial C}{\partial t}. \tag{9}$$

Return now to equation system (8) and apply this definition. With $x = vt + K$,

$$\frac{\partial C}{\partial t} + \frac{F}{A}\frac{\partial C}{\partial x} = 0 = \frac{dC}{dt}.$$

Thus along the lines $x = vt + K$, the pollutant concentration C is constant. This is a very powerful result (as will be seen below), and these lines are known as characteristics.

Next we establish the value of C on each characteristic (and thus at any specific time and location) using the boundary conditions. Consider the (x, t)-plane:

The particular line passing through $(a, 0)$ can be determined by setting $x = a$ and $t = 0$ in $x = vt + K$, and solving for K. The associated characteristic is

$$t = \frac{A}{F}(x - a).$$

Similarly a series of characteristics can be established. These are illustrated in Figure 12.4.2 in the positive (x, t)-plane.

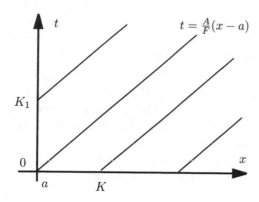

Figure 12.4.2: Diagram of characteristics.

Notice that when $t < A(x - a)/F$ then the characteristics cut the x-axis at $(K, 0)$, which implies that the value of C (which is constant along line $x = vt + K$) is $P(x) = P(K)$, from the second boundary condition in system (8). These values for $P(x)$ are the initial values of pollution at each x-location in the lake.

If $t > A(x - a)/F$ then $t = A(x - a)/F + K_1$ for some positive constant K_1. The point of intersection between this characteristic and the boundary $x = a$, is at $t = K_1$. On substitution into the first boundary condition of system (8),

$$C(a, t) = \frac{g(t - A(x - a)/F)}{F} = \frac{g(K_1)}{F}.$$

Thus we have as a solution for any location x, $a \le x \le b$, and with $v = F/A$,

$$C(x, t) = \begin{cases} g\left(t - \frac{A(x-a)}{F}\right)/F & \text{if } t > \frac{A(x-a)}{F} \\ P(x - \frac{Ft}{A}) & \text{if } t < \frac{A(x-a)}{F}. \end{cases} \tag{10}$$

We can now look at the cleaning time this model of a lake predicts, in the case where no further pollution enters the lake, and compare it with the cleaning time in our previous model from Section 2.5 where the automatic mixing of a pollutant was assumed. This is a particular case of the example above, when no pollution enters the lake, and thus the boundary condition at $x = a$ and $t > 0$ is $g(t) = 0$, so that $C(a,t) = 0$. The resulting solution is

$$C(x,t) = \begin{cases} 0 & \text{if} \quad t > \frac{x-a}{v} \\ P(x - vt) & \text{if} \quad t < \frac{x-a}{v} . \end{cases} \qquad (11)$$

Note that for $t > (b - a)/v$ (or $x > b$) there are no positive values for the pollution concentration in the lake, whence it is clean. The complete cleaning time is given by $(b - a)/v = A(b - a)/F$. Thus v is the velocity, in the x-direction, of the pollution through the lake.

wave motion Equation (11) describes a very important phenomenon: that of *wave motion*. Over time, the wave (of pollution) propagates in the x-direction, with velocity v. This notion can be understood by examining the equation solution as t increases. Notice that $P(a) = P(x - vt)$ for some x and t (from (11)), which implies that the pollution level at location $x - vt$ is the same as that at a. Alternatively stated, wherever $a = x - vt$ or $x = a + vt$. This latter equation implies that the pollution level at location a will have moved in the positive x-direction with velocity v in time t: the wave shifts to the right with velocity v. This moving wave phenomenon via advection is clarified in the sequence of diagrams of Figure 12.4.3.

The system is initialised with a 'wave' of pollution, $g(t)$, at the boundary $a = 0$, as in the first graph. For illustration the lake is assumed pollution free at this initial time. Each subsequent plot is the level of pollution in the lake for $a \leq x \leq b$, at a later time, with $t = t + nh$, where $n = 1, 2, \ldots$ and $h = \pi$ for the illustrated figure.

5% cleaning time In order to calculate the time required for a polluted lake, with fresh water flowing into it, to have its pollution level reduced to 5% of its current level, we apply (11), with $g(t) = 0$ and $P(x,0) \neq 0$ for some x, where $a \leq x \leq b$. To establish the time required for such a reduction in the amount of pollution in the entire lake, we convert concentration to mass $M(x,t)$, using $C = M/V(x)$, where $V = \Delta x \times$ width $\times$ depth is the volume of water at location x. Since width and depth are assumed constant, and $C\Delta x$ is a measure of mass, the solution of t in

$$\int_a^b C(x,t)dx = 0.05 \int_a^b P(x)dx,$$

Figure 12.4.3: Illustration of solution (11), with a wave of pollution $P(x) = \sin^2(x-vt)$ arriving at an unpolluted lake. For simplicity $v = 1$, and the 'rectangular' lake stretches from $a = 0$ to $b = 12$.

is the answer sought. This can be simplified to

$$\int_a^b P(x - vt)dx = 0.05 \int_a^b P(x)dx$$

$$\int_{a-vt}^{b-vt} P(s)ds = 0.05 \int_a^b P(x)dx.$$

Numerical solution

Maple can be used to solve the PDE (5), and obtain a general solution. Then, with specified boundary conditions, we can examine how the system changes with time and space.

```
> restart:with(PDEtools):
> pdelake := diff(C(x,t),t)+F/A*diff(C(x,t),x)=0;
> generalSolution:=pdsolve(pdelake);
Plotting the solution for given conditions
> PDEplot(pdelake,[y,5*(y-2),1+sin(5*(y-2))],y=0..50,orientation=[1,90]);
> PDEplot(pdelake,[y,5*(y-2),1+sin(5*(y-2))],y=0..50,orientation=[-90,90]);
```

The inclusion of orientation here allows you to plot t or x against pollution concentration. Alternatively, excluding orientation will result in a 3D plot, with a time axis, a spatial axis, and a pollution concentration axis.

Extensions and diffusion

diffusion

Although advection of pollution in the main central channel of a lake is the main contributer to the movement of pollution through it, there are many other processes at work. Chemical processes for example. Acting at a molecular level, certain nutrients react with chemicals to create further pollutants. Another process operating at the microscopic level is diffusion. Diffusion may be the main process for reducung pollution in sheltered bays, away from the main advection channel, and as such is essential to understand in order to ascertain safety levels for food sources from, or recreational sports in, these regions. Diffusion, working at the molecular level, operates in all directions. Thus the diffusion of chemicals is a slow process, compared with the advection modelled above, and occurring in all directions puts it beyond the scope of this book. However, we will develop the diffusion equation in a single direction to provide an understanding of how the process works.

the
diffusion
equation

For simplicity we will return to Figure 12.4.1, and imagine that diffusion may occur in the y direction only, at right angles to the direction of the advection in direction x. Using the same thin rectangular slice (in Figure 12.4.1) as before, let the face of the slice facing the y direction have area B. Thus $B = \Delta x \times$ depth and is comparable with A of the advection model. Let $\hat{J}(y, t)$ (comparable with $J(x, t)$) be the pollution flux through a diffusive process across a unit of area per unit of time. Then, in the same way as PDE (5) was developed for $C(x, t)$, with pollution flux $J(x, t)$ via advection in the x-direction across area A, we can develop a PDE for diffusion in the y direction across area B, to give

$$\frac{\partial C}{\partial t} + \frac{\partial \hat{J}}{\partial y} = 0. \qquad (12)$$

(Comparable with equation (3).) Since pollution moves from regions of high concentration to regions of low concentration, the flux $\hat{J}$ is proportional to the negative gradient of pollution concentration C. Thus we write

$$\hat{J} = -\gamma \frac{\partial C}{\partial y},$$

where γ is the constant of proportionality. Thus we have the diffusion equation in one direction,

$$\frac{\partial C}{\partial t} = \gamma \frac{\partial^2 C}{\partial y^2}. \qquad (13)$$

We will not solve this equation here, nor develop it to 3D space, but the interested reader will find details in any text on PDEs. However, the solution to the diffusion equation (13) will follow from the solution to the heat equation (see Section 12.2), which takes the same form, although with different boundary conditions. *analytical solution*

`Maple` can deal with functions of many variables, and finding their partial derivatives, although it (particularly Version 5 and previous versions) is not always able to solve PDEs without direct assistance concerning the form of the solution from the user. Thus, a good understanding of PDEs is required to make best use of the PDE facilities. Later versions come with a PDE toolbox, but it is left to the reader to explore this possibility. *numerical solution*

Skills developed here:

- Formulate a PDE to describe the advection of a pollutant from some source in a lake, such as a river.
- Understand the differences between modelling pollution spread via advection and diffusion.
- Distinguish cases where the use of an ODE is appropriate, and where a partial differential equation is appropriate.

12.5   Exercises for Chapter 12

1. Non-equilibrium temperature model. For Question 2, from Chapter 9, show the non-equilibrium temperature satisfies a PDE of the form

$$\frac{\partial U}{\partial t} = \alpha \frac{\partial^2 U}{\partial x^2} + \beta$$

where α and β are constants you must determine.

2. Non-equilibrium temperature for a cylinder. Extend the formulation in Question 6 of Chapter 9 to obtain a PDE for the non-equilibrium temperature $U(r,t)$.

3. Non-equilibrium temperature for a cylindrical heat fin. In Question 8 of Chapter 9 you found an ODE for the equilibrium temperature. Modify the derivation to obtain the following PDE for time-dependent temperature

$$c\rho \frac{\partial U}{\partial t} = k\frac{1}{r}\frac{\partial}{\partial r}\left(r\frac{\partial U}{\partial r} \right) - 2h(U - u_s),$$

where c is the specific heat, ρ is the density, k is the conductivity and h is the convective heat transfer coefficient.

4. Using Euler's identity. In Section 12.2 limits were required to establish values for the constants of integration, which led to equation (13). Given $b = a(1 + i)$ where a is a positive constant, what are

$$\lim_{x \to \infty} e^{bx} \qquad \text{and} \qquad \lim_{x \to \infty} e^{-bx} ?$$

What if $a < 0$? Show that your results lead to the solution given in (13). (Hint: Use Euler's identity.)

5. Oscillating surface temperature. Consider the PDE

$$\frac{\partial U}{\partial t} = \alpha \frac{\partial^2 U}{\partial x^2}$$

with the boundary conditions

$$U(0,t) = u_1 \cos(\omega t), \qquad -k\frac{\partial U}{\partial x} = 0, \quad \text{at } x = \infty.$$

These are the equations for the temperature in the ground due to a given oscillating temperature on the surface.

Solve these equations, leaving your answer as the real part of a complex valued temperature.

6. Amplitude of oscillating temperature. Find the amplitude of the complex valued temperature,

$$\hat{U}(x,t) = \frac{q}{kb}e^{-bx}e^{i\omega t}, \qquad b = \sqrt{\frac{\omega}{2\alpha}}(1+i), \quad i = \sqrt{-1}.$$

(Hint: Take the modulus of the complex valued temperature.)

7. Land mines. Solve the heat equation system from the land mine case study in the same way as the infinite soil problem from Section 12.2. The following differential equations and boundary conditions (equations (1) to (3) of Section 12.3 with boundary conditions (4) to (6)) are a model for the temperature of a land mine, $U_2(x,t)$, and the temperature of the soil, $U_1(x,t)$, above the land mine:

$$\frac{\partial U_1}{\partial t} = \alpha_1 \frac{\partial^2 U_1}{\partial x^2}, \quad -d < x < 0,$$

$$\frac{\partial U_2}{\partial t} = \alpha_2 \frac{\partial^2 U_2}{\partial x^2}, \quad x > 0,$$

$$-k\frac{\partial U_1}{\partial x} = q\cos(\omega t), \quad x = -d$$

$$U_1 = U_2, \quad x = 0$$

$$-k\frac{\partial U_1}{\partial x} = -k_1 \frac{\partial U_2}{\partial x}, \quad x = 0$$

$$-k_1\frac{\partial U_2}{\partial x} = 0, \quad x = \infty.$$

(a)
- Solve a modified problem, introducing complex valued functions by replacing $\cos(\omega t)$ with $e^{i\omega t}$.
- Assume a form of the solution $\hat{U}_1(x,t) = F_1(x)e^{i\omega t}$, and $\hat{U}_2(x,t) = F_2(x)e^{i\omega t}$.
- Substitute into the PDEs (partial differential equations) and obtain ODEs (ordinary differential equations) by using the fact that F_i is a function of x alone.
- Solve the ODEs to get a general complex valued solution, by assuming $F_1(x) = e^{\lambda_1 x}$ and $F_2(x) = e^{\lambda_2 x}$.
- Apply the boundary conditions to find the arbitrary constants.
- Hence obtain a solution for the soil surface temperature (at $x = -d$), given by

$$U_1 = \frac{q}{k_1 b_1}\left(\frac{(1+r)+(1-r)e^{-2b_1 d}}{(1+r)-(1-r)e^{-2b_1 d}}\right)e^{i\omega t}, \quad x = -d$$

$$r = \frac{k_2 b_2}{k_1 b_1}, \qquad b_1 = \sqrt{\frac{\omega}{\alpha_1}}\frac{(1+i)}{\sqrt{2}}, \qquad b_2 = \sqrt{\frac{\omega}{\alpha_2}}\frac{(1+i)}{\sqrt{2}}.$$

(b) We finally need to extract the real part of this complex valued expression to get an appropriate surface temperature. Use `Maple` to do this.

(c) Show that, in the limit as $d \to \infty$, this expression for the surface temperature reduces to the (complex valued) temperature for the surface temperature of a semi-infinite soil which was obtained directly in Section 12.2.

8. Depth of buried land mine. For the land mine case study, Section 12.3, use `Maple` to produce a graph of the amplitude of the soil surface temperature as a function of the depth the land mine is buried. Discuss the physical reasons for the behaviour in the graph.

9. Lake pollution in steady state. Suppose pollution has been entering a lake from a river at one end over a period of time, such that a stable steady state situation has been reached. Modelling the spread of pollution with (5), establish the level of pollution at any time and position (horizontal distance from the pollution source) in the lake.

10. Including a variable flow rate into the lake pollution model. In the case of a non-constant F, with $F(x,t)/A = v(x,t)$, velocity, show that equation (5) becomes

$$\frac{\partial C}{\partial t} + \frac{\partial}{\partial x}\left(C(x,t)\,v(x,t)\right) = 0\,.$$

11. Cleaning time. Suppose the pollution concentration throughout a lake is constant. If only fresh, unpolluted water flows into the lake, establish how long it will take for the amount of pollution already in the lake to reach 5% of its current level, in terms of the length of the lake and velocity of the water flow. Hence deduce a general expression for the time t_c it would take, for the pollution in a lake to reduce to a percent of its current level, if only fresh water flows into the lake.

Part V

Appendices

Appendix A

Differential equations

A.1 Properties of differential equations

A *differential equation* is an equation which involves the derivative(s) of some
unknown function. Such equations occur in many forms and thus specific
terminology, outlined below, is used to classify them. Solving a differential
equation involves finding an expression for the unknown function, for which
there are a variety of analytical techniques available. Since not all equations
can be solved analytically, numerical methods have been developed and are
widely used. Some analytical techniques relevant to this text are introduced
below, with a brief introduction to numerical methods covered in Chapter 4.

First some terminology relevant to the discussion of differential equations is
introduced:

- The *order* of a differential equation is the highest derivative occurring
  in it. Note that the order of the differential equation is the same as
  the number of arbitrary constants occurring in the general solution. For
  example, the differential equation

$$\frac{dy}{dt} - y = 0$$

  is first-order and has a general solution $y = C_1 e^t$, which contains only
  one arbitrary constant C_1. The differential equation

$$\frac{d^2y}{dt^2} - y = 0$$

  is second-order. Its general solution is $y = C_1 \cosh(t) + C_2 \sinh(t)$, which
  contains two arbitrary constants.

- A differential equation is *linear* if it can be expressed in the form

$$a_n(t)\frac{d^n y}{dt^n} + a_{n-1}(t)\frac{d^{n-1}y}{dt^{n-1}} + \ldots + a_1(t)\frac{dy}{dt} + a_0(t)y = F(t),$$

where the a_i and F are functions of t alone. Otherwise it is *nonlinear*. For example

$$\frac{dy}{dt} - y = 0 \qquad \text{and} \quad t\frac{dy}{dt} - y = \sin t$$

are linear, whereas

$$y^3\frac{dy}{dt} - y = 0 \qquad \text{and} \quad \frac{dy}{dt} - \sin y = 0$$

are nonlinear.

- An equation is said to be *homogeneous* when it involves only terms which are functions of the dependent variable and/or its derivatives. That is, no term must be a constant term or a function of the independent variable alone. The equation is *inhomogeneous* if a term involving the independent variable is included. Thus, for example,

$$\frac{dy}{dt} - y = 0$$

is homogeneous while

$$\frac{dy}{dt} - y = 2t$$

is inhomogeneous.

- A *system of differential equations* is a group of differential equations (more than one) which together describe some phenomenon. The equations are *coupled* if they depend on each other, such as in a predator–prey system

$$\frac{dx}{dt} = \beta x - c_1 xy, \qquad \frac{dy}{dt} = c_2 yx - \alpha y.$$

They *uncoupled* otherwise.

- The system is called a *dynamical system* when the independent variable is time and the solution used to predict future states. That is, the system describes how the modelled phenomenon changes with time. The coupled system above is thus a dynamical system.

- Differential equations are *ordinary differential equations* (ODE's) when there is only one independent variable, such as space or time. In the case of two independent variables, such as dependence on both space and time together, the equations are called *partial differential equations* (PDE's) in which case the derivatives are written with 'curly' d's, as for $y = y(x,t)$

$$\frac{\partial y}{\partial t} = 2\,x\,y\,t + \sin t.$$

A.2   Solution by inspection

Some differential equations are sufficiently simple for us to spot the general solution by inspection. This requires a good knowledge of the derivatives of the elementary functions such as polynomials, exponential functions, hyperbolic functions and trigonometric functions.

Example 1:   *Find a solution to the differential equation*

$$\frac{d^2y}{dt^2} = -b^2y,$$

where b is a non-zero constant (hence $-b^2$ is negative).

Solution: *We look for functions y with the property that their second derivative is a constant squared (b^2), times the function itself (y), and is negative. Functions with this property are the trigonometric functions $\cos(bt)$ and $\sin(bt)$. Hence the general solution is*

$$y = C_1 \cos(bt) + C_2 \sin(bt),$$

where C_1 and C_2 are arbitrary constants.

A few other common examples of differential equations whose solutions may be obtained by inspection include:

- $\dfrac{dy}{dt} = ay$, with solution $y = Ce^{at}$.

- $\dfrac{d^2y}{dt^2} = b^2y$, with solution $y = C_1e^{bt} + C_2e^{-bt}$ or, equivalently, with solution $y = c_1 \cosh(bt) + c_2 \sinh(bt)$.

- $\dfrac{dy}{dt} = at$, with solution $y = \dfrac{a}{2}t^2 + C$.

A.3   First-order separable equations

These are first-order differential equations which can be put in the form

$$\frac{dy}{dt} = f(y)g(t),$$

where $f(y)$ is a function of y only and $g(t)$ is a function of t only. We can solve these equations by 'separating the variables', that is, by dividing the differential equation through by $f(y)$ and then integrating both sides with respect to the independent variable t.

Example 1: *Solve the differential equation* $\dfrac{dy}{dt} = 2ty + 2t$ *using indefinite integrals.*

Solution: *We identify this differential equation as a separable differential equation by writing it in the form*

$$\frac{dy}{dt} = 2(y+1)t.$$

Thus, separating the variables, we obtain[1]

$$\frac{1}{y+1}\frac{dy}{dt} = 2t.$$

Integrating both sides with respect to t we obtain

$$\int \frac{1}{y+1}\frac{dy}{dt}\,dt = \int 2t\,dt.$$

On the LHS the change of variables rule for integration allows us to change the integration variable from t to y, giving,

$$\int \frac{1}{y+1}\,dy = \int 2t\,dt.$$

Carrying out the integrations we obtain

$$\ell n\,|y+1| = t^2 + C$$

where C is an arbitrary constant. Hence, solving for y,

$$|y+1| = e^{t^2 + C},$$

or

$$y(t) = Ae^{t^2} - 1,$$

where $A = \pm e^C$ is a convenient relabelling of the arbitrary constant.

Alternatively, the above example could be solved on some fixed interval using definite integrals, and thus not requiring the arbitrary constants, as follows.

---

[1]Many texts would separate the variables as

$$\frac{1}{y+1}\,dy = 2t\,dt.$$

The authors do not adopt this notation because the term $\frac{dy}{dt}$ is a derivative, not a fraction, and dy and dx make no sense on their own, outside an integral, unless one introduces more advanced mathematics such as the theory of differential forms.

Example 2:   Solve the differential equation $\dfrac{dy}{dt} = 2ty + 2t$ on the interval $[t_0, t]$.

Solution: As above, this differential equation as a separable differential equation and can be written in the form

$$\frac{dy}{dt} = 2(y + 1)t.$$

Thus, separating the variables, we obtain

$$\frac{1}{y+1}\frac{dy}{dt} = 2t.$$

Integrating both sides with respect to t

$$\int_{t_0}^{t} \frac{1}{y+1}\frac{dy}{dt}\, dt = \int_{t_0}^{t} 2t\, dt.$$

On the LHS the change of variables rule for integration allows us to change the integration variable from t to y, giving,

$$\int_{y_0}^{y} \frac{1}{y+1}\, dy = \int_{t_0}^{t} 2t\, dt$$

where $y(t_0) = y_0$. Carrying out the integrations we obtain

$$\ell n\, |y + 1| - \ell n\, |y_0 + 1| = t^2 - t_0^2,$$

and then

$$\ell n\left| \frac{y+1}{y_0+1} \right| = t^2 - t_0^2.$$

From here y can be solved for as before

$$\left(\frac{y+1}{y_0+1} \right) = \pm e^{t^2 - t_0^2}.$$

This can be written as

$$y(t) = Ae^{t^2} - 1,$$

where $A = \pm(y_0 + 1)e^{-t_0^2}$ is a convenient relabelling of the constant.

A.4   First-order linear equations

These are differential equations which can be put into the form

$$\frac{dy}{dt} + p(t)y = q(t)$$

where functions p and q are functions of t only. To solve these we find an *integrating factor* which helps us to integrate the equation directly. The integrating factor $R(t)$ is given by the formula

$$R(t) = e^{\int_a^t p(t)\,dt}.$$

If we then multiply throughout by this integrating factor, the differential equation can be converted into the form

$$\frac{d}{dt}\left(R(t)y(t)\right) = R(t)q(t),$$

and then the integral (with respect to t) of the LHS is simple to find, with the function on the right a function of t alone.

Example 1:   *Solve the differential equation* $\dfrac{dy}{dt} = 2t(y+1)$.

Solution: *First we see that the differential equation can be written in the appropriate form:*

$$\frac{dy}{dt} - 2ty = 2t.$$

Thus $p(t) = -2t$, $q(t) = 2t$ and the integrating factor $R(t)$ is given by

$$R(t) = e^{\int_a^t p(t)\,dt} = e^{\int_a^t -2t\,dt} = e^{-t^2 + a^2} = Ae^{-t^2},$$

where $A = e^{a^2}$ is constant. Hence, multiplying throughout by this factor, the differential equation becomes

$$\frac{d}{dt}\left(e^{-t^2}y\right) = 2te^{-t^2}.$$

Integrating both sides with respect to t we obtain

$$e^{-t^2}y = -e^{-t^2} + C$$

(where C is the arbitrary constant of integration) and so

$$y = -1 + Ce^{t^2}.$$

A.5   Homogeneous equations

We consider here *constant coefficient equations*. These are linear differential equations where the coefficient of each derivative, including that of zero order, are constants and the coefficient of y^0 is 0. This last condition, defines the equation as homogeneous. Thus they are a linear combination of derivatives of the dependent variable and include no functions of the independent variable alone.

For first-order equations they are of the general form

$$\frac{dy}{dt} + a_0 y = 0,$$

and for second-order equations they are of the general form

$$\frac{d^2y}{dt^2} + a_1\frac{dy}{dt} + a_0 y = 0,$$

where a_1 and a_0 are constants.

The characteristic equation

To solve such equations we use prior knowledge about their solutions: we look for a solution of the form $y(t) = e^{\lambda t}$, where the constant λ is to be determined. This always leads to an auxiliary equation, or characteristic equation, for λ. The general solution is then a linear combination of the solutions obtained.

Example 1: *Solve $\dfrac{d^2y}{dt^2} - \dfrac{dy}{dt} - 2y = 0$.*

Solution: *Substitute $y = e^{\lambda t}$ into the differential equation, to obtain*

$$\lambda^2 e^{\lambda t} - \lambda e^{\lambda t} - 2e^{\lambda t} = 0 \qquad so \qquad (\lambda^2 - \lambda - 2)e^{\lambda t} = 0.$$

Hence we obtain the characteristic (auxiliary) equation

$$\lambda^2 - \lambda - 2 = (\lambda + 1)(\lambda - 2) = 0.$$

This quadratic equation has roots $\lambda = -1$ and $\lambda = 2$ and thus the solutions are e^{-t} and e^{2t}. The general solution is a linear combination of e^{-t} and e^{2t},

$$y(t) = C_1 e^{-t} + C_2 e^{2t}.$$

Classification of roots

For second-order equations the characteristic equation is a quadratic equation and thus we do not always obtain distinct real roots. The form of the solution depends on the nature of the roots, for which there are three possible cases:

- Real roots: $\lambda = m_1$ and $\lambda = m_2$. Then

$$y = C_1 e^{m_1 t} + C_2 e^{m_2 t}.$$

- Complex roots $\lambda = \alpha \pm \beta i$. Then

$$y = e^{\alpha t} \left(C_1 \cos(\beta t) + C_2 \sin(\beta t) \right).$$

- Equal roots: $\lambda = m_1$. A second independent solution can be obtained by multiplying by t. Then

$$y = e^{m_1 t} \left(C_1 + C_2 t \right).$$

Two special cases of the above are worth noting:

- Real perfect squares: $\lambda = \pm m_1$. Then

$$y = C_1 e^{m_1 t} + C_2 e^{-m_1 t},$$

or an equivalent form is

$$y = c_1 \cosh(m_1 t) + c_2 \sinh(m_1 t),$$

where the arbitrary constants c_1 and c_2 are linearly related to the arbitrary constants C_1 and C_2.

- Purely imaginary roots: $\lambda = \pm m_1 i$. Then

$$y = C_1 \cos(m_1 t) + C_2 \sin(m_1 t).$$

A.6 Inhomogeneous equations

We consider again *constant coefficient equations*. This time we include a forcing term, that is, include a function dependent on t alone, or independent of the dependent variable. The inclusion of forcing defines the equation as inhomogeneous. For first-order equations the general form is,

$$\frac{dy}{dt} + a_0 y = F(t),$$

and for second-order equations

$$\frac{d^2y}{dt^2} + a_1\frac{dy}{dt} + a_0 y = F(t),$$

where a_i are constants and F is a non-zero function called the forcing term. (When $F(t) = 0$ the equation is homogeneous. Alternatively, when $F(t) \neq 0$ the equation is inhomogeneous.)

Splitting the solution

We solve this equation by splitting the solution into a homogeneous solution $y_h(t)$ and a *particular solution* $y_p(t)$, whence

$$y(t) = y_h(t) + y_p(t).$$

Here y_h is the *general solution* of the homogeneous equation (with $F(t) = 0$) and y_p is any particular solution satisfying the original equation.

Finding particular solutions

A particular solution of a differential equation is any function which satisfies the differential equation. If the RHS is a simple elementary function then it is usually possible to find a particular solution by making an educated guess based on the form of the RHS and its derivatives.

Example 1: Find a particular solution of $\dfrac{d^2y}{dt^2} - \dfrac{dy}{dt} - 2y = 3$.

Solution: Since the RHS is constant, we seek a solution of the form $y_p = A$ where A is a constant. Substituting into the differential equation we obtain $0 + 0 - 2A = 3$. Hence $A = 2/3$ and $y_p = 2/3$.

Example 2: Find a particular solution of $\dfrac{d^2y}{dt^2} - \dfrac{dy}{dt} - 2y = t^2$.

Solution: We look for a solution of the form

$$y_p = A + Bt + Ct^2,$$

where A, B and C are constants. Substituting this into the differential equation we get $2C - (2Ct + B) - 2(Ct^2 + Bt + A) = t^2$. Equating coefficients of the powers of t (noting that the RHS may be expressed as $t^2 + 0t + 0$) we obtain the set of simultaneous equations

$$-2C = 1,$$
$$-2C - 2B = 0,$$
$$2C - B - 2A = 0.$$

Solving these $C = -1/2$, $B = 1/2$ and $A = -3/4$. Hence

$$y_p = -\frac{1}{2}t^2 + \frac{1}{2}t - \frac{3}{4}.$$

Example 3:   Find a particular solution of $\dfrac{d^2y}{dt^2} - \dfrac{dy}{dt} - 2y = 2e^{5t}$.

Solution: We look for a solution of the form

$$y_p = Ae^{5t}.$$

Substituting into the differential equation we obtain

$$25Ae^{5t} - 5Ae^{5t} - 2Ae^{5t} = 2e^{5t}.$$

Hence
$$(25 - 5 - 2)A = 2$$

and $A = 1/9$. Thus

$$y_p = \frac{1}{9}e^{5t}.$$

Example 4:   Find a particular solution of $\dfrac{d^2y}{dt^2} - \dfrac{dy}{dt} - 2y = 2\sin(2t)$.

Solution: We try a solution in the form

$$y_p = A\cos(2t) + B\sin(2t).$$

Substituting this form into the differential equation we obtain

$$-4A\cos(2t) - 4B\sin(2t) - [-2A\sin(2t) + 2B\cos(2t)] - 2[A\cos(2t) + B\sin(2t)]$$
$$= 2\sin(2t) + 0\cos(2t).$$

Equating the coefficients of $\cos(2t)$ and $\sin(2t)$ we obtain

$$-6A - 2B = 0,$$
$$-6B + 2A = 2.$$

Solving for A and B gives $A = 1/10$ and $B = -3/10$ and hence

$$y_p = -\frac{1}{10}\cos(2t) + \frac{3}{10}\sin(2t).$$

Thus we have some general rules we can apply:

- For constant RHS try $y_p = A$, where A is a constant.

- For polynomial RHS try a general polynomial up to the order of the RHS polynomial.

- For exponential RHS try a general exponential with the same argument.

- For trigonometric RHS try a linear combination of sin and cos.

There are some exceptions which should be noted. When the particular solution is the same as the homogeneous solution then the usual approach for finding the general solution will fail. In the context of oscillations, this is called the resonance condition. In this case try the appropriate choice for a particular solution multiplied by t, as illustrated in the following example.

Example 5: *Find a particular solution of* $\dfrac{d^2y}{dt^2} - \dfrac{dy}{dt} - 2y = 3e^{2t}$.

Solution: *Note that the homogeneous solution is* $C_1 e^{2t} + C_2 e^{-t}$. *For the particular solution, rather than try* $y_p = Ae^{2t}$, *try* $y_p = Ate^{2t}$.

Substituting this into the differential equation yields

$$(4Ate^{2t} + 4Ae^{2t}) - (2Ate^{2t} + Ae^{2t}) - 2(Ate^{2t}) = 3e^{2t}.$$

Equating coefficients of te^{2t} *we gain no information* $(0 = 0)$. *However, from the coefficients of* e^{2t} *we obtain* $2A = 3$. *Hence a particular solution is*

$$y_p(t) = \frac{3}{2}te^{2t}$$

and the general solution is

$$y(t) = C_1 e2t + C_2 e^{-t} + \frac{3}{2}te^{2t}.$$

Appendix B

Further mathematics

B.1 Linear algebra

This appendix on linear algebra presents only the theory and concepts required to support the text. A general text on linear algebra, such as Lay (1994), is recommended for further background and a complete presentation of the theory.

Matrix notation and algebra

Square matrices, that is, matrices with the same number of rows and columns (in our case 2×2), can be added (subtracted), multiplied, or multiplied by a constant as follows:

- $\begin{bmatrix} a_1 & b_1 \\ a_2 & b_2 \end{bmatrix} + \begin{bmatrix} c_1 & d_1 \\ c_2 & d_2 \end{bmatrix} = \begin{bmatrix} a_1 + c_1 & b_1 + d_1 \\ a_2 + c_2 & b_2 + d_2 \end{bmatrix}$

- $\begin{bmatrix} a_1 & b_1 \\ a_2 & b_2 \end{bmatrix} \times \begin{bmatrix} c_1 & d_1 \\ c_2 & d_2 \end{bmatrix} = \begin{bmatrix} (a_1 c_1 + b_1 c_2) & (a_1 d_1 + b_1 d_2) \\ (a_2 c_1 + b_2 c_2) & (a_2 d_1 + b_2 d_2) \end{bmatrix}$

- $k \times \begin{bmatrix} a_1 & b_1 \\ a_2 & b_2 \end{bmatrix} = \begin{bmatrix} ka_1 & kb_1 \\ ka_2 & kb_2 \end{bmatrix}.$

They can also be post-multiplied by a vector with an equal number of rows (in our case 2):

- $\begin{bmatrix} a_1 & b_1 \\ a_2 & b_2 \end{bmatrix} \begin{bmatrix} x \\ y \end{bmatrix} = \begin{bmatrix} a_1 x & b_1 y \\ a_2 x & b_2 y \end{bmatrix}.$

For this last case we use the form

$$A\mathbf{x} = \begin{bmatrix} a_1 & b_1 \\ a_2 & b_2 \end{bmatrix} \begin{bmatrix} x \\ y \end{bmatrix}.$$

Matrices and vectors provide a very useful way of expressing systems of equations. With this above notation, the system

$$a_1 x + b_1 y = 0,$$
$$a_2 x + b_2 y = 0,$$

can be given as

$$A\mathbf{x} = \mathbf{0},$$

where

$$A = \begin{bmatrix} a_1 & b_1 \\ a_2 & b_2 \end{bmatrix}, \qquad \mathbf{x} = \begin{bmatrix} x \\ y \end{bmatrix}.$$

It should be noted that while the above notation is given for 2×2 matrices, the concepts can be extended to $n \times n$ matrices. There is a wealth of linear algebra theory which can then be applied to these systems, some of which is given below. (For a full understanding of the concepts a text on linear algebra, such as Lay (1994), is recommended.)

Determinants

Suppose A is the 2×2 matrix $\begin{bmatrix} a_1 & b_1 \\ a_2 & b_2 \end{bmatrix}$ and $\mathbf{x} = \begin{bmatrix} x \\ y \end{bmatrix}$ is a vector.

Then $A\mathbf{x} = \mathbf{0}$ is

$$\begin{bmatrix} a_1 & b_1 \\ a_2 & b_2 \end{bmatrix} \begin{bmatrix} x \\ y \end{bmatrix} = \begin{bmatrix} 0 \\ 0 \end{bmatrix} \qquad \text{or} \qquad \begin{aligned} a_1 x + b_1 y &= 0, \\ a_2 x + b_2 y &= 0. \end{aligned}$$

The *trivial solution* to this is $x = y = 0$. The *nontrivial solution* excludes this above possibility, and thus

$$x = -\frac{b_1}{a_1} y \qquad \text{if} \quad a_1 \neq 0.$$

Then

$$-\frac{a_2 b_1}{a_1} y + b_2 y = 0$$

so that

$$y \left(b_2 - \frac{a_2 b_1}{a_1} \right) = 0$$

and then, since $y \neq 0$, for a nontrivial solution to the vector equation (1) we must have

$$a_1 b_2 - a_2 b_1 = 0.$$

This value $(a_1 b_2 - a_2 b_1)$ is called the *determinant* of A, and is denoted $\det A$ or $|A|$. It is the difference between the diagonal products of A in the case of a 2×2 matrix. The important conclusion is that for a nontrivial solution to a vector equation $A\mathbf{x} = \mathbf{0}$, we require $|A| = 0$. (Such notions can be extended to $n \times n$ matrices.)

Eigenvectors and eigenvalues

An *eigenvector* of a 2×2 matrix A is a non-zero vector $\mathbf{x}$ such that $A\mathbf{x} = \lambda\mathbf{x}$ for some scalar λ. A scalar λ is an *eigenvalue* of A if there is a nontrivial solution $\mathbf{x}$ to $A\mathbf{x} = \lambda\mathbf{x}$: such an $\mathbf{x}$ is the eigenvector corresponding to the eigenvalue λ.

Note that, using the results of the previous section, the existence of a nontrivial (i.e. non-zero) solution $\mathbf{x}$ to

$$A\mathbf{x} = \lambda\mathbf{x} \quad \text{or} \quad A\mathbf{x} - \lambda\mathbf{x} = (A - \lambda I)\mathbf{x} = 0,$$

means that the determinant of $(A - \lambda I)$ is zero. That is, $|A - \lambda I| = 0$. (Here I is the identity matrix, and

$$I = \begin{bmatrix} 1 & 0 \\ 0 & 1 \end{bmatrix}$$

for 2×2 matrices.) Recall,

$$A - \lambda I = \begin{bmatrix} a_1 & b_1 \\ a_2 & b_2 \end{bmatrix} - \begin{bmatrix} \lambda & 0 \\ 0 & \lambda \end{bmatrix} = \begin{bmatrix} a_1 - \lambda & b_1 \\ a_2 & b_2 - \lambda \end{bmatrix}$$

and then

$$\begin{aligned} |A - \lambda I| &= (a_1 - \lambda)(b_2 - \lambda) - a_2 b_1 \\ &= a_1 b_2 - \lambda b_2 - \lambda a_1 + \lambda^2 - a_2 b_1 \\ &= \lambda^2 - \lambda(a_1 + b_2) + (a_1 b_2 - a_2 b_1). \end{aligned}$$

Notice that this is $\lambda^2 - \lambda(\text{trace} A) + (\det A)$, where trace A (or tr A) is the sum of the diagonal terms of A, and $\det A = |A| = a_1 b_2 - a_2 b_1$.

We call

$$|A - \lambda I| = \lambda^2 - \lambda(a_1 + b_2) + (a_1 b_2 - a_2 b_1)$$

the *characteristic polynomial* of A, and $|A - \lambda I| = 0$ is called the characteristic equation of A. This characteristic equation can be solved to find the eigenvalues. Clearly, in this case, there can be at most two distinct eigenvalues, with the possibility of complex valued eigenvalues.

Linear dependence and independence

Any point in the (x, y)-plane can be expressed as a linear combination of x and y, which have direction vectors $[1, 0]^T$ and $[0, 1]^T$ respectively.[1] We say

---

[1] The superscript T denotes the transpose. Thus

$$[1, 0]^T = \begin{bmatrix} 1 \\ 0 \end{bmatrix}.$$

that $[1, 0]^T$ and $[0, 1]^T$ *span* $\mathbb{R}^2$, or that they form a *basis* for $\mathbb{R}^2$: that is, every point in the plane can be expressed in terms of them. Further,

$$\begin{bmatrix} 1 \\ 0 \end{bmatrix} \neq c \begin{bmatrix} 0 \\ 1 \end{bmatrix}$$

and

$$\begin{bmatrix} 0 \\ 1 \end{bmatrix} \neq c \begin{bmatrix} 1 \\ 0 \end{bmatrix}$$

for any constant c. This property is called *linear independence* of the vectors. Thus we say that two vectors are linearly independent if neither can be expressed as a linear combination of the other. Another way to say this (the typical mathematical definition) is:

A set of vectors is linearly independent if the vector equation

$$c_1 \mathbf{v}_1 + c_2 \mathbf{v}_2 + \cdots + c_k \mathbf{v}_k = 0$$

has only the trivial solution $c_1 = c_2 = \cdots = c_k = 0$.

Theorem 4 (Linear independence of eigenvectors.) *Suppose* $\mathbf{v}_1, \ldots, \mathbf{v}_k$ *are eigenvectors that correspond to k distinct eigenvalues* $\lambda_1, \ldots, \lambda_k$ *of some matrix A. Then the set of eigenvectors* $\{\mathbf{v}_1, \ldots, \mathbf{v}_k\}$ *is linearly independent.*

Proof   *Suppose* $\{\mathbf{v}_1, \ldots, \mathbf{v}_k\}$ *is not linearly independent (i.e. linearly dependent), then there exists some p such that $\mathbf{v}_p$ is a linear combination of the eigenvectors* $\{\mathbf{v}_1, \ldots, \mathbf{v}_{p-1}\}$, *with constants c_i,*

$$\mathbf{v}_p = c_1 \mathbf{v}_1 + \cdots + c_{p-1} \mathbf{v}_{p-1}. \tag{2}$$

Now, multiplying by A,

$$A\mathbf{v}_p = c_1 A\mathbf{v}_1 + \cdots + c_{p-1} A\mathbf{v}_{p-1}$$

and then

$$\lambda_p \mathbf{v}_p = c_1 \lambda_1 \mathbf{v}_1 + \cdots + c_{p-1} \lambda_{p-1} \mathbf{v}_{p-1}, \tag{3}$$

since $A\mathbf{v}_i = \lambda_i \mathbf{v}_i$ by the definition of eigenvalues and eigenvectors. Also, multiplying (2) by λ_p gives

$$\lambda_p \mathbf{v}_p = c_1 \lambda_p \mathbf{v}_1 + \cdots + c_{p-1} \lambda_p \mathbf{v}_{p-1} \tag{4}$$

and then subtracting (4) from (3)

$$0 = c_1 (\lambda_1 - \lambda_p) \mathbf{v}_1 + \cdots + c_{p-1} (\lambda_{p-1} - \lambda_p) \mathbf{v}_{p-1}.$$

Since $\{\mathbf{v}_1, \ldots, \mathbf{v}_{p-1}\}$ are linearly independent and non-zero, the expressions $(\lambda_i - \lambda_p)$ for $i = 1, \ldots, p - 1$ are all zero. However, this is a contradiction because the eigenvalues are all distinct. This means that $\{\mathbf{v}_1, \ldots, \mathbf{v}_k\}$ are linearly independent.

What we have shown is that, in the case of A a 2×2 matrix with two distinct eigenvalues and two associated eigenvectors ($\mathbf{u}$ and $\mathbf{v}$), the eigenvectors are linearly independent. And if we have two linearly independent vectors in $\mathbb{R}^2$, they must necessarily span $\mathbb{R}^2$, that is, form a basis for $\mathbb{R}^2$. Thus x and y can be expressed as a linear combination of the eigenvectors

$$\mathbf{x} = z_1\mathbf{u} + z_2\mathbf{v}.$$

Matrix invertibility

An $n \times n$ matrix A is *invertible* if there exists a matrix A^{-1} such that $A^{-1}A = I = AA^{-1}$, where I is the identity matrix.

Theorem 5 (Matrix invertibility.) *Let A be an $n \times n$ matrix (a square matrix). Then the following are equivalent, that is, if any one of them is true then so are the rest.*

(a) *A is invertible*

(b) *the columns of A are linearly independent*

(c) *the columns form a basis for $\mathbb{R}^n$*

(d) *$|A| \neq 0$ (i.e. $\det A \neq 0$)*

(e) *$A\mathbf{x} = 0$ has no solutions other than the trivial one*

(f) *the rows of A are linearly independent*

(g) *the rows of A form a basis for $\mathbb{R}^n$.*

Proof. *The details are omitted, but for those who are interested, see any text on linear algebra, for example Lay (1994).*

B.2   Partial derivatives and Taylor expansions

Partial derivatives

Suppose z is dependent on two variables x and y. We say $z = f(x, y)$. If we keep y constant ($y = y_1$ say), then z can be considered as a function of x alone, and if the derivative of $f(x, y_1)$ exists then it is called the partial derivative of f with respect to x. Mathematically, it is described with 'curly' d's as

$$\frac{\partial f}{\partial x} = f_x \quad - \quad \text{a first-order partial derivative.}$$

Similarly, with x fixed ($x = x_1$ say) we have (if the derivative exists)

$$\frac{\partial f}{\partial y} = f_y \quad - \quad \text{the other first-order partial derivative.}$$

Example 1: $z = f(x,y) = x^2 y + x \sin y$

$$\frac{\partial f}{\partial x} = 2 x y + \sin y, \qquad \frac{\partial f}{\partial y} = x^2 + x \cos y,$$

and

$$\frac{\partial}{\partial y} \left(\frac{\partial f}{\partial x} \right) = 2 x + \cos y, \qquad \frac{\partial}{\partial x} \left(\frac{\partial f}{\partial y} \right) = 2 x + \cos y,$$

and

$$\frac{\partial}{\partial x} \left(\frac{\partial f}{\partial x} \right) = 2 y, \qquad \frac{\partial}{\partial y} \left(\frac{\partial f}{\partial y} \right) = -x \sin y .$$

These are the four second-order partial derivatives.

The partial derivatives represent slopes in the x- and y-directions respectively (see Figure B.2.1).

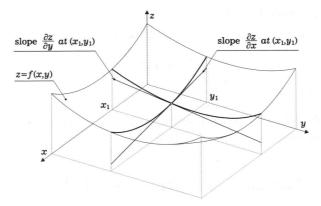

Figure B.2.1: Diagram showing the interpretation of the partial derivatives as the slopes of the graph of a function of two variables in the directions of the x and y axes.

The notation for these partial derivatives is given below, where it is worth noting the order of x and y in the subscripts.

$$\frac{\partial}{\partial y}\left(\frac{\partial f}{\partial x}\right) = \frac{\partial^2 f}{\partial y\, \partial x} = f_{xy},$$

$$\frac{\partial}{\partial x}\left(\frac{\partial f}{\partial y}\right) = \frac{\partial^2 f}{\partial x\, \partial y} = f_{yx},$$

$$\frac{\partial}{\partial y}\left(\frac{\partial f}{\partial y}\right) = \frac{\partial^2 f}{\partial y\, \partial y} = f_{yy},$$

$$\frac{\partial}{\partial x}\left(\frac{\partial f}{\partial x}\right) = \frac{\partial^2 f}{\partial x\, \partial x} = f_{xx}.$$

It is not always true that $f_{xy} = f_{yx}$. However, if all the derivatives concerned are continuous we do have $f_{xy} = f_{yx}$ and the order of differentiation does not matter. This was the case in Example 1 above.

For the partial derivatives then, y fixed at $y = y_1$ represents a vertical plane which intersects the surface $z = f(x, y)$ in a curve, and $\partial z/\partial x$ at the point (x_1, y_1) is the slope of the tangent to this curve at (x_1, y_1). (Similarly for x fixed at $x = x_1$.) This is illustrated in Figure B.2.1.

Taylor series expansions

We recall Taylor's Theorem for a single variable, and then extend the definition to functions of two variables.

Theorem 6 (Taylor's Theorem for a single variable) *If F and its first n derivatives F', F'', ..., $F^{(n)}$ are continuous on some interval $[a, b]$ and $F^{(n)}$ is differentiable on (a, b), then there exists a number c_{n+1}, between a and b, such that*

$$
\begin{aligned}
F(b) \quad = \quad & F(a) + F'(a)\,(b - a) + \frac{F''(a)}{2!}\,(b - a)^2 + \cdots \\
& + \frac{F^{(n)}(a)}{n!}\,(b - a)^n + \frac{F^{(n+1)}}{(n + 1)!}\,(c_{n+1})\,(b - a)^{n+1}.
\end{aligned}
$$

This equation, in the single variable case, provides an extremely accurate polynomial approximation for a large class of functions which have derivatives of all orders. With reference to Figure B.2.2, close to $x = \hat{x}$ we can approximate the curve $f(x)$ with the line tangent to the curve at $x = \hat{x}$.

For a linear approximation, $F(b) \approx F(a) + F'(a)\,(b - a)$ with an error $e(b)$ having $a < c < b$,

$$|e(b)| \leq \frac{\max|F''(c)|}{2}\,(b - a)^2.$$

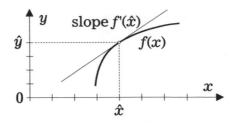

Figure B.2.2: Diagram showing the first-order Taylor expansion in the vicinity of the point $\hat{x}$.

For a quadratic approximation, $F(b) \approx F(a) + F'(a)\,(b-a) + \frac{F''(a)}{2}\,(b-a)^2$ with (again $a < c < b$)

$$|e(b)| \leq \frac{\max |F''(c)|}{6}\,|b-a|^3 \,.$$

Taylor's Theorem can be extended to the two variable case. As illustrated in Figure B.2.1, close to $(x,y) = (x_1, y_1)$ we can approximate the surface $z = f(x,y)$ with the tangent plane to the surface at $(x,y) = (x_1, y_1)$.

We extend the definition of Taylor's Theorem to functions of two variables in the following way. Let R be some region containing $P(a,b)$. Let $S(a + h, b + k)$ be in R such that PS is in R. This is shown in Figure B.2.3. Let f

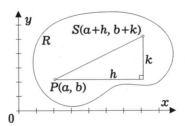

Figure B.2.3: Approximating a function near a point.

have continuous first- and second-order partial derivatives in R. Describe PS parametrically as $\quad x = a + t\,h, \quad y = b + t\,k$, with $0 \leq t \leq 1$. Now we study the values of $f(x,y)$ on PS by considering the function

$$F(t) = f(a + h\,t, b + k\,t).$$

We have

$$F'(t) = \frac{\partial f}{\partial x}\frac{\partial x}{\partial t} + \frac{\partial f}{\partial y}\frac{\partial y}{\partial t} = h\frac{\partial f}{\partial x} + k\frac{\partial f}{\partial y}$$

and F' is continuous and differentiable on $[0,1]$ because f_x and f_y are (by definition above).

Also

$$F''(t) = h\frac{\partial F'}{\partial x} + k\frac{\partial F'}{\partial y}$$

$$= h\left(h\frac{\partial^2 f}{\partial x^2} + k\frac{\partial^2 f}{\partial y\,\partial x}\right) + k\left(h\frac{\partial^2 f}{\partial x\,\partial y} + k\frac{\partial^2 f}{\partial y^2}\right)$$

$$= h^2\frac{\partial^2 f}{\partial x^2} + 2hk\frac{\partial^2 f}{\partial x\,\partial y} + k^2\frac{\partial^2 f}{\partial y^2}.$$

Since F has its first two derivatives continuous on the interval $[0,1]$ it satisfies the hypothesis for the Taylor series expression in a single variable with $n = 2$, so

$$F(1) = F(0) + F'(0)(1 - 0) + F''(c)\frac{(1-0)^2}{2!} \qquad (\text{for} \quad c \in [0,1])$$

$$= F(0) + F'(0) + \frac{1}{2}F''(c). \tag{1}$$

But $F(1) = f(a + h, b + k)$ and $F(0) = f(a, b)$ by definition. Also $F'(0) = h\,f_x(a, b) + k\,f_y(a, b)$ and $F''(c) = (h^2\,f_{xx} + 2hk\,f_{xy} + k^2\,f_{yy})$ evaluated at $(a + ch, b + ck)$.

Now substituting these expressions into (1) above

$$f(a + h, b + k) = f(a, b) + h\,f_x(a, b) + k\,f_y(a, b) + (\text{terms of higher order}),$$

and the *tangent-plane approximation* is

$$f(a + h, b + k) \approx f(a, b) + h\,f_x(a, b) + k\,f_y(a, b).$$

B.3 Review of complex numbers

Complex numbers have the form $z = a + ib$ where $i = \sqrt{-1}$. The number a is called the real part of z, $R(z)$, and the number b is called the imaginary part of z, $I(z)$.

Complex numbers can be added, subtracted, multiplied and divided, as shown by the following examples:

- $(7 + 3i) + (-3 + 2i) = (7 - 3) + (3 + 2)i = 4 + 5i$

- $(2 + 3i)(3 - i) = 2 \times 3 - 2i + 3 \times 3i - 3i^2 = 6 + 7i - 3 \times (-1) = 9 + 7i$

- $\dfrac{2 + i}{1 + i} = \dfrac{2 + i}{1 + i} \times \dfrac{1 - i}{1 - i} = \dfrac{2 - 2i + i - i^2}{1 - i^2} = \dfrac{2 - i + 1}{1 + 1} = \dfrac{3 - i}{2}.$

Also useful when dealing with complex numbers is *Euler's identity*,

$$e^{i\theta} = \cos\theta + i\sin\theta, \tag{1}$$

which implies that

$$e^{-i\theta} = \cos\theta - i\sin\theta. \tag{2}$$

When solving ODEs, the function $e^{-i\theta}$ is often more convenient to deal with than $\cos\theta$ and $\sin\theta$. Furthermore, the identity is useful for obtaining roots of complex numbers. For example, for $\theta = \pi/2$,

$$e^{i\pi/2} = \cos\left(\frac{\pi}{2}\right) + i\sin\left(\frac{\pi}{2}\right) = i.$$

So

$$i^{1/2} = (e^{i\pi/2})^{1/2} = e^{i\pi/4}$$

and from (1),

$$i^{1/2} = e^{i\pi/4} = \cos\frac{\pi}{4} + i\sin\frac{\pi}{4} = \frac{1}{\sqrt{2}}(1+i).$$

Note that $i = \cos\left(\frac{3\pi}{2}\right) + i\sin\left(\frac{3\pi}{2}\right)$, since $\cos\left(\frac{3\pi}{2}\right) = 0$ and $\sin\left(\frac{3\pi}{2}\right) = 1$, and thus we also have

$$i^{1/2} = \cos\left(\frac{3\pi}{4}\right) + i\sin\left(\frac{3\pi}{4}\right) = -\frac{1}{\sqrt{2}}(1+i).$$

B.4   *Hyperbolic functions*

From Euler's identity (equation (1), of B.3) we can define $\sin x$ and $\cos x$ as

$$\sin x = \frac{e^{ix} - e^{-ix}}{2i}, \qquad \cos x = \frac{e^{ix} + e^{-ix}}{2}. \tag{1}$$

And recall that $\sin^2 x + \cos^2 x = 1$ indicates that in the (x, y)-plane, the point $(\sin x, \cos x)$ lies on the unit circle and the functions $\sin x$ and $\cos x$ are called circular functions. In the same way we can define the *hyperbolic functions*, where $(\sinh x, \cosh x)$ lies on the rectangular hyperbola with equation $\sinh^2 x - \cosh^2 x = 1$.

Formally they are defined in terms of e^x as

$$\sinh x = \frac{e^x - e^{-x}}{2}, \qquad \cosh x = \frac{e^x + e^{-x}}{2}, \tag{2}$$

whence

$$\tanh x = \frac{e^x - e^{-x}}{e^x + e^{-x}}.$$

These functions are commonly used to simplify expressions of exponentials and thus often appear in the integral solutions of the text.

To get an idea of how they behave they are graphed below using `Maple` in Figures B.4.1 to B.4.3.

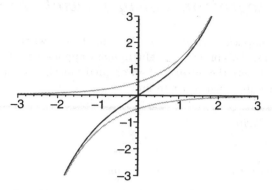

Figure B.4.1: Graph of sinh x (black) with $e^{-x}/2$ (grey) and $-e^{-x}/2$ (grey).

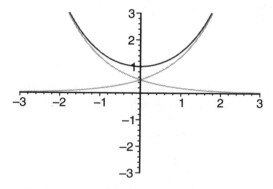

Figure B.4.2: Graph of cosh x (black) with $e^{-x}/2$ (grey) and $e^{x}/2$ (grey).

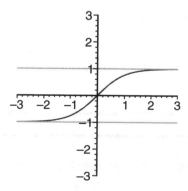

Figure B.4.3: Graph of tanh x (black) with upper and lower bounds at ± 1 (grey).

B.5   Integration using partial fractions

Integrating a complicated fraction may be simplified by writing it as the sum of simpler fractions. This process is useful in many applications, but in particular for integration where the integral of an original function becomes the sum of integrals, which are much easier to find.

Example 1:   Write
$$\frac{x+4}{x^2 - 5x + 6}$$
as a sum of simpler fractions and hence find its integral.

Solution: The denominator of the function can be factorised and written as two partial fractions
$$\frac{x+4}{x^2 - 5x + 6} = \frac{x+4}{(x-2)(x-3)} = \frac{A}{x-2} + \frac{B}{x-3}.$$

Now
$$\frac{A}{x-2} + \frac{B}{x-3} = \frac{A(x-3) + B(x-2)}{(x-3)(x-2)}$$

so that equating the coefficients in the numerator, with those of the original function, gives two simultaneous equations, $A + B = 1$ and $-3A - 2B = 4$. Solving these we have $A = -6$ and $B = 7$. Thus the original fraction may be written as the sum of partial fractions
$$\frac{x+4}{x^2 - 5x + 6} = \frac{-6}{x-2} + \frac{7}{x-3}.$$

While integration of the original function is clearly not simple, the latter expression is easy to integrate:
$$\int \frac{-6}{x-2}dx = -6\,\ell n\,|x-2| + C_1 \quad \text{and} \quad \int \frac{7}{x-3}dx = 7\,\ell n\,|x-3| + C_2.$$

Thus
$$\int \frac{x+4}{x^2 - 5x + 6}dx = -6\,\ell n\,|x-2| + 7\,\ell n\,|x-3| + C_3$$
where $C_3 = C_1 + C_2$.

Example 2:   Write the fraction
$$\frac{1}{x^2 - a^2}$$
as a sum of simpler fractions.

Solution: *Firstly, the denominator of the original fraction can be factorised*

$$\frac{1}{x^2 - a^2} = \frac{1}{(x - a)(x + a)}$$

and we would like to write this as

$$\frac{A}{x - a} + \frac{B}{x + a}.$$

But

$$\frac{A}{x - a} + \frac{B}{x + a} = \frac{A(x + a) + B(x - a)}{(x - a)(x + a)}$$

from which, by equating the numerators,

$$1 = Ax + Aa + Bx - Ba.$$

Thus $A + B = 0$ (equating the coefficients of x) and $a(A - B) = 1$ (equating the constants) which implies that $A = 1/2a$ and $B = -1/2a$. So the original fraction can be written as

$$\frac{1}{x^2 - a^2} = \frac{1}{2a(x - a)} - \frac{1}{2a(x + a)}.$$

In general, fractions can be separated into partial fractions as follows:

$$\frac{f(x)}{x(x + a)(x - a)} = \frac{A}{x} + \frac{B}{x + a} + \frac{C}{x - a},$$
$$\frac{f(x)}{x^2 + a} = \frac{A}{x} + \frac{Bx + C}{x^2 + a},$$
$$\frac{f(x)}{x(x - a)^2} = \frac{A}{x} + \frac{B}{x - a} + \frac{C}{(x - a)^2}.$$

Note that when splitting a fraction into partial fractions, the numerator of each partial fraction is a polynomial with degree one less than the denominator. (For further details see, for example, Adams (1995).)

Appendix C

Notes on Maple

C.1 Brief introduction

Introduction

`Maple` is an extensive software package, able to carry out many sophisticated calculations in many areas of mathematics. As an extension to the calculator, it is able to cope with extremely long numbers, up to about 500 000 digits, solving differential equations, and partial differential equations, graphing numerical solutions, complex algebra, and much more. Since this book is concerned with differential equations, and their solution, not all of which are solvable analytically, `Maple` (or some other equivalent software package) is essential for a complete understanding of the processes developed and discussed in the text.

This brief introduction is far from adequate for those who are new to such packages, but it serves to illustrate the `Maple` environment, the manner in which statements and functions operate, and how to get help on any topic.

Basics

Below (Figure C.1.1) is a `Maple` screen dump of some basics in value assignments and algebra. Note the use of semicolons at the end of each statement: a semicolon allows the results to be displayed on the worksheet, while a colon suppresses information to the screen, although it exists and can be accessed at any time by typing the variable name. Note too, that `restart` resets all variables to blanks, and it is good practice to start each independent code segment with this command.

Figure C.1.1: `Maple` screen dump illustrating some basic assignment statements, and introducing the `Maple` environment.

Getting help

At any time, a question mark in front of a command will provide full information on that topic, such as, `?plot` for information, options and examples on plotting functions. Information on the commands required for a variety of functions pertaining to equations, plotting procedures and solution finding, typically used in this text, can also be found under '`Maple` code', in the text index.

'Further information' (below) suggests a website for help on particular topics, as well as suggesting a basic text (of which there are many) for those requiring a detailed introduction to `Maple`.

C.2 Using `Maple` to solve DEs

This book is concerned with both the numerical and symbolic solution of differential equations, and the plotting of these results. Thus the use of `Maple` is limited to the appropriate functions and plotting routines.

Figure C.2.1, illustrates how to find the analytical and numerical solutions to a differential equation, as well as how to use the plotting and display functions to graph the results. These routines are typical of those which required repeatedly throughout the book. Note the use of `with plots` in the opening line to access routines in the plotting library of `Maple`. (This code segment is taken from Section 2.5.)

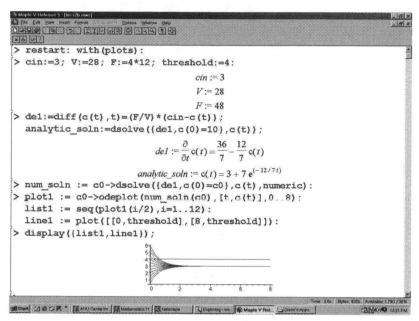

Figure C.2.1: `Maple` screen dump illustrating the solution to a differential equation, and a plotting routine.

Further information

There are any number of excellent books on introducing `Maple`, such as Holmes et al. (1993), and also information and online introductions from the web, such as at `http://www.maplesoft.com`. More advanced texts are also available, such as Heck (1996).

Appendix D

SI units

Fundamental Units of Primary Quantities

Primary Quantity	SI Unit	cgs Unit
mass	kilogram, kg	gram, g
length	metre, m	centimetre, cm
time	second, s	second, s
temperature	Kelvin, K	Kelvin, K

Secondary units comprised of fundamental units

Quantity	SI Units
density ρ	$\mathrm{kg\,m^{-3}}$
velocity v	$\mathrm{m\,s^{-1}}$
acceleration a	$\mathrm{m\,s^{-2}}$
force F	Newtons $\mathrm{N = kg\,m\,s^{-2}}$
pressure p	Pascal $\mathrm{Pa = N\,m^{-2} = kg\,m^{-1}\,s^{-2}}$
energy E	Joule $\mathrm{J = kg\,m^2\,s^{-2}}$
power $\dot{E}$	Watt $\mathrm{W = J\,s^{-1}}$
heat flux J	$\mathrm{W\,m^{-2}}$
thermal conductivity k	$\mathrm{W\,m^{-1}\,K^{-1}}$
specific heat c	$\mathrm{J\,kg^{-1}\,K^{-1}}$
thermal diffusivity α	$\mathrm{m^2\,s^{-1}}$
Newton cooling coefficient h	$\mathrm{W\,m^{-2}\,K^{-1}}$

Bibliography

Adams, R. A. (1995). *Calculus: A Complete Course* (3rd ed.). Ontario: Addison-Wesley.

Anderson, A. (1996). Was *rattus exulans* in New Zealand 2000 years ago? AMS radiocarbon ages from the Shag River Mouth. *Archaeol. Oceania. 31*, 178–184.

Banks, R. (1994). *Growth and Diffusion Phenomena: Mathematical Frameworks and Applications*. Berlin: Springer-Verlag.

Beck, M., A. Jakeman and M. McAleer (1993). *Construction and Evaluation of Models of Environmental Systems*, Chapter 1, pp. 3–35. New York: Wiley.

Beddington, J., C. Free and J. Lawton (1978). Characteristics of successful natural enemies in models of biological control of insect pests. *Nature 273*, 513–519.

Begon, M., J. Harper and C. Townsend (1990). *Ecology, Individuals, Populations and Communities* (2nd ed.). Oxford: Blackwell.

Borelli, R. and C. Coleman (1996). *Differential Equations; A Modelling Perspective.* New York: Wiley.

Braun, M. (1979). *Differential Equations and Their Applications* (2nd ed.). Berlin: Springer-Verlag.

Burgess, J. and L. Olive (1975). Bacterial pollution in lake burley griffin. *Water (Australian Water and Wastewater Association) 2*, 17–19.

Caulkins, J., R. Marrett and A. Yates (1985). Peruvian anchovy population dynamics. *U.M.A.P. Journal 6*(3), 1–23.

Costantino, R., J. Cushing, B. Dennis and R. Desharnis (1995). Experimentally induced transitions in the dynamic behaviour of insect populations. *Nature 375*, 227–375.

Costantino, R., R. Desharnais, J. Cushing and B. Dennis (1997). Chaotic dynamics in an insect population. *Science 275*, 389–391.

Cramer and R. May (1972). Interspecific competition predation and species diversity: A comment. *Journal of Theoretical Biology 34*, 289–293.

Cyrix (1998). Technical documentation for Cyrix 6x86-P-166 processor.

Daley, D. and J. Gani (1999). *Epidemic Modelling. An Introduction.* Cambridge: Cambridge University Press.

Davies, P. (1994). *War of the Mines: Cambodia, Landmines and the Impoverishment of a Nation.* London: Pluto Press.

Dekker, H. (1975). A simple mathematical model of rodent population cycles. *Journal Mathematical Biology 2*, 57–67.

Dodd, A. (1940). The biological campaign against prickly-pear. Technical report, Brisbane Government Printer, Australia.

Edelstein-Keshet, L. (1988). *Mathematical Models in Biology.* New York: Random House.

Fulford, G. R. and P. Broadbridge (2000). *Industrial Mathematics: Case Studies in Heat and Mass Transport.* Cambridge: Cambridge University Press.

Fulford, G. R., P. Forrester and A. Jones (1997). *Modelling with Differential and Difference Equations.* Cambridge: Cambridge University Press.

Gani, J. (1972). Model-building in probability and statistics. In O. Sheynin (Ed.), *The Rules of the Game*, pp. 72–84. London: Tavistock Publications.

Gani, J. (1980). The role of mathematics in society. *Mathematical Scientist 5*, 67–77.

Garret, L. (1995). *The Coming Plague: Newly Emerging Diseases in a World Out of Balance.* New York: Penguin.

Gotelli, N. J. (1995). *A Primer of Ecology.* Sunderland, MA: Sinauer Associates.

Greenhalgh, D. and R. Das (1993). An SIRS epidemic model with a contact rate depending on population density. In O. Arino, D. Axelrod, M. Kimmel and M. Langlais (Eds.), *Mathematical Population Dynamics: Analysis of Heterogeneity*, pp. 79–102. Winnipeg: Wuerz Publications.

Grenfell, B. and A. Dobson (1995). *Ecology of Infectious Diseases in Natural Populations.* Cambridge: Cambridge University Press.

Hassell, M. (1978). *The Dynamics of Arthropod Predator–Prey Systems.* Princeton, N.J.: Princeton University Press.

Hassell, M. P. (1976). *The Dynamics of Competition and Predation.* London: Edward Arnold.

Hearn, C. (1998). Private communication.

Heck, A. (1996). *An Introduction to Maple* (2nd ed.). New York: Springer-Verlag.

Holman, J. (1981). *Heat Transfer.* New York: McGraw-Hill.

Holmes, M., J. Ecker, W. Boyce and W. Siegmann (1993). *Exploring Calculus with Maple.* New York: Addison-Wesley.

Hurewicz, W. (1990). *Lectures on Ordinary Differential Equations*. New York: Dover.

Jones, J. C. (1993). *Combustion Science. Principals and Practice*. Brisbane: Milennium Books.

Keeton, W. T. (1972). *Biological Science* (2nd ed.). New York: W W Norton.

Kermack, W. and A. McKendrick (1927). A contribution to the mathematical theory of epidemics. *Proceedings of the Royal Society. A 115*, 700–721.

Kincaid, D. and W. Cheney (1991). *Numerical Analysis*. Belmont, CA.: Brooks Cole.

Kormondy, E. K. (1976). *Concepts of Ecology* (2nd ed.). Englewood Cliffs, NJ: Prentice-Hall.

Krebs, C., M. Gaines, B. Keller, J. Myers and R. Tamarin (1973). Population cycles in small rodents. *Science 179*, 35–41.

Lay, D. C. (1994). *Linear Algebra and its Applications*. Reading, MA: Addison-Wesley.

Lewin, R. (1983a). Predators and hurricanes change ecology. *Science 221*, 737–740.

Lewin, R. (1983b). Santa rosalia was a goat. *Science 221*, 636–639.

Lignola, P. and F. D. Maio (1990). Some remarks on modelling CSTR combustion processes. *Combustion and Flame 80*, 256–263.

Louie, K., M. Roberts and G. Wake (1993). Thresholds and stability analysis of models for the spacial spread of a fatal disease. *IMA Journal of Mathematics Applied to Medicine and Biology 10*, 207–226.

Loyn, R., R. Runnalls, G. Forward and J. Tyers (1983, Sep). Territorial bell miners and other birds affecting populations of insect prey. *Science*, 1411–1413.

MacArthur, R. and J. Connell (1966). *The Biology of Populations*. New York: Wiley.

May, R. (1981). *Theoretical Ecology: Principles and Applications* (2nd ed.). Oxford: Blackwell Scientific Publications.

Merzhanov, A. G., V. Barzykin, V. Abramov and F. Dubovitskii (1961). Thermal explosion in the liquid phase with heat transfer by convection only. *Russian Journal of Physical Chemistry 35*, 1024–1027.

Mesterton-Gibbons, M. (1989). *A Concrete Approach to Mathematical Modelling*. Reading, MA: Addison-Wesley.

Michaelis, L. and M. Menten (1913). Die kinetik der invertinwirkung. *Biochem. Z. 49*, 333–369.

Microsoft (1995). Microsoft encarta.

Miller, T. (1995). Mathematical model of heat conduction in soil for landmine detection. Technical Report DMS-C95/95, CSIRO Division of Mathematics and Statistics.

Monro, J. (1967). The exploitation and conservation of resources by populations of insects. *Journal of Animal Ecology 36*, 531–547.

Murray, J. (1990). *Mathematical Biology*. New York: Springer-Verlag.

Myers, J. and C. Krebs (1974). Population cycles in rodents. *Scientific American 230*, 38–46.

NHMRC (1997). Staying healthy in childcare. Technical Report Oct., National Health and Medical Research Council, Australia.

Oakley, R. and C. Ksir (1993). *Drugs, Society and Human Behaviour* (6th ed.). St Louis, HO: Mosbey.

Parrish, J. and S. Saila (1970). Interspecific competition, predation and species diversity. *Journal Theoretical Biology 27*, 207–220.

Przemieniecki, J. (1994). *Mathematical Methods in Defence Analysis*. Washington DC: American Institute of Aeronautics and Astronautics.

Quammen, D. (1997). *The Song of the Dodo: Island Biogeography in an Age of Extinctions*. London: Pimlico.

Renshaw, E. (1991). *Modelling Biological Populations in Space and Time*. Cambridge: Cambridge University Press.

Rivers, C., G. Wake and X. Chen (1996). The role of milk powder in the spontaneous ignition of moist milk powder. *Mathematics and Engineering in Industry 6*, 1–14.

Roberts, M. (1992). The dynamics and control of bovine tuberculosis in possums. *IMA Journal of Mathematics Applied to Medicine and Biology 9*, 19–28.

Roberts, M. (1996). The dynamics of bovine tuberculosis in possum populations, and its eradication by culling or vaccination. *Journal of Animal Ecology 65*, 451–464.

Roberts, M. and J. Heesterbeek (1993). Bluff your way in epidemic models. *Trends in Microbiology 1*, 343–348.

Rolls, E. (1969). *They All Ran Wild: A Story of Pests on the Land in Australia*. Melbourne: Angus and Robertson.

Sansgiry, P. and C. Edwards (1996). A home heating model for calculus students. *College Mathematics Journal 27*(5), 394–397.

SBS (1998). *SBS World Guide*. Sydney: Australian Broadcasting Commission.

Smedley, S. I. and G. C. Wake (1987). Spontaneous ignition: Assessment of cause. *Annual Meeting of the Institute of Loss Adjusters of NZ. Palmerston North 10–12 June*, 1–14.

Stewart, I. (1989). Portraits of chaos. *New Scientist 1689*(4 Nov), 22–27.

Taylor, J. G. (1980). *Force-on-Force Attrition Modelling.* Virginia: Operations Research Society of America.

van de Koppel, J. Huisman, J. van der Wal and H. Olff (1996). Patterns of herbivory along a productivity gradient: An empirical and theoretical investigation. *Ecology* 77(3), 736–745.

Vivaldi, F. (1989). An experiment with mathematics. *New Scientist 1689*(28 Oct), 30–33.

Wake, G. and M. Roberts (1995). Percy possum plunders. *CODEE Winter*, 13–14.

Wilmott, P. (1998). *Derivatives. The Theory and Practice of Financial Engineering.* Chichester: Wiley.

Index